CATHODOLUMINESCENCE
OF
GEOLOGICAL MATERIALS

TITLES OF RELATED INTEREST

Boninites
A. J. Crawford (ed.)

Carbonatites
K. Bell (ed.)

The dark side of the Earth
R. Muir Wood

Experiments in physical sedimentology
J. R. L. Allen

Image interpretation in geology
S. A. Drury

The inaccessible Earth
G. C. Brown & A. E. Mussett

The interpretation of igneous rocks
K. G. Cox *et al.*

An introduction to X-ray spectrometry
K. L. Williams

Komatiites
N. T. Arndt & E. G. Nisbet (eds)

Mathematics in geology
J. Ferguson

Metamorphic geology
C. Gillen

Metamorphic processes
R. H. Vernon

Metamorphism and metamorphic belts
A. Miyashiro

Petroleum geology
F. K. North

Petrology of the igneous rocks
F. H. Hatch *et al.*

Petrology of the metamorphic rocks
R. Mason

Petrology of the sedimentary rocks
J. T. Greensmith

A practical approach to sedimentology
R. Lindholm

A practical introduction to optical mineralogy
C. D. Gribble & A. J. Hall

Principles of physical sedimentology
J. R. L. Allen

Rutley's elements of mineralogy
H. H. Read

Sedimentary structures
J. D. Collinson & D. B. Thompson

Sedimentology
M. R. Leeder

Simulating the Earth
J. Holloway & B. Wood

Statistical methods in geology
R. F. Cheeney

Volcanic successions: modern and ancient
R. A. F. Cas & J. V. Wright

The young Earth
E. G. Nisbet

Cathodoluminescence OF GEOLOGICAL MATERIALS

D. J. MARSHALL

Nuclide Corporation, Acton, Massachusetts

with a chapter contributed by
Anthony N. Mariano

Consultant, Carlisle, Massachusetts

Boston
UNWIN HYMAN
London Sydney Wellington

Allen & Unwin Inc.,
8 Winchester Place, Winchester, Mass. 01890, USA

Published by the Academic Division of
Unwin Hyman Ltd
15–17 Broadwick Street,
London W1V 1FP, UK

Allen & Unwin (Australia) Ltd,
8 Napier Street, North Sydney, NSW 2060, Australia

Allen & Unwin (New Zealand) Ltd in association with the Port Nicholson Press Ltd, 60 Cambridge Terrace, Wellington, New Zealand

First published in 1988

Library of Congress Cataloging-in-Publication Data

Marshall, Donald J.
Cathodoluminescence of geological materials:
Bibliography: p.
Includes index.
1. Petrology. 2. Luminescence. I. Mariano, Anthony N.
I. Title.
QE431.5.M365 1987 552′.06 87–12160
ISBN 0–04–552026–7 (alk. paper)

British Library Cataloguing in Publication Data

Marshall, Donald J.
Cathodoluminescence of geological materials:
an introduction.
1. Optical mineralogy 2. Microscope and microscopy
I. Title II. Mariano, Anthony N.
549′.125 QE369.D6
ISBN 0–04–522026–7

Typeset in 9 on 11 point Concorde by Phoenix Photosetting, Chatham and printed in Great Britain by The University Printing House, Oxford

Preface

When energetic electrons bombard the surface of certain materials, visible radiation is emitted. The process is called cathodoluminescence (CL), and is the basis of a technique which is now widely used for the petrographic examination of geological and other materials. In some instances, important information can be obtained by cathodoluminescence which is not obtainable by other readily available techniques.

Major articles pointing out the use of cathodoluminescence as a petrographic tool were first published in 1965 (Smith & Stenstrom 1965, Long & Agrell 1965). Since that time, CL has progressed from the status of a research technique, available only to those laboratories equipped with an electron probe or a specially built electron beam system, to the status of a routine laboratory technique, in use even at the undergraduate level in some universities. In addition to its rôle in teaching and research at universities, many petroleum research and field laboratories and other industrial and government research centers are using the CL technique as one of their routine investigative tools.

Many investigators continue to use the electron probe and SEM for CL observations but a significant number are now using a cathodoluminescence attachment which can be mounted on the stage of a petrographic microscope.

The CL observations supplement the petrographic observations made with transmitted light and polarized light. CL complements these more established techniques in some very important ways. For example, the diagnosis of authigenic material, especially silica and feldspar, is frequently done more easily with CL. The recognition that carbonate cements may have complex zonation, and the delineation of the successive stages of this complex cement formation, is facilitated. Fine-grained mineralization is often more easily detected. The identification of impurities present in certain minerals, and even the determination of their ionic charge states, can sometimes be done.

To date, the CL technique has most often been used for the examination of sedimentary samples. However, some very interesting work has recently been done on igneous rocks. Also limited work with metamorphic rock samples indicates that many of the metamorphic minerals luminesce and their luminescence characteristics may provide useful information.

The fact that certain minerals exhibit CL was established in the late 1800s. Sporadic references to the CL technique's application in geological studies appeared prior to the definitive papers in 1965 by Smith and Stenstrom and by Long and Agrell. Despite the interval of 20 years, the technique still remains relatively unknown to many who could use it for the examination of geological materials. Two principal reasons for this are:

(a) Some investigators are using CL as a routine tool just as they would use a petrographic microscope. They do not always make specific comments about the CL observations in publications.

(b) Luminescence results are certainly best shown in color, rather than in black and white. However, it is difficult, primarily because of the expense, to publish color photomicrographs in most journals.

This publication contains a number of color plates which illustrate the value of the CL technique. These may encourage those who are not familiar with its potential to begin to use it. The examples have been collected from many sources and represent a time span of over 20 years – hence the quality varies. During the time that these examples have been collected, the author has been with Nuclide Corporation, the designer and manufacturer of the Luminoscope®, and the examples are weighted towards work done with this instrument. However, there are many examples which were done with other types of instrument.

As a supplement to the color plates included in this text, we also list here a number of references which include one or more color plates.

Reference	Subject
Amieux (1982)	general
Carrasco (1973)	dolomite
Dickson (1985)	carbonates
Ebers and Kopp (1979)	carbonates
Gaal (1976–77)	gems
Goebel and Patzelt (1976)	general
Goni and Rémond (1969)	sphalerite
Grögler and Liener (1968)	meteorites
Hanley *et al.* (1977)	diamond
Iijima *et al.* (1983)	quartz, feldspar
James and Klappa (1983)	carbonates
Kopp (1981)	general
Lang (1979)	diamond
Le Poole *et al.* (1968)	cassiterite, scheelite
Mariano (1978)	carbonatites and general
Matter and Ramseyer (1985)	sandstones
Mendelssohn *et al.* (1979)	diamond
Olaussen *et al.* (1984)	quartz
Ponahlo and Koroschetz (1985)	natural and synthetic gems
Reeder and Prosky (1986)	dolomite
Richter and Zinkernagel (1975)	sandstones, perthites, carbonates
Richter and Zinkernagel (1981)	calcite, aragonite, dolomite, oolites, fossils
Sippel (1971)	lunar samples
Smith and Stenstrom (1965)	carbonates, feldspars, quartz, fossils
ten Have and Heijnen (1985)	synthetic calcite
Zinkernagel (1978)	quartz, general

This text will concentrate on CL although numerous pieces of information developed with other forms of excitation will be introduced from time to time. There is some confusion in the literature because many publications use the general term 'luminescence' without specifying the excitation source. The collection of definitions of the more important terms in Chapter 1 may help to alleviate some of this confusion.

There is a great deal of CL information in the literature of the phosphor industry and some of this is referenced in this text. Since much of this information is for synthetic materials, it is potentially very useful in understanding CL mechanisms, because of the better material definition. The application of the results to geological cases requires care because natural samples are not usually so well defined. The phosphor industry literature can certainly serve as a useful guide to those who may choose to go on from this introductory text into research on the CL of geological materials.

It would be useful to present for each mineral a list of the activators and quenchers, a set of illustrations showing how CL intensity varies with activator identity and concentration, and an example of the emission spectrum for each activator. Unfortunately, it is the rare case where enough information exists to do this. This is partly due to the still-limited input from the solid-state physicist on materials of geological interest and partly to the complex nature of luminescence itself. It is difficult to isolate the CL characteristics of even a simple synthetic system, much less the complex systems in natural minerals. Until this information is developed it seems best to present the various bits and pieces, gleaned from thermoluminescence, X-ray excited luminescence, and elsewhere, that are available, even though they may not always represent a complete and consistent picture. Perhaps gaps in information that are revealed herein will help to stimulate interest in the necessary research.

Similarly, one would like to be able to present the detailed theoretical background for the CL properties of the various minerals. This theoretical treatment is very involved and has been carried out for only a relatively few materials. An outline of some of the theoretical approaches used is presented in Chapter 1 and the interested reader will find here many references which lead to this literature.

A few publications approach the desired goal of comprehensive treatments of the CL of geological materials. Medlin's series of papers on the thermoluminescence of carbonates and quartz are examples of these. Mariano, Ito and Ring and Geake, Walker and coworkers have done important work with feldspars. Hall and Ribbe have done a thorough study of cassiterite. Sommer has made a comprehensive study of several carbonate minerals using CL, X-ray excited luminescence, spectral scans, electron probe analysis, and X-ray diffraction of natural and synthetic samples. These are all excellent examples of the direction of research required for the other minerals.

A textbook, especially an introductory textbook, is usually not the place to put forth new, possibly controversial, subject material. That is more normally the subject for refereed journal articles. However, so much of the material on CL fits into this category that it is difficult to avoid some interpretations which may be controversial. All are advised to read the material with an open, questioning attitude for it is only by the input from many people of diverse viewpoints that consensus will be reached.

Some of the included material, quoted from various published sources, is not consistent. To illustrate some of the inconsistencies in the published literature, consider the seemingly routine problem of determining the emission spectrum for calcite. The range of published values (Table 7.2) for the position of the CL emission peak maximum extends from 578 to 605 nm with no clear understanding yet as to whether the differences are instrumental or real and, if they are real, what impurity, excitation, crystallographic, or other factors play important rôles in creating these differences.

Unfortunately, if we were to wait until the subject of Cathodoluminescence of Geological Materials had reached a higher degree of maturity, an introductory text might be considerably delayed. We believe that the probability of inspiring additional investigations is worth the risk of including some controversial material.

Even though there are a number of demonstrated uses of CL, a great deal remains to be learned. There is ample opportunity for further research. New questions are being asked as rapidly as old ones are being answered. Some questions are very difficult to answer conclusively. To realize the full potential of CL in geological applications, much more research must be done. This will need to involve the combined talents of the earth scientist, the instrumental analyst, and the solid-state physicist.

D. J. Marshall

Acknowledgments

Many people have contributed to the author's knowledge of the applications of cathodoluminescence in the study of geological and other materials. Some of these associates have provided color plates or other information which is acknowledged in the appropriate locations. However, I would like to give special acknowledgment to the following individuals with whom I have had important contacts over a number of years:

Anthony Dickson, *University of Cambridge*
Otto Kopp, *University of Tennessee*
William Meyers, *SUNY, Stony Brook*
John Miller, *University of Edinburgh*
Michael Owen, *St Lawrence University*
John B. Thomas, *Amoco, Denver*

Special acknowledgment is also due to Dr Anthony N. Mariano. He is the actual author of the material in Chapter 8 but his influence and advice is imbedded throughout the other chapters as well. His continuing support has been very important.

An early draft of the manuscript was reviewed in all or in part by James Marshall, William McQuown, Michael Owen and Paul Ring. Many of their constructive comments have been incorporated.

Many of my associates at Nuclide Corporation have, each in their own way, contributed to the development of instruments and techniques for CL studies and to my understanding of these.

When reading acknowledgments in other books, one sometimes tends to look upon the acknowledgments to editors, publishers, and copy editors as a formality. After completing the process of preparing this manuscript, I realize that the acknowledgments to these professionals are well-earned.

Most importantly, the value of the continuing support of my wife, Barbara, through many evenings and weekends of absence or neglect, cannot be measured.

I would like to thank those who have given permission for the reproduction of color plates – full source details are given in the plate legends. I would also like to thank the following individuals and organizations who have kindly given permission for the reproduction of other copyright material:

Text material: extract in Chapter 5 from R. F. Sippel & A. B. Spencer (1970), *Proc. Apollo 11 Lunar Science Conf.* **3**, 2,413–26, copyright © 1970 Pergamon Press Ltd, reproduced with permission. Table 9.1 NASA.

Line illustrations: D. R. Beaman, J. A. Isasi, and the American Society for Testing and Materials (1.2); Figure 1.4 reproduced from Figure 1 of K. Nassau (1978), *American Mineralogist* **63**, 219–29, copyright © 1978 Mineralogical Society of America; Figure 1.6 reproduced from H. W. Leverenz (1968), *An introduction to luminescence of solids*, by permission of the author and Dover Publications Inc.; W. L. Medlin and Academic Press Inc. (1.7a); L. E. Orgel and the American Institute of Physics (1.7b); R. F. Sippel and the American Institute of Physics (3.1); Nuclide Corporation (3.2); the Editor, *Geologische Rundschau* (3.5a & 6.2); K. Ramseyer (3.5b); (3.9a); Dynasil Corporation of America (3.9b); Harshaw/Filtrol Partnership and Polaroid Corporation (3.9c,e); Kaufman Glass Co. (3.9d); Figure 4.1 reprinted by permission of the publisher, The Electrochemical Society, Inc., and F. A. Hummel; Figures 4.2 & 4.3 reproduced from Figures 5 & 7 of M. R. Hall & P. H. Ribbe (1971), *American Mineralogist* **56**, 31–44, copyright © 1971 Mineralogical Society of America; Figures 4.4, 4.5a,b, 4.6, 4.7 & 5.7 reprinted with permission from A. N. Mariano & P. J. Ring (1975), Europium-activated cathodoluminescence in minerals, *Geochimica et Cosmochimica Acta* **39**, 649–60, copyright © 1975 Pergamon Press Ltd; Divisão de Geologia e Mineralogia, Direccion Nacional da Producão Minerão, Brasil, and A. N. Mariano (4.5b, 5.1–3); Figure 4.8 reproduced from M. Ross *et al.* (1968), *Science* **159**, 1,099–102, copyright © 1968 American Association for the Advancement of Science, with the permission of M. Ross; Figure 4.9 reprinted by permission from *Nature* **203**, 134–6, copyright © 1964 Macmillan Journals Ltd; Figure 5.5 reprinted with permission from J. E. Geake *et al.* (1973), Luminescence of lunar, terrestrial, and synthesized plagioclase, caused by Mn^{2+} and Fe^{3+}, *Proc. Fourth Lunar Science Conf., Geochimica et Cosmochimica Acta* Suppl. 4 **3**, 3,181–9, copyright © 1973 Pergamon Press Ltd; Gordon and Breach (5.6a,b); Figure 5.8 reproduced from R. F. Sippel & A. B. Spencer (1970), *Proc. Apollo 11 Lunar Conf.* **3**, 2,413–26, copyright © 1970 American Association for the Advancement of Science; W. L. Medlin and the American Institute of Physics (6.1, 7.1, 7.2); E. Schweizerbart'sche Verlagsbuchlandlung (6.2); W. M. Hanusiak (6.3–5); The Mineralogical Society and J. V. P. Long (7.3); Figures 7.4–6 reproduced from *Chemical Geology* **9**, 257–73, by permission of Elsevier Science Publishers and S. A. Sommer; Society of Economic Paleontologists and Mineralogists (7.8) and J. R. Frank *et al.* (1982); B. J. Pierson and Blackwell Scientific Publications (7.9); I. J. Fairchild and Blackwell Scientific Publications (7.10, 7.11); Figure 9.1 reproduced from D. B. Nash (1973), *Journal of Geophysical Research* **78**, 3,512–14, copyright © 1973 American Geophysical Union; Academic Press Inc. (9.2).

Contents

List of tables

List of color plates

Note: for most of the plates, the scale refers to the original 35mm transparency. The horizontal dimension of the photograph corresponds to the 35mm dimension of the transparency and the scales can be obtained accordingly. For example, the horizontal dimension of Plate 12a represents 3.8mm on the original thin section and the longest dimension of the original actual apatite grain is about 1.7mm.

Abbreviations

Å	Ångström: 1 ångstrom = 10^{-8}cm
AC	Alternating current electrical power supply
CL	Cathodoluminescence
cm	Centimeter
CMA	Cathodoluminescence microscope attachment
CRT	Cathode-ray tube
DC	Direct current electrical power supply
EDS	Energy-dispersive spectroscopy
EPR	Electron paramagnetic resonance
eV	Electron volt, a unit of energy
FWHM	Full width at half maximum, a measure of peak width
HREE	Heavy rare earth elements
IR	Infra-red
keV	Kilo-electron volts
kV	Kilovolts
LES	Luminescence excitation spectroscopy
LREE	Light rare earth elements
M	Magnification factor of a lens
mA	Milliampere
meV	Million electron volts
mTorr	MilliTorr, a unit of pressure: 1 milliTorr = 0.133 pascal
mm	Millimeter
NA	Numerical aperture of an optical lens
nm	Nanometer: 1 nanometer = 10^{-9} meters
Pascal	SI system unit of pressure. 1 atmosphere = 1.01×10^{-5} pascals
PhL	Photoluminescence, i.e. luminescence excited by UV
PM	Photomultiplier
PoL	Polarized light
ppma	Parts per million by number of atoms
ppmw	Parts per million by weight
REE	Rare earth elements
SEM	Scanning electron microscope
ThL	Terminology used for *thermoluminescence* in this text to distinguish from *transmitted light*
TL	Term used for *thermoluminescence* by specialists in the field of thermoluminescence
TrL	Transmitted light
μm	Micron, a unit of length: 1 μm = 10^{-6} meters
UV	Ultraviolet
Vacuum UV	Vacuum ultraviolet, the wavelength range from 2,000 to 100Å.
W	Watt
WD	Working distance of an objective lens
XL	X-ray excited luminescence

$CaSiO_3$:Mn (e.g.) = $CaSiO_3$ with added Mn

$CaSiO_3$:(0.01 Mn) (e.g.) = $CaSiO_3$ with 0.01% added Mn: weight or mole percent should be specified

Epigraph

'629. One of the most striking of the phenomena attending this research has been the remarkable power which the molecular rays in a high vacuum possess of causing phosphorescence in bodies on which they fall. Substances known to be phosphorescent under ordinary circumstances shine with great splendour when subjected to the negative discharge in a high vacuum. Thus, a preparation of sulphide of calcium, much used now in Paris for coating clock faces, which remain luminous after dark, is invaluable in these researches for the preparation of phosphorescent screens whereon to trace the paths and trajectories of the molecules. It shines with a bright blue-violet light, and, when on a surface of several square inches, is sufficient to light up a room. Modifications of these phosphorescent sulphides shine with a yellow, orange, and green light . . .' (from Crookes 1879, p. 659).

Chapter 1

Introduction to luminescence phenomena

Conventional petrographic examination of geological samples makes use of the interaction of light with the mineral, either transmission through the mineral or reflection from the polished surface. Other somewhat more specialized petrographic techniques can also be used. Cathodoluminescence (CL) is one of these. CL is an additional way of looking at thin sections and other types of geological samples. The sample is bombarded by an energetic electron beam and often responds by emitting light of various wavelengths. The wavelength and intensity of the light emission characterize the mineral and the distribution of certain impurities within it. These distinctive patterns provide new information, much of which may be unobtainable, or certainly less readily obtainable, by any other presently available method. Many observers, when they first see a sample under CL, are astonished at the crispness of detail, the features not previously seen, and, often, just the spectacular beauty of the sample. No photographic copy can do full justice to the view of CL that is seen through the microscope.

The color plates in this book illustrate the sort of information revealed by CL. The discussion of the geological significance of some of the features revealed is presented in Chapter 4 *et seq.* The first three chapters provide a definition of terms and a discussion of the basic physics of luminescence, a bit of history, and a discussion of instrumentation. The terminology used is the same as Leverenz (1968), which is based on the experience of many years of phosphor research for the cathode-ray tube and television industries. In later chapters, we will present what is known about the CL of the more common minerals. Finally, we will look more closely at the fields where CL is presently being applied, including the study of feldspars, quartz, carbonates, and mineral deposits, particularly those associated with the rare earth elements.

1.1 Definition of terms

When energy is directed at the surface of some materials, nonthermal radiation may be emitted. The emitted radiation is most frequently observed in the visible range, but ultraviolet (UV) and infra-red (IR) emission may be found in some instances. This physical process is called luminescence (Fig. 1.1).

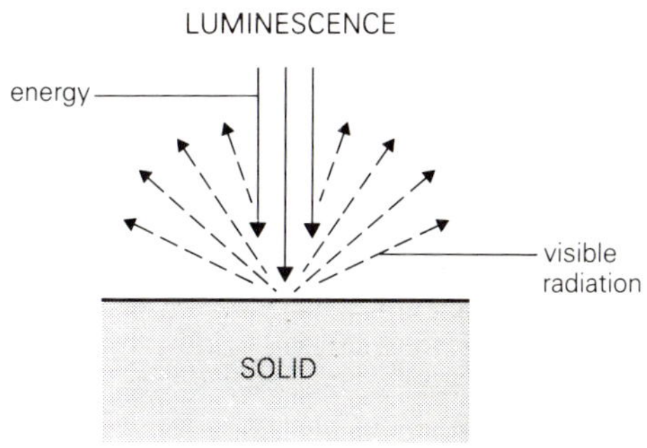

ultraviolet and visible light:	1.8 to 4.9 eV	=	photoluminescence
electron impact:	0 to 50,000 eV	=	cathodoluminescence

Figure 1.1 The nature of the luminescence process. Incident energy, in any of several possible forms (see text), causes visible radiation emission from a luminescent solid. The range of readily available energies for two of the more commonly used forms of incident energy, light beams and electron beams, are indicated.

The general phenomenon of luminescence includes both *phosphorescence* and *fluorescence*. Phosphorescence is the correct term for those processes where the radiation persists after the incident energy source is turned off. Fluorescence is the correct term for those processes where the emitted radiation ceases as soon as the excitation ceases. The actual formal definition for fluorescence (Leverenz 1968) is that the radiation ceases within 10^{-8} seconds after the excitation has been removed.

The energy directed at the surface can be in many forms: for example, the surface can be bombarded by UV photons, ions, X-rays, or electrons; or it can be mechanically disturbed. Each of these sources of energy can produce luminescence in certain materials. Several forms of luminescence are recognized, depending upon the form of the primary excitation (Leverenz 1968).

(a) *Chemiluminescence* refers to luminescence developed as a result of a chemical reaction.
(b) *Bioluminescence* is chemiluminescence in living organisms, for example in fire-flies.
(c) *Photoluminescence* is the correct term when the primary excitation is photon impact, for example the familiar 'black light' used as a source of UV in mineral studies.
(d) *Roentgenoluminescence* refers to the excitation of luminescence by X-ray bombardment. Röntgen discovered the X-ray in 1895.
(e) *Ionoluminescence* occurs when certain materials are bombarded by an energetic ion beam.
(f) *Cathodoluminescence* refers to the process in which the primary excitation is by a beam of electrons, the cathode being the source of the electrons. Cathodoluminescence is the process that occurs in the cathode-ray tube used in oscilloscopes and in black & white and color television sets.

The term 'electron-excited fluorescence' is occasionally used, and is a specific indicator of both the type of excitation used and the type of emission observed. A number of other terms are in use in the literature and sometimes one encounters an article in which only the term 'luminescence', undefined, is used. This can be misleading. We will strive to specify the type of luminescence being discussed throughout this text. When the general term 'luminescence' does appear, it will invariably mean cathodoluminescence unless otherwise defined.

One insight into a possible advantage of cathodoluminescence (CL), as contrasted to photoluminescence (UV excitation), is obtained from consideration of the range of energies available. The electron energy can be as high as 50 keV in instruments such as the electron probe. This is to be contrasted with the very low energies of the photons from a UV source, typically 1.8–4.9 eV. Whereas only a limited number of minerals will exhibit photoluminescence, most of the common rock-forming minerals which are not conductors and not opaque will show CL under some circumstances when subjected to the higher energy excitation available in CL studies. This more general excitation capability may not always be an advantage because selection of the wavelength of excitation from a UV source, and spectral analysis of the luminescence, can sometimes provide a means of discriminating between different activators (Gies 1975). There has been a renewal of interest in photoluminescence.

Finally, we should note that the process of *incandescence* occurs when materials are heated to a sufficiently high temperature. This heating may take place inadvertently in some samples as they are being excited by the electron beam used to generate cathodoluminescence. Experimental conditions are usually selected to avoid this.

Even though most of our comments will be specific to excitation by electron bombardment, it is of interest to note that the nature of the emission, at least above a certain threshold energy, may be relatively independent of the type of excitation. Thus Derham *et al.* (1964) noted that the color and relative efficiency of enstatite luminescence appeared to be about the same for excitation by low-energy protons (5–100 keV), 2 meV (million electron volts) and 4 meV protons, 5–40 keV electrons, and X-rays (40 kV, copper anode). Of course such things as absolute efficiency and damage effects do differ among these different excitation techniques. The different excitation sources produce similar spectra because, ultimately, the spectra arise from transitions of electrons in the material between various energy levels. A portion of the energy introduced by any of the various forms of bombardment ends up exciting electrons within the material. For many of the minerals, the only luminescence information available is from excitation by UV, X-ray or ions. We will introduce some of this information on the assumption that CL results will be analogous. This assumption is an extrapolation from the work of Derham *et al.* and a few other authors who have compared CL with other forms of luminescence.

The subject of this text is the cathodoluminescence of solid inorganic materials. Many organic materials (including fingerprints on thin sections) will also luminesce under UV excitation. These organic materials often show CL, although they may be rapidly evaporated or degraded by the electron beam. In connection with the luminescence of organic materials, we note that Winkler (1983) has reported on a procedure for applying fluorescent organic dyes to geological samples and studying them under UV excitation. He found interesting features revealed but the information has yet to be interpreted.

1.2 Thermoluminescence

The phenomenon of *thermoluminescence* has found applications in geological and archeological studies and there is still frequently a confusion between thermoluminescence and cathodoluminescence. Materials which exhibit thermoluminescence (ThL*) contain electrons excited to metastable energy states of long lifetime. The excitation took place at some earlier time by some form of ionizing radiation, e.g., alpha particles emitted from radioactive elements intimately associated with the thermoluminescent materials (Aitken 1974). The material will emit visible radiation, i.e., exhibit ThL, when these excited states decay. The rate at which this decay occurs is very slow at ambient temperatures but increases rapidly when the material is heated. (However, it should be noted that the heat does not cause the luminescence.) When the material is heated in the laboratory under controlled conditions, a glow curve can be obtained. This is a curve of the emission intensity as a function of sample temperature for a selected heating rate. Using the glow curve along with other information, the ThL 'age' of the sample can be estimated. ThL 'age' is not the geological age of the material but refers only to the time period that the material has been associated with the radioactive elements since the last firing or last time at high temperature.

Usually the total intensity is measured by a photometer although spectral analysis may be done. The sample is usually in a powdered form so there is no information on

* Specialists in thermoluminescence use TL as an abbreviation for thermoluminescence. Optical microscopists sometimes use TL for transmitted light. Since we refer frequently to both terms, we will use ThL for thermoluminescence and TrL for transmitted light to minimize confusion.

mineralogical distribution within the sample. Charlet (1970), however, designed an instrument for ThL which permits measurements on an individual mineral grain and is more akin to the instrumentation used for CL studies.

Many investigators were involved in the study of the geological applications of ThL in the 1950–70 period. The hoped-for applications as a routine geological dating tool have not materialized but it is an important routine tool for radiation dosimetry and for archeological dating (Aitken 1974) up to ages of about 500,000 to 1 million years. Presumably dating of geological events within such a time span is also possible (Dreimanis *et al.* 1978). Charlet (1971) has also demonstrated its usefulness in provenance studies, making use of the ThL characteristics of various minerals as source rock indicators.

There is a great deal of published data on activators and quenchers of ThL in geological materials, due especially to the work of Medlin (1959, 1961a, b, 1963a, b). In a few instances he obtained X-ray excited luminescence (XL) information also and concluded that the same activators and quenchers were responsible for XL and ThL and that the emission spectra were similar. Many of the observations of the physical bases of ThL parallel those of CL, as would be expected, and the work of Medlin and others on the physical bases for ThL has proved very useful in understanding CL.

There are also some closely related observations of CL and ThL in geological materials. The ThL of feldspars, for example, is much brighter than that of quartz. The ThL of authigenic quartz is duller than that of detrital quartz (Charlet 1971). Both of these observations parallel those of CL.

1.3 Causes of luminescence

When an energetic electron beam impinges on a solid surface, a number of physical processes take place (Fig. 1.2), including the generation of X-rays, secondary and backscattered electrons, and cathodoluminescence. Much of the incident beam energy also goes into heating the sample. The beam energy is a maximum at the point of entry and decreases with distance within the solid as energy is lost by various mechanisms. At a certain distance within the solid, the energy will be in the optimum range for each of the processes mentioned; X-rays will be most efficiently generated when the beam energy is still high, while the energies involved in the efficient production of cathodoluminescence are on the low end of the energy scale as portrayed in the figure, near the limit of the beam penetration. The extent to which any of the products is observed externally will depend also on the absorption which takes place as they exit from the solid.

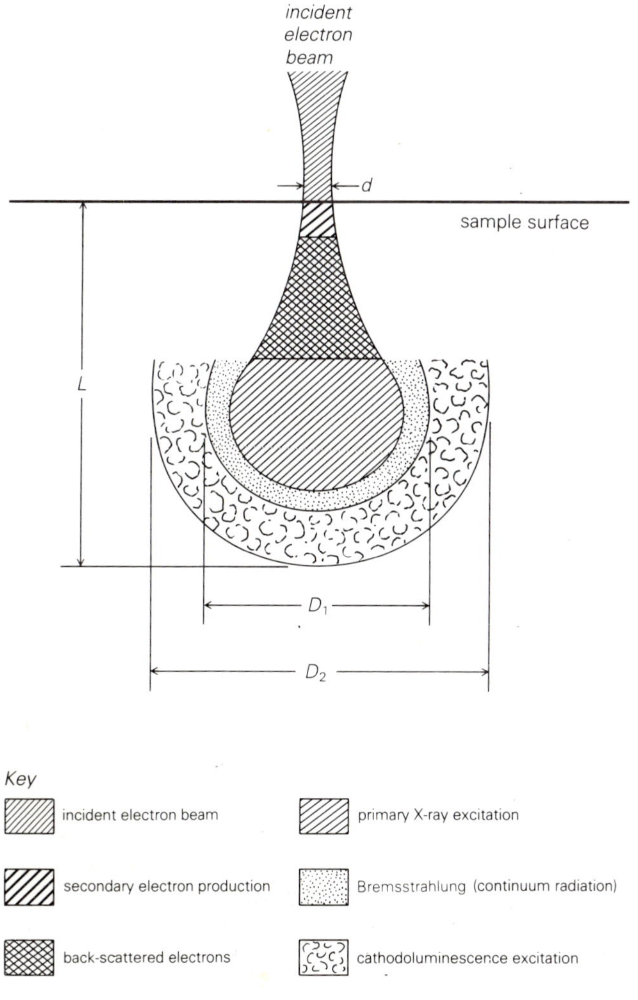

Figure 1.2 A qualitative picture of the interaction of an electron beam with a solid. The beam is focused to a diameter, *d*, at the surface. Elastic and inelastic interactions take place within the solid and include such processes as excitation, scattering, absorption, and diffusion. These lead to a steady reduction in the energy of the electrons with depth and an enlargement of the beam. Various secondary emissions are induced throughout the enlarged volume provided the electron energy remains high enough. The various radiations are emitted over a 4π solid angle but only those exiting from one of the sample surfaces are normally observed. The intensity of the emission observed from the surface depends not only on the volume over which it is formed but also on the subsequent interactions within the material. For charged particles the interactions are strong and only those produced very near the surface are emitted. Thus secondary electrons are produced throughout the volume, but only those formed very near the surface point of incidence will be able to escape from the solid. The diameter over which they are produced is only slightly larger than the beam diameter itself. Primary X-rays are produced over the entire region where the electron beam energies remain above the appropriate level, and some of these X-rays will exit, even from their deepest limit of formation. This enlarged diameter, D_1, over which secondary X-rays are observed is one of the limits to resolution in analyzers that make use of them. Cathodoluminescence is similarly generated over the entire region where the electrons retain greater than a few eV energy, and this region, of diameter D_2, is larger than that over which characteristic X-rays are generated and observed. CL emission is not completely absorbed within the solid and some of that generated even at the greatest depths will be emitted from the surface. For CL on a CMA or SEM, the conclusion to be drawn is that the generation of CL is not only at the surface but extends a (small) distance, *L*, inside – a distance often estimated to be in the 2–8 μm range for nonmetallic minerals. The enlarged diameter, D_2, over which CL is produced is sometimes a limitation to the resolving power available for CL observations on instruments like the SEM. The shape of the enlarged volume is only qualitatively portrayed and some authors will draw it as a series of concentric hemispheres, instead of the teardrop shape shown. It would be interesting to know more about the details of the shape but the significant point is that CL is produced over an enlarged volume and the shape and size of this volume will change with the electron energy. (Modified after Beaman and Isasi 1972.)

The details of the interaction of the beam with the solid are obviously complex and difficult to study experimentally. Few studies exist for geological materials. The penetration depth increases with the energy of the beam and may increase as the square of the energy. It is calculable from 'range-energy' equations for some materials. For beams in the 10–20 kV range, the depths are usually estimated to be 1–2 μm. See Ehrenberg and Franks (1953) and Leverenz (1968) for additional discussions on this subject.

1.3.1 *Energy levels in atoms, molecules, and solids*

The basic process of luminescence involves the excitation of an electron to a state of higher energy. External energy is supplied in the excitation process by photons, by electron impact, by X-rays, or by other forms of energy. The electron remains in the excited level only briefly and then returns to a state of lower energy, emitting a photon in the process. The electron may return directly to the ground state or may possibly return via one or more intermediate states (Fig. 1.3). The wavelength of the emitted photon corresponds to the energy level difference between the particular states involved, as specified by the Bohr frequency condition. To understand the details of CL, we must understand the details of the energy levels of the electrons in the mineral. This is a very complex subject and it will only be possible to outline it in this introductory text.

Throughout the discussion, the units used for energy include electron volts (eV) and the wavelength of the emitted light in ångstroms (Å) or nanometers (nm). The relationships between the various energy units are:

1 nanometer (nm) = 10 ångstroms (Å)

wavenumber (cm^{-1}) = 10^{-7}/(wavelength in nm)

energy in electron volts (eV) = 12,398/(wavelength in Å)

energy in electron volts (eV) = 1,239.8/(wavelength in nm)

Visible radiation is in the range of wavelengths between 400 and 700 nm, corresponding to energies between 1.77 and 3.10 electron volts. These energy and wavelength relationships are summarized in Figure 1.4.

First let us consider an isolated atom in free space. The state of lowest energy of the atom is called the ground state. In this state, as in all others, the electrons occupy a set of definite energy levels which can, in principle, be obtained from solutions of the equations of quantum mechanics. The solution leads to the electron configuration which is the number of electrons in each of the available energy levels.

As an example, consider the case of a neutral sodium atom. The configuration of the ground state, in the standard quantum mechanical notation, is $1s^2 2s^2 2p^6 3s$, i.e., there are two electrons in s orbitals in the first shell, two electrons in s orbitals in the second shell, six electrons in p orbitals in the second shell and a single electron in the third shell. The first and second shells are filled with their maximum allowable number of electrons. This configuration is sometimes written as (neon core) 3s, i.e., it is the neon core plus a single 3s electron.

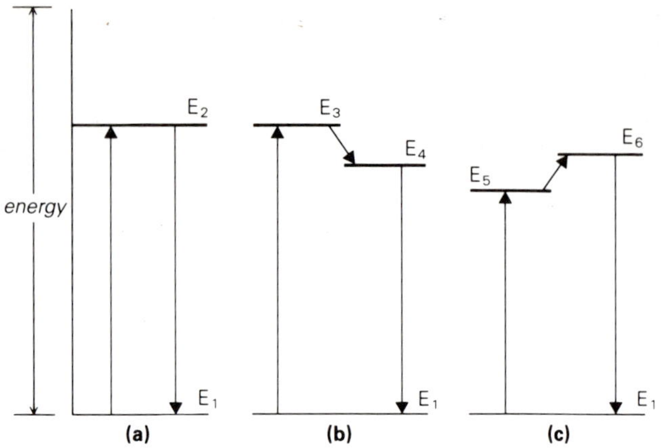

Figure 1.3 The basic process of luminescence. In (a) an electron associated with a neutral or ionic species is excited to a higher energy level, E_2, and subsequently returns from this level to the ground state, E_1, which is usually the zero energy reference. A photon of wavelength corresponding to the energy difference $E_2 - E_1$ is emitted. In (b) an electron is excited to the level E_3 from where it returns first to E_4 and then to the ground state. A photon of wavelength corresponding to $E_3 - E_4$ is emitted and also one corresponding to $E_4 - E_1$. In (c) we indicate a situation where there is a trap. The electron is excited to the energy level E_5, but the return to the ground state is a forbidden transition. However, if the electron receives a small amount of additional energy from an external source, it is excited to energy level E_6, and the transition from this state to the ground state is allowed. This is one of the bases for the phenomenon of thermoluminescence.

If this neutral atom is raised to a state of higher energy, the possible levels are derived, in principle, from further solutions of the equations of quantum mechanics. In practice, the possible higher energy levels are determined from a

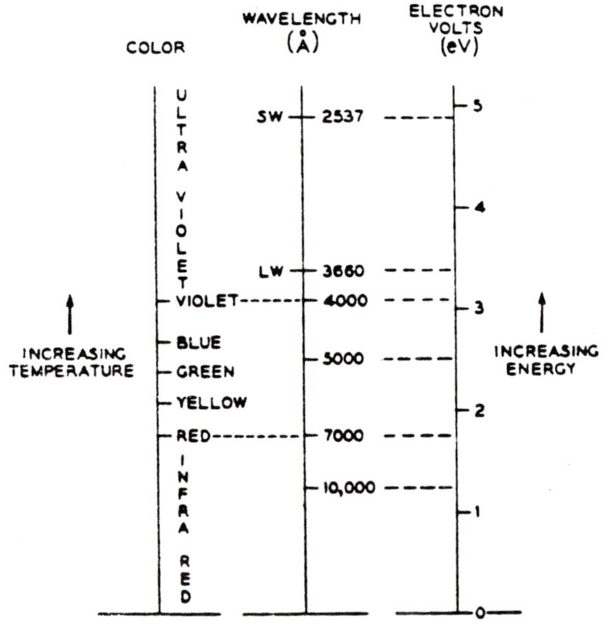

Figure 1.4 The relationship between the color of emitted light, and the wavelength and energy in eV of the photons, for spectrally pure light. The wavelengths of ultraviolet components of the ordinary UV light source at 2,537 and 3,660 Å (253.7 and 366 nm) are indicated. (Modified after Nassau 1978.)

Table 1.1 Energy level differences for the first few excited states of Na and Ca atoms and ions (values converted from Moore's (1949, 1952) tables of wavenumbers).

Levels	Wavelength of emitted photon (nm)*	Energy difference from ground state (eV)
Sodium atom: configuration = $1s^22s^22p^63s$ = (neon core) 3s		
3s	ground state	0
3p (doublet)	589.6	2.102
	589.0	2.104
4s	388	3.19
3d	343	3.62
.†		
.		
.		
ionization	241.3	5.14
Sodium ion: configuration = $1s^22s^22p^6$ = (neon core)		
$2p^6$	ground state	
3s (doublet)	37.7	32.84
.	37.6	32.93
.		
.		
Calcium atom: configuration = (argon core) $4s^2$		
$4s^2$	ground state	
4p (triplet)	660	1.88
	658	1.88
	654	1.90
3d (doublet)	491.8	2.52
.	491.4	2.52
.		
.		
ionization	202.8	6.11
Calcium ion: configuration = (argon core)		
$3p^6$	ground state	
3d	49.1	25.27
.		
.		

* This is the wavelength of the photon emitted when the excited atom or ion returns directly from the stated level to the ground state.

† The series of dots in the table indicate that there are many additional unlisted levels, e.g., between the 3d level and ionization for the sodium atom.

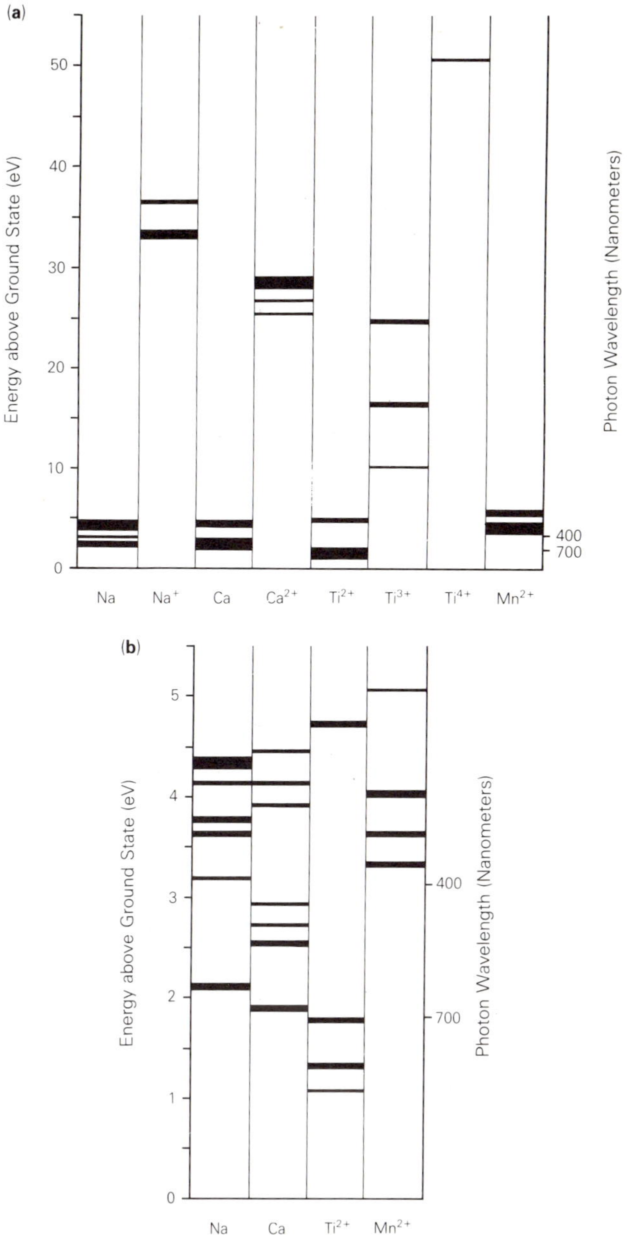

Figure 1.5 (a) A portion of the energy level diagram of several free atoms and ions. (b) The same with an expanded scale. The boundary lines corresponding to the visible range are indicated on the right. The neutral Na and Ca atoms and Ti^{2+} have levels from which an electron may return to the ground state and emit a photon in the visible range in the process. The Na^+, Ca^{2+} and Ti^{4+} ions, however, have their lowest energy level in the vacuum ultraviolet, well outside the visible range. Mn^{2+} has several levels which are near the visible range and a slight perturbation of these levels when the Mn^{2+} is incorporated into a solid may shift the levels sufficiently to produce visible emission.

combination of theory and experiment. An important point is that there is a discrete number of such levels, not a continuum.

For the neutral Na atom, the electron most likely to be excited is a 3s electron. The energy level scheme for the first few excited states of neutral sodium is shown in Table 1.1 and in Figure 1.5. If this atom is impacted, the excited electron can only assume one of the discrete levels shown. After excitation it may return directly to the ground state or may pass through an intermediate energy level first, depending on the level it is excited to and also on certain selection rules which apply to such transitions.

The excited state of lowest energy for Na is a 3p state and is a doublet; that is, it has two closely spaced lines separated by only 0.002 eV. Return to the ground state from the 3p levels results in the emission of yellow light at 589.6 and 589.0 nm, the familiar sodium D lines.

The terminology used to designate the various energy levels has been established in the field of spectroscopy and is an indication of the angular momentum of the state. The ground state of the sodium atom is described as ^2S. The general form of notation for the levels is ^rX, where r is the electron spin multiplicity and X is the spectroscopic designation for the states of differing total orbital angular momentum (S, P, D, G, . . .). See the text by Burns (1970) for further discussion of the designations.

When all of the excited levels that have been observed for sodium are listed, they extend to the limiting energy of

5.14 eV. At this high energy level, the electron is effectively completely removed from the Na atom. This corresponds to the first ionization potential of sodium. The first ionization potential is the energy required to remove an electron to the highest energy level of the neutral atom, i.e., to infinity.

In a similar fashion we can discuss the energy level diagram for the sodium ion. This can be constructed in exactly the same manner as for the neutral atom except that the ground state is now that of the singly charged ion which has two electrons in s orbitals and six electrons in p orbitals in the second shell, i.e., the second shell is filled. The ground state of the ion is designated as ^1S. For the neutral sodium atom the excited electron came from an unfilled s shell and excitation required a relatively small energy. For the sodium ion, the excited electron will come from a filled shell. The energy required to excite an electron in a filled shell is much greater than that required to excite an electron in an unfilled shell and the first available energy level to which one of the electrons in the Na^+ ion can be excited is at the much higher energy of 32.8 eV above the ground state of the ion (Table 1.1). This level is also a doublet. Emission resulting from excitation to this lowest energy level of the sodium ion is outside the visible range, in the vacuum ultraviolet.

Similar considerations would apply for any of the other alkali elements, i.e., visible emission may result from excitation of the neutral atom but emission resulting from excitation of the singly charged ion is at much higher energy, well outside the visible range.

In a similar fashion, we can review the energy levels for an alkaline earth element, calcium. The energy levels for the calcium atom and the normal valence state of Ca^{2+} are shown in Table 1.1 also. The ground state of the atom is ^1S. The emissions from the lowest level triplet state and doublet states of the excited Ca atom are in the visible region. For Ca^{2+} we note that the first available energy level is at the energy of 25.27 eV, corresponding to 49 nm wavelength of emission, and any emission involving transitions between this level and the ground state is outside the visible and in the vacuum ultraviolet.

ENERGY LEVELS OF MOLECULES

In considering the luminescence of any solid material, we must be concerned not just with the energy levels of the individual ions and their surrounding electron cloud but also with the energy levels of the entire lattice. To introduce some of the concepts which arise, let us first consider the case of a diatomic molecule made up of two identical atoms.

The simple picture of the diatomic molecule is that of two balls on a spring. The spacing between the two atoms in the molecule oscillates about an average separation. The amplitude of the oscillation increases as the temperature of the system is increased. For a given state of electron excitation, we can draw the envelope of the two extremes of the oscillation for different temperatures and the result is a curve of the form of curve 'E_1' in Figure 1.6. This could be referred to as the ground state of the diatomic molecule. Actually, the amplitude of oscillation does not increase continuously with temperature but must change by definite amounts because of the quantization of energy. Each of the discrete levels so produced is called a vibrational energy level.

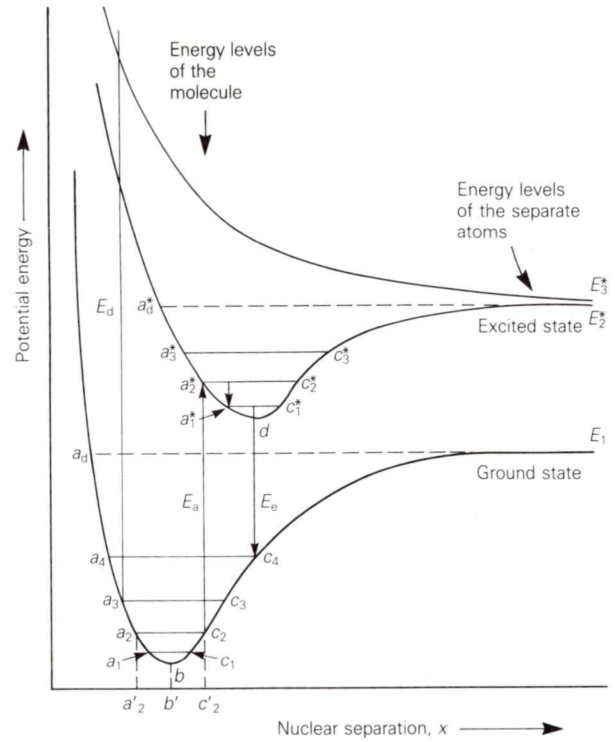

Figure 1.6 The energy diagram for a hypothetical free molecule made up of two atoms of the same species. The ground state can have several different levels of vibrational energy and the envelopes of the amplitude of oscillation of each of these defines the potential energy curve for the ground state, E_1. Thermal agitation, or small amounts of external energy input, can cause transfer between vibrational energy levels in the ground state, e.g., a_1–c_1 to a_2–c_2. Absorption of larger amounts of external excitation energy, e.g., E_a, from an energetic electron or photon, can excite the system to the vibrational level a_2^*–c_2^* on energy state, E_2^*. From this level, by small energy losses it can change to a_1^*–c_1^*. Finally, the excited system can return to the ground state at vibrational level a_4–c_4, emitting a photon (luminescence) of energy, E_e, in the process. The energy loss in the small energy changes, e.g., from a_2^*–c_2^* to a_1^*–c_1^*, can be emitted as a low-energy (infra-red) photon or, in actual solids, be exchanged with the lattice in a temperature equilibration process. Because of the many closely spaced vibrational energy levels, the line spectra of an isolated atom or ion are replaced by band spectra, i.e., there are closely spaced lines, corresponding to the various vibrational levels. Excessive excitation of the hypothetical molecule, e.g., by input of an amount of energy, E_d, leads to the state defined by the curve, E_3^*, in which the molecule dissociates in a single vibration. (Modified after Leverenz 1968, p. 30.)

When the diatomic molecule is excited by one of the means already referred to, the state of electron excitation can change to a higher level. This level also has a multiplicity of vibrational energy levels and it too can be represented by another curve of the same form as curve 'E_1', e.g., by curve 'E_2' of Figure 1.6.

In the simple 'balls on a spring' picture of the diatomic molecule, the total energy of the system is constant but the relative amounts of potential and kinetic energy change. The kinetic energy is a minimum at the ends of the path where the relative velocity of the balls is slowing to zero and reversing direction. The kinetic energy is a maximum at the 'center' of the path, i.e., at the average spacing. The system spends the majority of time in the state where its velocity is a minimum, that is at one extreme or another of the interatomic spacing. Hence electron excitation after external energy input is most likely to take place from one of the end-points. The time required for the electron excitation is short in comparison with the period of the interatomic oscillation and hence excitation is represented by a very nearly vertical line on Figure 1.6. If such a vertical line from one or the other of the end-points on curve 'E_1' intercepts one of the higher electron energy states such as curve 'E_2', then the transition is a possible one.

ENERGY LEVELS IN A SOLID: CONFIGURATIONAL COORDINATES

The concept of the diatomic molecule is most useful in the study of, for example, gas molecules, N_2 and O_2. But in a mineral solid the situation is more complex and, instead of a diatomic molecule, each atom or ion is surrounded by several 'near neighbors,' the number and identity of which differ for each composition and depend upon the symmetry. Therefore, we must think about a given atom or ion undergoing a fluctuation in its position with respect to all neighbors. The concept of a configurational coordinate has been introduced to furnish a qualitative description of the situation. The configurational coordinate may be thought of, roughly, as an average of the interatomic spacings for a large number of the particular species of ion. With the use of the configurational coordinate concept, the same form of energy level diagram used to discuss a diatomic molecule can be used to discuss the energy levels of a particular species of atom or ion in a solid (Klick & Schulman 1957, Vlam 1954).

One of the additional features which is introduced in the case of a solid is that within a given electron energy state there can be changes from one vibrational level to another by the exchange of energy with the remainder of the material it is associated with. The energy released in the change from one vibrational level to another is not emitted as visible radiation, and this type of transition is called a radiationless transition. In some materials these transitions predominate over those which produce luminescence.

Let us return now to the discussion of the sodium atom. When Na is incorporated into a mineral such as halite, then we must be concerned with the energy levels of the ion – not the neutral atom. As we have seen, all of the levels for the ion are in the vacuum ultraviolet or even higher energies, and simple excitation of the Na^+ electrons in halite cannot produce visible CL. However, in halite, as in any compound, it is not necessarily only the energy levels of the free ion that are important, it is the energy levels of the NaCl crystal as expressed by the configurational coordinate diagram for this situation. One can speak of the influence of the remainder of the Cl^- and Na^+ ions in the crystal on the energy levels of the free Na^+ ion and vice versa. In some cases, the interaction between the ions in the lattice is quite important. In other cases, the crystal environment has only a slight effect on the energy levels of the free ion. For the case of Na^+ in halite, the effects of the remainder of the ions in the crystal do not alter the picture, as far as visible CL is concerned, and a very pure halite crystal shows little or no CL (Murata & Smith 1946, Teegarden 1966). In particular, halite does not show CL of the yellow color that we would normally ascribe to atomic Na emission.

Similarly, we can conclude that the CL of calcite cannot be simply associated with the excited states of the Ca^{2+} ions, but must involve the effective electron energy levels in the entire lattice.

For many minerals, the situation is the same as for halite and calcite, i.e., the CL of the highly pure mineral is very weak. This raises the question of whether an ideal, absolutely pure crystal of any substance will exhibit CL. Intuitively, it would seem that there are many cases (many substances) where at least some of the available energy levels in the ideal crystal would be in the range to produce visible light emission. However, many very pure samples do not exhibit CL. For example, authigenic feldspar shows either very weak or no CL, whereas feldspars formed at higher temperatures almost always show CL. It is presumed that the authigenic feldspar is much purer than the higher-temperature variety (Sec. 5.3) and that its lattice is more highly ordered. Similarly, apatite is an excellent host for a variety of impurities that cause it to be luminescent; but very pure apatite, formed under oxidizing conditions which disallow the introduction of Mn^{2+} and other impurities, is non-luminescent (Ch. 8). And there are other examples which will be referred to in due course.

The question raised in the previous paragraph is, of course, moot because the ideal situation of a pure mineral with a perfect lattice does not exist. All minerals, synthetic or natural, have defects and depart more or less from the ideal. There are different types of defects and different possible impurities and each has its own effect on the creation of CL conditions. Broadly speaking, the resulting luminescence can be placed into one of two categories, intrinsic or impurity-activated.

1.3.2 Intrinsic luminescence

The type of CL which is found in an apparently very pure phosphor or natural or synthetic mineral is usually called intrinsic. A definition of 'intrinsic' is 'belonging to the real nature of a thing; not dependent on external circumstances; essential; inherent.' Thus the designation of luminescence as 'intrinsic' implies that it is always present in a given mineral species, though, as we shall see, this is rarely the case for minerals.

Many authors (e.g., Randall 1939, Hall & Ribbe 1971, Sippel 1971) have commented on the intrinsic CL of silicates. They have stated that it is common to nearly all silicates and appears in the blue region as a broad band. Indeed, one often sees broad-band blue CL in the feldspars and in quartz, as well as in other minerals. However, there are many examples revealing that the CL of authigenic quartz and authigenic feldspar are both very weak (see Chs. 5 & 6). These authigenic minerals are formed under conditions of low temperature and slow crystallization, conditions which favor the production of a more perfect lattice with low impurity levels. This evidence argues against the terminology of 'intrinsic' CL because, to be intrinsic, it must always occur. The conclusion which might be drawn in these cases is that the ideal, pure crystal does not luminesce. However, when there is crystal imperfection, due to any of several causes, the distorted lattice has energy levels which will produce emission in the blue region of the spectrum.

Intrinsic luminescence can be enhanced by nonstoichiometry, by lattice damage, and by other means of increasing defects which are shown in the following outline:

(a) nonstoichiometry
(b) structural imperfections

(i) poorly ordered
(ii) radiation damage
(iii) shock damage

(c) impurities (nonactivators), substitutional or interstitial, which distort the lattice

NONSTOICHIOMETRY

An absolutely pure mineral which does not have exactly the correct elemental proportions is called nonstoichiometric. The absolute purity condition is not realizable but it is convenient to isolate the condition of nonstoichiometry from that of impurities for the present discussion. Consider the case of willemite, Zn_2SiO_4 which Bhalla and White (1971) have investigated (Ch. 4). They found 'weak pale violet-blue' luminescence in very pure samples of willemite. By heating the samples, which possibly allowed zinc to evaporate preferentially, they created a condition of nonstoichiometry and found that the CL was enhanced. Clearly, since no additions were made to the system, there had to be some compensatory change in the ionization state of the Si or O to maintain the charge neutrality. Furthermore, the structure (lattice) is modified when the zinc is removed, and adjustments are required to compensate for this also. The exact details of what actually happens are difficult to obtain. Ions on lattice sites may move to interstitial positions and become neutral atoms, etc. To the external observer, one of the obvious results is an enhanced CL. The CL spectrum is a broad band in the blue region of the spectrum. The broad-band nature of the CL spectrum probably reflects the fact that there are several different types of compensation and each has a slightly different effect on the energy levels – hence the broad-band distribution of energy levels in this case.

So, for the case of willemite, intrinsic CL can apparently be associated with nonstoichiometry and the purer the willemite, the weaker the CL.

STRUCTURAL IMPERFECTIONS

Defect structures

There is evidence for many minerals that intrinsic luminescence is associated with defects, and crystals which form at higher temperatures and have more defects are more brightly luminescent than crystals of the same mineral, formed slowly at low temperatures with lower defect concentrations. The examples already mentioned of authigenic quartz and feldspar are illustrations of this.

Radiation and shock damage

As noted in previous paragraphs, defects can influence the CL of minerals. In addition to the defects associated with nonstoichiometry, or with rapid crystallization at higher temperatures, at least two other forms of damage have been recognized as influencing CL. The first of these is radiation damage. Alpha particles bombarding quartz produce a region of damage and the CL is changed from that of the remainder of the quartz grain (Smith & Stenstrom 1965, Morton 1978). An example is discussed in Section 6.2 and shown in Plates 5a and b. That this is damage is supported by the observation that the CL characteristics can be restored to those of the original grain by annealing. A similar changed CL was seen in radiation haloes in diamonds (Mendelssohn *et al.* 1979) and has also been reported in kyanite (Chinner *et al.* 1969).

The other form of damage which has been shown to influence CL is shock damage. Sippel and Spencer (1970b) observed that the CL of shock-damaged lunar feldspars was shifted in wavelength and the CL emission peak was broader than in the undamaged counterpart. They also found that the polarization of the CL, usually easily noticeable in feldspars, disappeared. These observations are discussed in more detail in Section 5.4.

These two pieces of evidence strongly support the argument that structurally damaged areas in minerals can have CL characteristics drastically different from those of the undamaged areas.

Impurities and intrinsic luminescence

There is evidence that some types of impurities can enhance intrinsic CL. The most important evidence in this direction is for the case of titanium impurity in the feldspars and quartz. When titanium is added as a dopant to feldspars (Sec. 5.1), there is an enhancement in the blue region of the spectrum and the emission is broad-band. Many authors believe that the titanium is not serving as the source of the CL, i.e., as an activator (see Sec. 1.3.3), but is causing structural defects which lead to an enhancement of the intrinsic CL of the feldspars.

Similarly, there is an enhancement of the ThL of quartz with the addition of Ti (Medlin 1963a) which may have a similar explanation. The added Ti was the Ti^{4+} ion in solution but, as Medlin points out, it is possible that the valency changed during the crystallization. What the effective charge state of the Ti is, whether it is introduced in lattice sites or interstitially, what other impurities, if any, are introduced simultaneously to maintain neutrality – these are all questions which remain unanswered. The description of the CL of minerals with titanium impurities has involved both the use of the term 'activator' and the term 'intrinsic.' More work is needed to resolve the issue.

The term 'intrinsic' is firmly established in the literature of luminescence and we will continue to use it, with the understanding that it implies elemental purity but not structural perfection or stoichiometry. The terminology of self-activated is also encountered frequently and refers to intrinsic luminescence, luminescence not caused by impurities serving as activators.

1.3.3 Impurities

The most important type of defect in a practical sense is the presence of impurities. Many different impurities in minerals can serve as centers for the production of CL, and for many of these the CL can be directly associated with the energy levels of the impurity ion in the crystal environment. Impurities which function in this manner are called 'activators.' Activator impurities are usually believed to be substitutional, but occasionally CL effects have been attributed to interstitial impurities. There is also the possibility, very common in geological samples, of more than one activator impurity being present.

ACTIVATORS

The terminology used in the phosphor industry to identify the impurity content is to write either the weight or the mole percent and the chemical symbol for the impurity after the formula for the compound (Leverenz 1968). Thus synthetic calcite with a 1 percent Mn impurity would be written as $CaCO_3$: (0.01Mn) with some accompanying clarification as to whether it is mole or weight percent.

To discuss the rôle of impurities as activators we will return to the discussion of energy levels for ions in a solid. The energy level diagrams for sodium and calcium, presented earlier, are useful when discussing the principles, but these elements are not known to be important as activators for CL in minerals. So now let us consider the case of manganese, an element which is very important as an activator of CL in many of the common minerals.

The doubly charged state of Mn is apparently the ionization state most important in the CL of minerals. This ion substitutes for, for example, Ca^{2+} and Mg^{2+} in many carbonates as well as other minerals. The configuration of the neutral Mn atom is

$$(\text{Ar core})\ 3d^5 4s^2$$

The doubly charged ion is formed by the loss of the two 4s electrons. Unlike Na^+ and Ca^{2+} which have filled shells, Mn^{2+} has five electrons in a half-filled D shell. The electrons in this unfilled shell are easier to excite (lower energies required) than those in a filled shell, but the resultant spectra can be quite complex. A portion of the energy level diagram for Mn^{2+} is shown in Figure 1.5. The ground state is ^6S and the first few excited states are ^4G, ^4P, ^4D, and ^4F. These are for the isolated, doubly charged ion in free space. What happens when this ion is actually incorporated into the structure of a mineral?

When the manganous ion is incorporated into a crystal structure, the available electron energy levels are still obtained from a solution of the equations of quantum mechanics. But now the formulation of the problem must also take into account the effects of the other ions in the crystal surrounding the manganese ion, i.e., it must take into account the effects of the crystal field. For the case of many elements including manganese, the other transition elements, and the rare earth elements, an approach which is available is the use of crystal field theory (Orgel 1966, White & Keester 1966, Burns 1970). The influence of the crystal field is to change the magnitude of the energy levels and also to cause additional splitting of closely spaced levels. By application of the principles of crystal field theory, one can derive information on the magnitude of the energy levels and their splitting. The influence of the crystal field is related to the field strength, which is a measure of the charge and spacing of the other cations and anions in the structure, nearest neighbors being the most important. It is also related to the symmetry of the site or the coordination number for the ion.

If the crystal structure and crystal field were known exactly, and the complex calculations could be done precisely and completely, the result would be an energy level diagram of the same form as shown in Figures 1.5a and 1.6. Since this is not the case, the approach is to make the approximate calculations as a function of Dq, a parameter which defines the strength of the crystal field. Dq is a function of the spacing, charge, and nature of the anions surrounding the activator ion (Orgel 1955a, 1955b).

In order to proceed with the general calculation, one must know the environment of the manganese ion. Let us consider the case when Mn^{2+} is present as a trace impurity in calcite, substitutional on a site normally occupied by Ca^{2+}. Then one has a clear definition of the ions surrounding the Mn^{2+} ion and the symmetry is trigonal. However, according to Medlin (1963a, 1968), the more easily calculated cubic symmetry approximation is quite good. The energy level diagram for cubic symmetry assumes the form shown in Figure 1.7b. For this symmetry, crystal field theory predicts that the four closely spaced ^4G levels are split into states, designated as $^4T_{1g}$, $^4T_{2g}$, ^4E_g, and $^4A_{1g}$. (These various symbols are group theory designations, the terminology for which is summarized, for example, in the 1970 text by Burns.) The significant point is that the closely spaced multiplet states of the free gaseous ion are separated into states with appreciable energy differences because of the influence of the crystal field. The energy levels and the energy level differences vary with Dq and in some cases the relative positions of the energy levels actually change, e.g., the ^4P level crosses the curve for the ^4D level at large values of Dq. If the value of the parameter Dq for Mn^{2+} in calcite is known, it is possible to draw a vertical line on the diagram and define the energy level arrangement and energy level differences. If the crystal field parameter is not known, another approach is necessary. One such approach is described in detail by Medlin (1963a, 1968) who did extensive investigation of Mn^{2+} activated ThL in calcite.

Medlin (1963a, 1968) made use of the fact that the energy level diagram must also describe the situation for light absorption which also involves excitation of an electron. Not all of the energy levels need necessarily be good absorbers but whatever wavelengths of light are absorbed must be represented by one of the levels in the diagram. The absorption spectrum for Mn in calcite is weak and difficult to measure so, instead, he measured the absorption spectrum for pure $MnCO_3$ which, he says, is closely related to that for Mn in $CaCO_3$. The best fit to the absorption spectrum was used to estimate the value of Dq and from this he concluded that the CL emission arises in a transition from the $^4G(T_{1g})$ state to the ^6S ground state.

An extension of this analysis to other carbonate minerals with different structures and different Dq values leads to

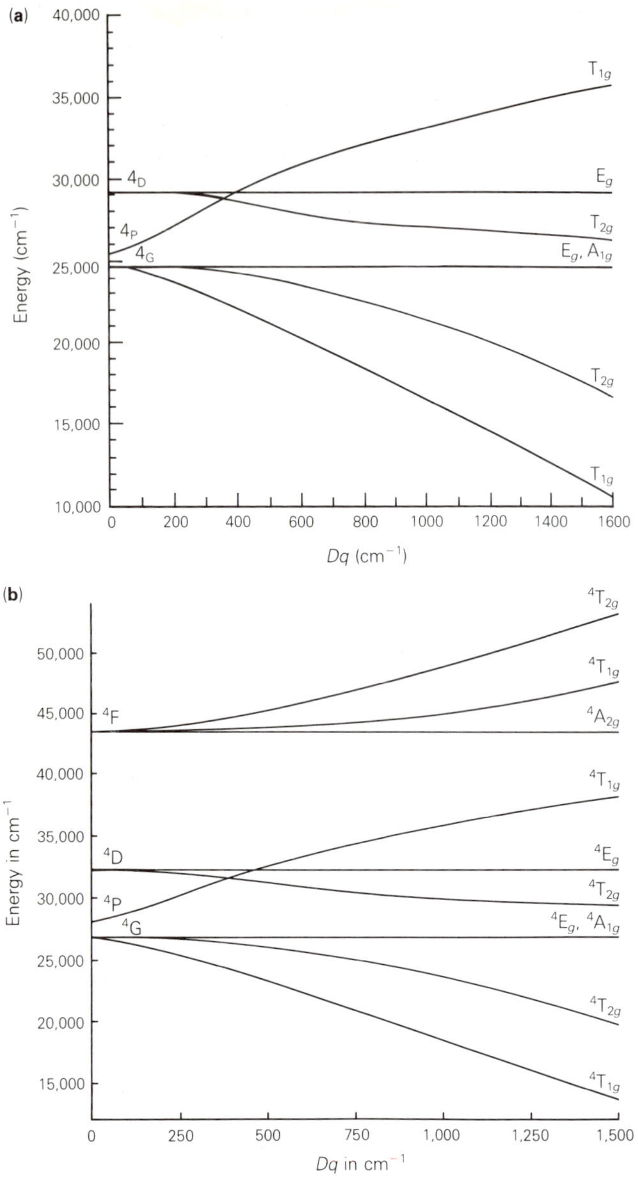

Figure 1.7 (a) An example of the portrayal of energy levels of an isolated ion, in this case the Mn^{2+} ion. The energy above the ^6S ground state can be given in terms of wavelength of the photon emitted when an electron returns from a particular level to the ground state. It can also be given in terms of the energy in electron volts required to excite an electron from one level to another. S, P, D, G, F are the spectroscopic designations used for different levels. (Modified after Medlin 1968.) (b) When the ion is incorporated into a solid, the energy level scheme is changed. This figure shows the energy level diagram when the Mn^{2+} ion is incorporated into a location in which the crystal field has cubic symmetry. The effect is to split the levels of (a) into sublevels. Then the energy level spacings depend upon the parameter *Dq*, which is a measure of the crystal field strength for the particular crystal environment. (Modified after Orgel 1955b, 1966. See also Medlin 1968 and Burns 1970, Chs. 2 and 3.)

prediction of the change of CL wavelength as one goes from calcite to magnesite to dolomite, etc. (Sommer 1972a). This variation in wavelength is a characteristic of ions such as Mn^{2+}, with unfilled outer shells. The energy levels are influenced by the crystal field and the CL emission spectrum is not a characteristic property of the ion – it is a property of the material.

It is important to note that a diagram of the form of Figure 1.7b can be drawn for a particular Mn^{2+} ion with the simplifying assumptions as discussed. In a real calcite crystal, one has many Mn ions and each of these can be represented by a similar, but not identical, diagram, because of slight differences in crystal environment. These slight differences arise from such things as imperfection in the host calcite, statistical variations in the distribution of the Mn impurity itself, possible random distribution of other impurities, etc. These effects will lead to slightly different *Dq* values for each ion. The net result is that there will be a distribution of *Dq* values and a distribution of the wavelength of the CL emission, leading to the typical broad-band CL spectrum for calcite which is shown in Figure 7.1. This, then, is another characteristic of an ion which is appreciably influenced by its environment – the emission peak will be broad.

When the same approach is used for an element such as Eu which has the configuration

$$\text{(krypton core)}4d^{10}4f^{6}5s^{2}5p^{6}5d^{1}6s^{2}$$

then a slightly different set of conclusions is drawn. Eu can substitute in various minerals in either the doubly or triply charged states, and in some cases both valences may be present in the same mineral. The electron configurations of these two states are

$$Eu^{2+} \quad \text{(krypton core) } 4d^{10}4f^{6}5s^{2}5p^{6}5d^{1}$$
$$Eu^{3+} \quad \text{(krypton core) } 4d^{10}4f^{6}5s^{2}5p^{6}$$

In the case of the doubly charged ion, the electron most probably excited is the only electron in the 5d shell. The energy levels of this unshielded electron are influenced by the environment. Therefore, for Eu^{2+} the peak position is variable and the peak is broad. For the triply charged ion, the excited electrons are in the unfilled 4f shell. These electrons are well shielded by the filled 5s and 5p shells and are not appreciably influenced by the environment. Therefore, for Eu^{3+} the emission is a characteristic of the ion, not of the material. The peak-maximum position is nearly the same for many minerals and the peaks are narrow. The CL emission spectra for most of the other trivalent REE are narrow bands also and the CL emission peak locations in various matrices are even more constant than that of Eu^{3+} (Larach 1968).

Many other elements besides Mn and Eu can serve as activators in the appropriate host mineral. Others frequently encountered include Cr^{3+}, Fe^{3+}, the other rare earths, Pb^{2+}, and others. Ti^{4+} has also been shown to be effective in enhancing the CL of some minerals but it is not certain whether it serves as an activator on a substitutional site, or possibly in some other manner or in some other valence state. (Leverenz (1968, p. 239) uses the terminology of 'intensifier activator' for an element which intensifies existing emission lines, such as Ti apparently does. He uses the term 'originative activator' for an element which produces new emission lines.) Which of these elements can be incorporated into a given mineral is governed by the familiar geochemical rules, based on ionic charge, size, coordination number, etc. Which of these elements, if incorporated, can be effective as an activator depends upon the complex considerations of the solid-state physics involved which we have attempted to outline. A study of the electron configuration will furnish important qualitative clues to the functioning of these as activators in the possible valence states.

Natural minerals often contain more than one activator. The locations of the CL emission spectrum peaks due to one impurity are changed slightly, if at all, by the presence of another impurity, at least at low concentrations. The spectra are superimposed. However, the relative intensities of the various emission peaks may change because the excitation energy is coupled to the different activators with widely different efficiencies. Examples of superposition of REE spectra are shown in Figure 4.5b (strontianite). Figure 8.7 (apatite) contains an example of superposition of REE and Mn^{2+} spectra, and Figure 4.10 (zircon) is an example of superposition of REE and intrinsic spectra.

The construction of energy level diagrams is a very involved process and knowledge is far from complete for most minerals. The information required to construct them is derived from a combination of theory and experiments. In addition to the light absorption experiments mentioned, CL, PhL, and other luminescence observations are an important part of the experimental input.

Another source of information is the technique of luminescence excitation spectroscopy (LES) (Medlin 1963a, Telfer & Walker 1975, Marfunin 1979, Walker 1983). In this technique a monochromatic source of light of variable wavelength is used to excite the sample, and the intensity of the photoluminescence emission at a particular wavelength is monitored as the wavelength of the excitation source is varied. The luminescence emission wavelength being monitored can be selected by a suitable filter or by a monochromator. The result is a spectrum in which the wavelengths of the peaks correspond to some of the energy levels which are present and from this many of the details of the energy level diagram can be constructed. LES is particularly useful for several reasons, as pointed out by Walker (1983). For example, sample preparation is minimal. Also, it can be used even when the activator impurity being studied is present in small amounts, because it is a direct measurement, not a difference measurement, as is the case for absorption measurements.

Some of the cases where the solid-state physics of mineral luminescence have been discussed in the literature include willemite (Bhalla & White 1970), calcite (Medlin 1959, 1968), and the feldspars (Geake *et al.* 1973, Telfer & Walker 1975, 1978). Leverenz's text (1968) also contains a discussion of the subject. Marfunin (1979) also has an extensive discussion of energy level schemes. His text is more of a discussion of the physics of CL of an isolated mineral and does not include color plates. It is a good introduction to the extensive literature in Russian on CL. The article by Walker (1985) is a thorough discussion of the physics of luminescence with many specific applications to minerals.

Besides crystal field theory there are other approaches to the study of energy levels in a solid, including molecular orbital and band theory formalism. Whereas crystal field theory associates the excited electron with a particular ion, molecular orbital formalism is more appropriate when bonding is predominantly covalent and electrons are in more widespread orbits or are associated with groups of atoms. This approach has been used to describe intrinsic luminescence in scheelite, due to charge transfer within the WO_4^{2-} group (Nassau 1978). In the band theory approach, the electrons are associated with the whole mineral. This approach is especially applicable to conductors and semiconductors. Sphalerite is the best example of a mineral whose luminescence spectrum is interpreted using band theory (Marfunin 1979). For a comparison of these formalisms with crystal field theory, the paper by Nassau (1978) is recommended. It will lead the interested reader to many additional references.

Though there are many instances where a discussion of emission wavelength as a function of site coordination, etc., can be found, a great deal of this discussion still remains qualitative and not yet at a level of understanding where it can be used in a predictive manner. This will not always be the case and perhaps the glimpse of applications presented by this text will help to foster additional interest in this field.

The structure of glasses is not so well defined as for the common minerals but glasses can exhibit CL and the influence of activators and quenchers can be noted. Rindone (1966) presented a comprehensive discussion of the subject and showed examples of many different impurities which cause luminescence in glasses.

COACTIVATORS OR SENSITIZERS

There are a few important examples of materials which show luminescence (and this is particularly true of photoluminescence) only when two different impurities are simultaneously present. The terminology of 'coactivators' or 'sensitizers' is used to describe this situation.

The description of the action of a coactivator for the case of photoluminescence is presented in Robbins (1983) for the case of Pb and Mn in minerals like calcite. Incident UV is very inefficiently coupled to the Mn ion but very efficiently coupled to the Pb ion. By a process called resonant transfer the energy acquired by the Pb ion is transferred (possibly over many atomic spacings) to the Mn ion which then, finally, releases it in the form of the familiar reddish-orange-yellow luminescence of calcite. All this takes place, of course, in less than 10^{-8} seconds for fluorescence. Thus calcite must contain both Mn and Pb (or other coactivator) to show photoluminescence. However, when the energy input is an electron beam, then calcite need contain only Mn to show CL. The energy from the electron beam can be coupled efficiently to the Mn without the need for an intermediate step. The author is unaware of any documentation of coactivators for CL but this is not to say that it may not sometimes play a rôle.

QUENCHERS

A few elements can perform a rôle opposite to that of the activators; that is, they modify the energy level arrangement so that the luminescence process does not operate or is very inefficient. These are called quenchers. The most important quenching ion in minerals is Fe^{2+}. Carbonates, for example, are activated by Mn^{2+} but if a sufficient amount of Fe^{2+} is also present, there is no net CL. Apparently the presence of the quencher causes new closely spaced energy levels to be set up and the electron can easily return to the ground state with the emission of a succession of low-energy photons (IR) or by losing energy to the lattice as heat.

Quenchers are sometimes also referred to as 'poisons', 'killers', or 'inhibitors'.

GENERAL COMMENTS ON ACTIVATORS AND QUENCHERS

In order to be effective the activator (or quencher) must be incorporated in the solid as a substitutional or interstitial impurity, not in a coexisting mineral phase. Thus we may have a natural carbonate sample exhibiting ferruginous staining on the surface and in cracks but unless the Fe is also incorporated in the lattice, its presence has no bearing on the luminescence properties of the carbonate. (However, some of the CL emission from the carbonate can be absorbed if the viewing path passes through the iron oxide phase.)

The actual amounts of activator and quencher required will vary, depending upon the mineral, but they are quite low in most samples of interest. Typically, one is dealing with activator impurity levels of less than 1,000 ppm and there are some examples where only a few ppm will suffice, if the concentration of quencher is low.

One of the difficulties in establishing activator and quencher identities, or demonstrating intrinsic CL, is the difficulty of analysis. The electron probe, a natural choice for low-level analysis of impurities in small sample areas, has a detection limit for many elements at levels of 100–1,000 ppm and this is the *high* end of the concentration range of interest for CL. CL emission spectra can be detected in some cases from impurities present at the 50 ppb level, i.e., Pr in pure La_2O_3 – an ideal host material (Kniseley *et al.* 1969). Many of these analytical problems remain to be solved, possibly with the ion probe or laser probe type of mass spectrometer (Hutcheon *et al.* 1978, Conzemius & Svec 1978, Shankai *et al.* 1984) or with the techniques of proton induced X-ray emission (PIXE) (Blank *et al.* 1984). These instruments have detection limits in the ppm range or lower and sampled areas may be as small as a few microns in diameter.

Determination of the amounts of impurities is difficult but the location of the impurity in the crystal structure, and its charge state, are sometimes even more difficult to establish. Techniques such as electron paramagnetic resonance (Dieleman 1966, Marfunin 1979, Wildeman 1969/70) and the variation of X-ray emission with the direction of an exciting electron beam (Taftø & Spence 1982) are available. Laser interference microscopy (Pearce 1984, Pearce & Thomas 1984) may play a rôle. Luminescence excitation spectra (Telfer & Walker 1975, Walker 1983) have also been used to predict site locations of impurities.

1.3.4 CL intensity versus activator concentration

The CL intensity generally increases with the concentration of the activator to a maximum and then decreases. The decrease of CL intensity at higher concentrations of activator is sometimes referred to as self-quenching or concentration quenching. Picture the situation where transitions in the neighborhood of one activator ion result in the emission of CL. The emitted photons may excite another activator ion in the vicinity, rather than exit from the crystal. The process of CL is taking place, but much of it is internal and not visible to the external observer. It is of interest to estimate the concentration conditions under which this might be expected to happen, i.e., the conditions under which we expect these activator ions to behave independently rather than interacting with each other. Such a calculation is given by Leverenz (1968) and it is summarized here for the case of calcite.

Consider the case of calcite. The density is 2.7 g/cm^3 and the molecular weight is 100 g. From this we can calculate (Leverenz 1968, p. 479) that each molecule occupies an average volume of 6.15×10^{-23} cm^3.

If, as a first approximation, we consider the volume of each $CaCO_3$ molecule to be a cube, then the dimensions of the cube are 3.94 Å on each side. Each Ca^{2+} ion would, on average, be spaced about 3.94 Å from the next nearest Ca^{2+} in calcite, i.e., it would be at the center of a cube of volume $D_{Ca}{}^3$ where D_{Ca} is 3.94 Å.

Now consider the situation for an Mn^{2+} impurity, substituting for the Ca^{2+}. If the Mn^{2+} concentration is 1 ppma, then the average volume of the cube surrounding each Mn^{2+} ion is a million times larger than the average volume occupied by each Ca^{2+} ion. The cube edge dimension, D_{Mn}, is obtained as

$$D_{Mn}{}^3 = 6.15 \times 10^{-23} \times 10^6 \text{ cm}^3$$
$$D_{Mn} = 394 \text{ Å}$$

and the average spacing between the Mn activator ions is 394 Å. The average spacing decreases with increasing Mn concentration as follows:

Mn concentration	Average spacing between Mn ions (Å)
1 ppm	394
1,000 ppm (0.1%)	39.4
1%	18
10%	8.5

The maximum CL intensity, as a function of the concentration of the activator ion, often occurs at about the 0.1–1 percent concentration level (Leverenz 1968, p. 330) where the activator impurity separation is a few '10s' of ångstroms. Telfer and Walker (1978), for example, reported that the CL intensity in doped feldspars occurs at a dopant concentration of about 1.5 percent for Fe or Mn (see Fig. 5.6 in Ch. 5). Medlin (1961b) observed a maximum (ThL) at about 0.2 percent for Mn in dolomite and at about 0.3 mole percent for Mn in calcite (Medlin 1960).

1.3.5 Other forms of excitation

Several authors have commented on the essential independence of luminescence spectra on the method of excitation (Marfunin 1979, pp. 148, 190–1, Williams 1966, pp. 35, 41). A few authors have provided experimental confirmation of this (Medlin 1963a, Derham *et al.* 1964).

For a given mineral sample, the energy levels available for excitation are an inherent property of the sample and do not depend on the excitation process. However, whether a given level is excited may depend on the excitation process, e.g.

whether the excitation energy is high enough. A given transition may be allowable under one type of excitation and forbidden under another. All of the peaks in the CL spectrum may be produced by some form of excitation whereas only a portion of them may be produced by another form. Electron beams, in particular, have a large amount of energy, excess to that required for actual excitation and they tend to excite all of the available levels. Ultraviolet light sources are more selective in the levels that they can excite and the process of sensitization or coactivation (Section 1.3.3) is more important.

Different excitation methods can have widely differing incident 'beam' intensities (Section 3.4) and different efficiencies for excitation. One of the practical consequences of this will be widely different luminescence intensities, even if the spectra are qualitatively similar.

Finally, it should be pointed out that some excitation processes, especially ion beams and intense electron beams, can actually cause disruption within the mineral structure by such processes as heating, sputtering, or ion implantation. These disruptions can lead to modified spectra and sometimes to rapid decreases in intensity with time of bombardment.

Even though the underlying physics is the same for the various excitation methods, there are many practical differences such as sample preparation requirements, depth of excitation, etc., which are quite important. These are discussed in Section 3.4 and summarized in Table 3.1.

Before continuing with the discussion of CL, a few comments on other techniques using light excitation may be useful. In one technique (Yanguas & Dravis 1985) a fluorescent dye is mixed with the thin section epoxy cement and the sample is excited by blue light. This technique enhances macro- and micro-porosity. The luminescence observed is from the dye material. Winkler (1983) has also described a technique in which an organic dye is applied to the sample. In another technique (Dravis & Yurewicz 1985), the sample is excited by UV or blue light and fluorescence emission from the sample minerals at longer wavelengths is observed. The causes of the fluorescence in the latter case are not certain – these authors suggest trace element activators or perhaps residual preserved organic matter. The latter technique has proved useful in carbonates for grain and texture recognition in dolomites, diagenetic fabrics, cement zonation, etc. The further developments in the use of the latter technique, and understanding of the causes, may also lead to a better understanding of CL in carbonates.

1.4 The luminescence spectrum

The 'color' of the luminescence of a solid is the perception by the observer of what may actually be a complex spectrum. The wavelength of the components of this spectrum depends on the magnitude of the energy level differences through the relationships given in Section 1.3.1:

$$\text{wavelength (nm)} = 1{,}239.8/E$$

where E is the energy level difference in electron volts. The spectrum is a complex function of the solid, the activators, and the quenchers. It should also be emphasized that the actual emission spectrum of intensity versus wavelength is not a simple line spectrum but may be very complex. Occasional samples will show a single narrow peak, some show a broad band (see Fig. 7.1), and others show one or more broad bands with narrow peaks superimposed (Fig. 4.6a), indicating more than one activator or that different charge states for the same activator are present.

A given mineral may accept more than one possible activator and exhibit a different luminescent color for each activator. Feldspars (see Ch. 5) can accept Eu^{2+}, Mn^{2+}, and Fe^{3+} as activators and they will luminesce blue, greenish-yellow, and red, respectively, for these different activators. A given activator may be accepted by several different minerals and exhibit a different luminescence color in each of the minerals. The activator Mn^{2+} can be incorporated in feldspars, apatite, and carbonates, to mention only a few (see Ch. 4). The luminescence colors of these three Mn-activated minerals are greenish-yellow, yellow, and reddish-orange respectively.

Several of the REE activators (Dy^{3+}, Sm^{3+}, Eu^{3+}, for example) produce 'line' spectra whose wavelength is almost independent of the host crystal, and CL can be used to qualitatively analyze for these elements in some cases.

We are concerned here, primarily, with visible range emission but there are also important cases of UV and IR emission, subjects which will almost certainly grow in interest.

The present level of sophistication of interpretation of CL emission spectra makes use only of the major peaks. In many cases, however, the spectra have minor peaks present also and repeated scans at expanding sensitivities might well open up even more areas of information.

1.5 Temperature effects

The intensity of CL is a function, often strong, of sample temperature. Leverenz (1968, p. 342) discusses this and states that the effect of increasing temperature can be in either direction, i.e., it may either increase or decrease CL intensity, depending on the composition of the phosphor. For most phosphors, luminescence efficiencies decrease with increasing temperature, a process which can be called thermoquenching. Conversely, these same samples will show an increase in intensity with decreasing temperature. This increase is dramatic in some cases, e.g., quartz (Plates 5c & d), and represents a potentially valuable additional source of information. The CL spectral peaks are also somewhat narrower at lower temperatures. Examples of CL temperature dependence will be given in Chapters 4 and 6.

1.6 Characterization of luminescence

Characterization of CL from a given mineral can be done in terms of the intensity, spectrum, polarization, phosphorescence, and possible transient nature of the emission. The variables in the exciting electron beam are the beam energy, beam current density, and angle of incidence. Specimen temperature may also be a variable. The orientation of the crystal structure of the mineral, relative to the direction of the electron beam and the direction of observation, may also have an effect in some minerals.

1.6.1 *Intensity*

In general, the intensity increases as either the electron beam energy or the electron beam current density increases. The intensity reaches a maximum or plateau for many luminescent minerals and may even decrease at higher power densities (Coy-Yll 1969/70, Leverenz 1968).

There is a definite effect associated with beam energy. If, for instance, beam power density ($V \times I$/area) is kept constant, and the energy is varied, the CL intensity shows a definite increase with V. At low energies, no matter how high the current (within reason!), no CL is observed. Above some threshold value (often called the 'dead voltage' in the phosphor industry literature; see Curie 1963) the CL intensity will increase with V at constant beam power density. This implies that there is some effect associated with the surface layer and penetration of this layer is required before the representative CL of the sample is developed. The magnitude of the required voltage is as low as a few kV for many carbonates and bright phosphors such as synthetic ZnS. It is several kV for quartz.

The surface of a polished thin section is a region which has undergone severe mechanical manipulation during sawing, grinding, and polishing operations, and this may be at least part of the reason for the higher values of electron beam voltage required to excite CL effectively in some minerals, especially those whose CL depends upon lattice damage as well as activators. It is also tempting to speculate that there are some cases of 'poisoning' of the near surface layers by oxygen, i.e., a change in the charge state of the activator.

In some instruments, the CL is viewed from the same side as the surface that is being excited. In other cases, emission is observed from the opposite surface and the observed CL must pass through the sample. In the latter case, there may be lower intensity and spectral modifications due to absorption of the CL. This is particularly important when looking at large crystals, e.g., diamonds.

1.6.2 *Color and the spectral analysis of luminescence*

For most situations of geological interest, it is observed that the 'color' of a given luminescing mineral sample is the same to the unaided* eye for all values of the voltage and current density above a certain minimum or threshold value. However, sometimes the unaided eye may interpret increased intensity as an apparent color shift. A few minerals, most importantly quartz, may show a gradual color shift with time during bombardment.

When the subjective observations of the unaided eye are replaced by measurements with a monochromator, a wealth of additional information is obtained (Geake *et al.* 1973, Mariano & Ring 1975). Both broad-band and narrow-band spectra are observed and often one can identify the activators responsible, and sometimes even their charge states, from the CL emission spectra. This subject is discussed further in Chapter 8.

* That is, to the observer simply looking at the sample through the microscope without the use of optical instrumentation like a spectrophotometer and without the use of color photomicrography.

1.6.3 *Phosphorescence*

Phosphorescence is observed in some minerals. Occasionally the decay times are long enough to be observed with the unaided eye but more often appropriate instrumentation is required to observe it (Lebiedzik *et al.* 1974). Coy-Yll (1969/70) shows curves of intensity versus time, after cessation of the excitation, for several mineral samples. From these we can estimate that the times for the intensity to fall to $1/e$ (0.36) of its initial value, for apatite, calcite, and quartz, are about 20, 15, and 10 msec, respectively. Hall and Ribbe (1971) measured a decay time of 0.17 msec in cassiterite. Görz *et al.* (1970) measured decay times (and rise times also) for many silicates. For most of their samples, decay times were 0.3 msec or less but some were as high as 40 msec. The decay time is probably a function of the activator(s) and the host mineral, not of the excitation intensity, but very little information has been published on this subject for geological materials.

1.6.4 *Polarization*

Polarization of the emitted radiation is sometimes observed but as yet this property has been studied for only a very few minerals. Görz *et al.* (1970) have compiled some of this information for many silicates. Bhalla and White (1970) have studied in detail the polarization of CL emission from willemite. The CL from most feldspars exhibits polarization effects. Information on this characteristic is provided in Chapter 4 for each individual mineral for which it is available and in Chapter 5 for the feldspars.

The polarization effect is essentially a measure of the differences in intensity of the CL when the polarization analyzer is in the respective position for maximum and minimum intensities. The magnitude of the polarization effect will depend upon the crystal plane exposed to the bombarding beam and for some planes will be very small or actually zero. (See section on willemite in Ch. 4.) In an actual natural sample, there will be many different grains of the same mineral species. These grains may have differing orientations and therefore they will show different magnitudes of polarization effect.

Occasionally, the polarization effect is so great that there is actually a color change, depending upon the orientation of the polarizer with respect to the crystal axes (Görz *et al.* 1970, Sippel 1971). Kiflawi and Lang (1974) have reported polarized emission in certain types of diamonds which arises from linear defects, and in this case the defect orientation, not the crystal axis orientation, determines the degree of polarization effect.

1.6.5 *Transient and time-dependent luminescence*

The CL of phosphors can deteriorate with time under electron bombardment. This phenomenon apparently occurs to at least some extent in all phosphors and is referred to as 'electron beam aging' in the phosphor industry. Gilliland (1971) has a discussion of this with particular reference to the compound $BaSi_2O_5$:Pb*. He mentions some of the responsible mechanisms in synthetic phosphors as:

(a) precipitation of metal from the host causing darkening;
(b) thermally induced destruction of luminescence centers;
(c) loss of oxygen in oxygen-dominated phosphor material.

For the particular case of $BaSi_2O_5$:Pb, Gilliland found a 50 percent reduction after about 5 minutes bombardment at $30\,\mu A/cm^2$ beam current density and accelerating voltages of 10–25 kV. (This is a power density of only $6\times10^{-1}\,W/cm^2$ which is relatively low compared to those used for geological samples. See Table 3.2.)

The CL properties of most geological samples appear to be steady-state, reproducible properties, at least with the power densities and viewing times normally employed. Electron beam aging over normal viewing times is not significant. A notable exception to this is some quartz which may show a color change, typically from blue to red, under normal excitation. This change takes place over a period of 'tens of seconds' and is sometimes not recoverable. The process is not well understood or well documented. Prolonged observation of other minerals, e.g., feldspars, apatite, and anhydrite, may also lead to a noticeable reduction in luminescence intensity but the effect is much less than that in quartz. And it is not the case for all samples of these minerals, only those with a particular constitution. Geake *et al.* (1977) mention that the electron beam may reduce ferric iron to ferrous in feldspars, causing a reduction in the intensity of the CL emission. Other forms of excitation produce different aging effects. Ion beams, in particular, produce rapid aging (see Sec. 3.8.2).

1.7 Observational techniques

CL can be observed on a wide variety of electron beam instrumentation, as discussed in Chapter 3. However, the emphasis in this text will be on the observations made with relatively simple electron beam units which can be mounted directly on the stage of the standard petrographic microscope, and the observation of the CL is done with the optical microscope. Some minerals luminesce so brightly that their CL can be observed through the microscope even in a well lit room. More typically, CL observations are made in an actual dark room. The wealth of additional detail often visible when the eye is accommodated to total darkness is well worth the slight inconvenience.

The ability to alternate between CL and transmitted light (TrL) observations is an important one. The standard microscope illuminator can be connected so that it can be switched off when the CL is switched on, and vice versa, making it possible to alternate rapidly between CL and TrL. Some investigators prefer to control the TrL intensity with a separate control and to be able to blend in a low intensity of TrL with the CL for enhancement of certain features (Kopp 1980). In either mode of operation, the TrL intensity should be kept low to improve the accommodation of the eye to the CL.

When the intensities of TrL are too low in normal petrographic work, one is naturally accustomed to 'cranking up' the illuminator intensity for better viewing. For natural samples, the CL intensity reaches a practical maximum and additional excitation intensity will not increase the CL – indeed it may even lead to incandescence. Additional 'cranking up' of the CL cannot be done. This finite limit on CL intensity imposes a restriction on the maximum magnification which can be used and also poses the need for the best possible microscope transmission. Note that the intensity per unit area of observation, or of recording film area, decreases with the square of the magnification. The practical limit for brightly luminescing samples is a magnification of about 200×, although this may improve with improvements in microscope transmission and in film speed. Sometimes CL features which are too weak to be seen by eye under the maximum excitation are recordable by very long photographic exposures.

At this time, most investigators will describe the CL observations in terms of colors as seen with the (subjective) unaided eye and by color photomicrography. The latter presentation is also variable because of differences in film types, development processes, etc. The description of the color is facilitated with the use of a system like the Munsell color charts (Munsell 1975, Greer *et al.* 1968), although these charts do not exhibit the vivid colors seen through the microscope. A few investigators (Mariano & Ring 1975, Geake *et al.* 1973, Sippel & Spencer 1970b, Greer *et al.* 1968) have attached spectrophotometers to the microscope and obtained CL spectra. Both wide-band and narrow-band spectra are observed, depending upon the host mineral and the activator(s). A wealth of additional data on identities of activators and even their charge states can be derived from the spectra (Mariano & Ring 1975).

1.8 Applications of cathodoluminescence in petrographic studies

There are several general areas of application for CL which we will briefly summarize here. Each of these will be expanded upon in the later chapters.

(a) The distribution of different minerals within a sample (thin section, usually, but not necessarily) can often be visualized quickly. Calcite and dolomite, for example, usually show different luminescence and stand out from one another clearly under CL when they coexist (Plates 7c, d, & 10a). Calcite or dolomite cement often stands out clearly from the other minerals in a sedimentary sample (Plate 3a). The distribution of two or more feldspar phases in a granite will often show up vividly under CL (Plate 2a).
(b) Luminescent minerals present as fine grains (e.g., apatite in a granite) or as a narrow vein filling (e.g., calcite) are often immediately visible by the CL technique. These very fine features might go unnoticed, or at

* A terminology used in the phosphor industry to indicate the presence of an activator in a material is to write the activator symbol after the formula for the pure material, e.g., $BaSi_2O_5$:Pb.

least be very difficult to locate and identify, by conventional techniques (Plates 2e & f).

(c) Variations in the CL emission from different portions of the same mineral can indicate different generations with different conditions of formation, etc. The revelation of multiple stages of carbonate cements is the most dramatic example of this application. The distinction between authigenic and detrital feldspar and authigenic and detrital quartz is another important example of this useful application.

(d) The 'color' of the luminescence of a mineral can be an indicator of the trace element content of the mineral. Apatite, as one example, is usually a bright yellow when activated by Mn^{2+} but the CL colors of apatite activated with REE include blue (Eu^{2+} activation), blue-violet to violet (Eu^{2+} and Eu^{3+}), and white (Sm^{3+} and Dy^{3+} activators) (Mariano & Ring 1975).

(e) Usually the same mineral species, formed under different conditions or derived from different sources, will show different CL properties. Zircon is an example of a sedimentary rock component with varying CL properties; quartz and feldspar are others. The analysis of sediment–source rock (provenance) relationships can sometimes be assisted by making use of the varying CL properties of the same mineral from different sources. This particular application will be even more important when we have a better understanding of all the reasons for CL in minerals such as quartz.

(f) CL can be a useful tool to monitor mineral separation processes. CL observations are not restricted to thin sections or slabs. Loose grains can be examined, non-destructively. If a mineral separation scheme involves two or more minerals with different CL properties, it is convenient to remove a sample from the separator from time to time, inspect it, and return the material to the separator, if necessary.

As an example of this application, consider apatite and feldspar from granitic rocks. Apatite (typically bright yellow CL) and feldspar (typically blue or greenish-yellow CL) would not be confused and a single grain of one in the midst of 'thousands' of the other would stand out clearly.

1.9 Summary

A thorough understanding of the wide range of luminescence phenomena seen in geological samples requires a thorough analysis in terms of solid-state physics, an analysis which has been done for only a few minerals. However, even without the development of this knowledge, the observational data have produced a tool of great value. Before we investigate this tool in more detail, we will consider some of the historical development of the method.

Chapter 2

History of cathodoluminescence observations

As Leverenz (1968) pointed out, if we consider luminescence in all its possible forms, as enumerated in the previous chapter, one can find references dating back not only to Aristotle (384–322 BC) but even earlier to the Chinese in 1500–1000 BC. If, however, we restrict ourselves to the luminescence of solids, apparently the first reported observation was by Vincenzo Cascariolo, of Bologna, in 1602–3. He heated powdered barite with coal and obtained a material which glowed in the dark.

Fortunio Liceti wrote about the material created by Cascariolo in 1640 and called it litheophosphorus. The word phosphorus means 'light bearer.' The element phosphorus was not discovered until about 30 years later and was named after Cascariolo's luminescent material, because of the light given off when it was oxidized. The term 'phosphorescence' came to be applied to any material which showed persistent emission.

Stokes demonstrated in 1852 that the emission from a solution of an aqueous wood extract absorbed incident light of one wavelength and emitted light of longer wavelength. He demonstrated also that the emission in this case stopped the moment the incident light was removed. He noted that the emission from the mineral fluorspar behaved in the same fashion and therefore he coined the word 'fluorescence' from the mineral fluorspar.

The above information was garnered from the excellent text by Leverenz (1968) which presents an extensive discussion of the early history of the subject.

Crookes (1879), in his extensive research on molecular physics and discharge phenomena in high vacua, observed luminescence excited by charged particles. His observations on specific minerals will be included in later chapters but some of his comments are appropriate here also. Thus, to quote him directly (p. 659)

> One of the most striking of the phenomena attending this research has been the remarkable power which the molecular rays in a high vacuum possess of causing phosphorescence in bodies on which they fall. Substances known to be phosphorescent under ordinary circumstances shine with great splendour when subjected to the negative discharge in a high vacuum.

Crookes made observations on diamonds, rubies, emeralds, and many other minerals. He observed polarization of luminescence for certain minerals and even did spectral analysis of the luminescence in some cases.

Nichols, Merritt, Howes, and Wilber investigated luminescence phenomena in the early 1900s and their work appeared in several papers and in Publication 384 of the Carnegie Institute of Washington in 1928 (Nichols & Merritt 1909, Nichols *et al.* 1928). This early work and that of others at this same time are discussed in detail in the thesis by Everett Glover (1977). As Glover notes, by 1923 there were over 1,300 references to some facet of luminescence listed in a publication by Merritt *et al.* (1923). Glover further notes that several correct conclusions were drawn during this period, including the rôle of manganese in activation of calcite; but he further states:

> The understanding of the theory was in a state of ferment in this period and awaited one main development: the application of the modern conception of the atom with transition of electrons between energy levels giving rise to radiation.

Early references to apparatus for CL studies of geological materials, including devices for visual observation or attachments to be used with the microscope, include those of Nichols *et al.* (1928) and Gallup (1936).

When the electron probe came into use as a means of analyzing minerals, investigators noted the luminescence of some minerals (Long 1963, Stenstrom & Smith 1964). It was not until 1965, however, that a systematic discussion of these observations was presented in the published literature. Smith and Stenstrom (1965) showed a large number of examples including iron-quenching and zonation in dolomites, differences in detrital and secondary quartz, zonation in luminescing feldspars, and others. Their publication in the *Journal of Geology* was especially valuable because it contained color plates.

Long and Agrell also published results from their work in cathodoluminescence in 1965. They used both an electron probe and also a smaller, simpler unit which could be mounted directly on the petrographic microscope. They reported on observations in several minerals including calcite, dolomite, wollastonite, quartz, and perthite. They made a combined CL and electron probe study of a zoned calcite and demonstrated the correlation between luminescence intensity and the concentrations of the activator, Mn, and the quencher, Fe. Sippel and Glover (1965) demonstrated that cathodoluminescence revealed an amazing complexity in some carbonate cements and published photographs showing examples of multiple zoning in carbonate cements (up to '50 distinguishable growth bands' in one instance). They noted that CL also occasionally revealed the presence of completely unsuspected features. One such example, a vein of calcite cutting across, and in optical continuity with, a large single crystal of calcite, appeared in the cover photograph of *Science* magazine. Then, in 1968, Sippel published a classic paper on the use of luminescence petrography in the study of sandstones. He demonstrated its use to distinguish between primary and secondary quartz in sandstones and to distinguish between solution and overgrowth phenomena. Fracture events and recrystallization processes were also demonstrated to be more readily visible.

In the late 1960s and early 1970s, the field of luminescence benefited from the emphasis on space-related research as attempts were made to understand more fully the transient luminescence phenomena observed on the Moon (Greer *et al.* 1968, Weber & White 1970, Geake *et al.* 1970).

In numerous publications, spanning the time period from the early 1960s to the present, Geake and Walker and coworkers have described their work on the luminescence of rocks and minerals. They have used both ion excitation and also low-voltage electron beam excitation. Their work was primarily concerned with lunar samples and meteorites but it also included many examples of terrestrial minerals for comparison. The early results they presented (Geake & Walker 1966) were for unseparated samples, but they quickly turned their interest to the individual minerals. Much of their interest has been devoted to determining the reasons for the luminescence of feldspars.

A great deal of experimental and theoretical work in CL has been done because of the interest in development of new phosphors for the cathode-ray tube industry, and for the color television industry in particular. The textbook by Leverenz (1968) is a valuable reference to this work which is strongly oriented to synthetic materials.

There are a number of more recent publications on geological applications of CL. The comprehensive articles by Nickel (1978), Mariano (1978), Kopp (1981), and Amieux (1982) deserve particular mention. Extensive bibliographies on CL have been prepared by Bröcker and Pfefferkorn (1978, 1980).

One of the most unusual references to early applications of luminescence which the author has encountered is in the paper by Weber (1963). I will quote him directly, in part (p. 92).

> The history of phosphorescence is much older. In the Far East, it was utilized by Chinese painters in the 10th century. They devised phosphorescent pigments from oyster shells which they used to introduce mysterious elements into their painting, a magic ox, for instance, invisible during daylight but appearing in a pasture during the twilight hours. This material, obtained by calcining oyster shells with sulfur, was rediscovered by the Englishman, Canton, in 1768 and called Canton's phosphor . . .

(The author has tried for a number of years to locate such a painting still in existence but without success!)

Thus we can note in summary that luminescence phenomena have been observed for many centuries and cathodoluminescence has been an applied technique for over 100 years, but the real development of its geological applications has taken place only in the past 20 years or so. As we shall see, there remains a great deal to be done to derive full value from the information it provides.

Chapter 3

Instrumentation for cathodoluminescence studies

3.1 Introduction

The cathodoluminescence observations by Stenstrom and Smith (1964) and Smith and Stenstrom (1965) and the early work by Long (1963) and Long and Agrell (1965) were accomplished with an electron probe as the excitation source. The electron probe provides the advantages of performing compositional analysis, its primary function, as well as a limited capability for direct cathodoluminescence observations. The electron probe is a more expensive and more elaborate piece of instrumentation than is required for routine CL observations and some designs may not offer the advantage of permitting rapid interchange between conventional transmitted light observations and CL.

As soon as it was realized that the CL observations were potentially very useful in petrographic studies, suggestions for excitation sources, simpler than the electron probe, were made. Long (1963) had demonstrated a cathodoluminescence microscope attachment using a cold cathode electron gun. This instrument design was referred to in the article by Smith and Stenstrom (1965). Sippel (1965) also pursued this approach and published a detailed description of a unit which could be mounted directly on the stage of a standard petrographic microscope. Long and Agrell (1965) included in their paper a description of a microscope-mounted unit with a hot filament. Herzog *et al.* (1970) also described a commercial version of a microscope-mounted attachment for cathodoluminescence studies, the Luminoscope®*.

An advantage of most of the cathodoluminescence microscope attachments (CMAs) is that they provide the capability to perform transmitted light (TrL), polarized light (PoL), or CL observations consecutively or simultaneously on the identical area of a given specimen. An interesting feature located with one technique can immediately be studied with the other technique(s). Another advantage is that the specimen can usually be a standard thin section, as prepared for conventional petrographic studies, without the need for the conductive coating required by the electron probe. The CMAs can also excite a relatively large sample area at any one time since the current available is many times greater than the beam current available in the electron probe. Hence the CMA beam can be defocused to cover a larger area while still maintaining power density, adequate to develop visible luminescence.

Both the electron probe and the CMAs continue to be important instruments for CL observations.

The electron probe and the CMA must be operated under vacuum conditions. The vacuum is required only to generate and transmit the electron beam and not to facilitate the CL observations.

3.2 Electron probes

There are thousands of electron probes in use throughout the world and there are many different designs included in this group. Typically, they utilize an electron beam with energy variable from about 20 keV to as much as 50 keV, beam current to the target ranging from 0.001 to 0.02 microamperes (μA), and a spot diameter in the range of 1 to 50 microns (μm). A summary of the operating parameters of several units is included as part of Table 3.2 in Section 3.5. For very small sample areas, the electron probe can be operated to provide a very high current density. In fact, this can easily be carried to the extreme where the sample is damaged or incandescence is caused.

The maximum beam current is well under 1 μA in the electron probe and for spot diameters larger than about 0.01 mm the power density available is less than that available in the CMAs.

CL observations are frequently done with the electron probe (Reid & Smith 1969). Probes are often equipped for relatively high magnification of the sample area but may not always provide the best quality optics for polarized light and transmitted light observations. A conductive coating is required on the samples and this reduces the intensity of the CL reaching the observer.

*Luminoscope® is a registered trademark of the Nuclide Corporation.

3.3 Cathodoluminescence microscope attachments

A number of cathodoluminescence arrangements have been designed to mount directly on the stage of an optical microscope. They usually offer the advantages of quickly alternating between transmitted light viewing and cathodoluminescence, and they are of simpler instrumentation than the electron probe. Some of the designs which have been published in the literature are summarized here. Some instruments of unpublished design are also in use in various laboratories.

3.3.1 *CMA design of Long and Agrell*

Long and Agrell (1965) described a CMA which utilizes a heated filament electron gun. The electron gun is mounted on a sample chamber which is attached directly to a microscope stage. An electron beam energy of up to 9 keV is provided and current densities are of the order of 0.1 $\mu A/mm^2$ (see Table 3.2). They suggested that a cold cathode electron gun would also be a good choice for this application.

3.3.2 *CMA design of Sippel*

Sippel (1965), following Long's suggestion (above), designed a CMA which utilizes a cold cathode electron gun. The cold cathode gun is a simpler and more rugged source of electrons than the heated filament gun. This feature makes it especially useful in petrographic applications, for several reasons which will be discussed in a later section.

Sippel's original design provided for an alternating current power supply and a beam energy of about 10 keV. Spot size was about 1 × 3 cm and beam currents were not stated. He designed the sample chamber to accept standard thin sections. The electron gun is mounted at an angle to the chamber to provide an intersection of the electron beam and the thin section surface on the optical axis of the microscope (Fig. 3.1). The luminescence is viewed after it passes through the thin section. Subsequently, Sippel and Spencer (1970b) reported modifications to this design including the use of a DC power supply and also focusing of the electron beam. Some of the important work done with it will be discussed in later chapters, especially Chapter 6.

3.3.3 *The Luminoscope®*

The Luminoscope® (Herzog *et al.* 1970) is a commercial design for a CMA. It also uses a cold cathode electron gun. The beam energy is variable up to 15–16 keV DC. Beam current is variable up to 1 mA although it is usually restricted by the operator to a maximum of 0.5–0.7 mA. The beam spot diameter on the sample is adjustable from a minimum of about 1 mm to a maximum of over 2 cm by means of a magnetic lens (focus coil).

The relationship between the electron gun, the sample surface, and the viewing optics in the Luminoscope® is shown in Figure 3.2.

The Luminoscope® is designed to accept a single thin section of a size up to 50 mm × 75 mm (2 in × 3 in), or two or more smaller thin sections. Any portion of the entire 2 in × 3 in sample can be brought beneath the excitation beam which intersects the sample surface on the optical axis of the microscope on which the Luminoscope® is mounted. A pair of windows are provided on either side of the sample chamber (Fig. 3.2) so that the specimen can be viewed with transmitted light, interchangeably with the cathodoluminescence observation, at exactly the same spot on the specimen, and within a second or less. The CL is viewed from the same side as the electron beam and hence it does not undergo absorption by passing through the thin section.

The Luminoscope® electron gun is not mounted at an angle to the chamber. Instead, the electron beam is formed and directed initially in a horizontal path, parallel to and above the sample surface. After leaving the cold cathode electron gun and passing through the focus coil, the electron beam comes under the influence of a deflecting magnet assembly (Fig. 3.2) which deflects the electron beam on to

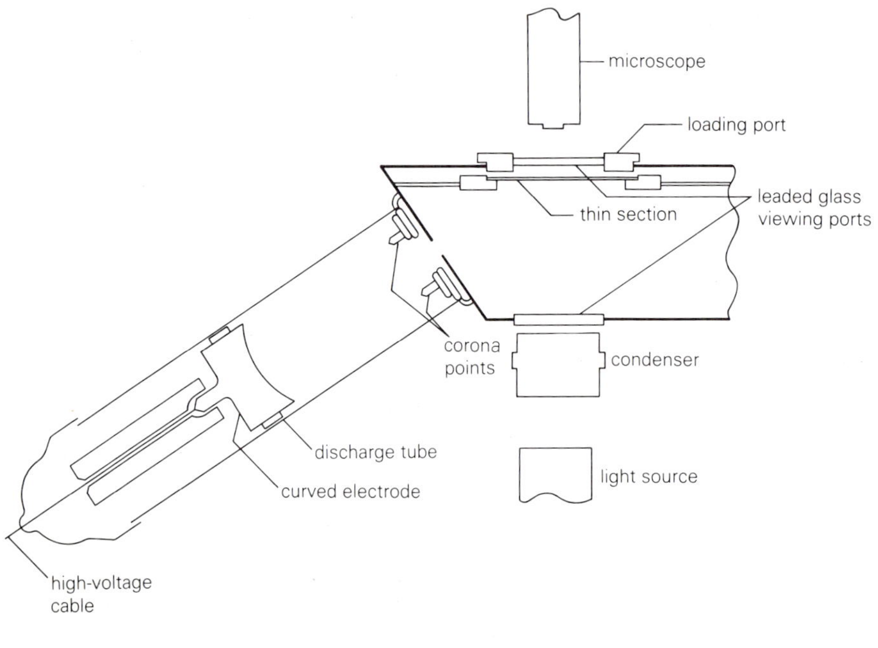

Figure 3.1 The cathodoluminescence microscope attachment designed by Sippel (1965). The electron beam from a cold cathode electron gun strikes the (inverted) thin section at the point where the thin section intersects the microscope optical axis. The luminescence is viewed from the 'back' side of the thin section. The initial design had up to 10 keV (AC) beam energy. Subsequently he modified the design to provide for a beam focusing element and a higher voltage DC power supply.

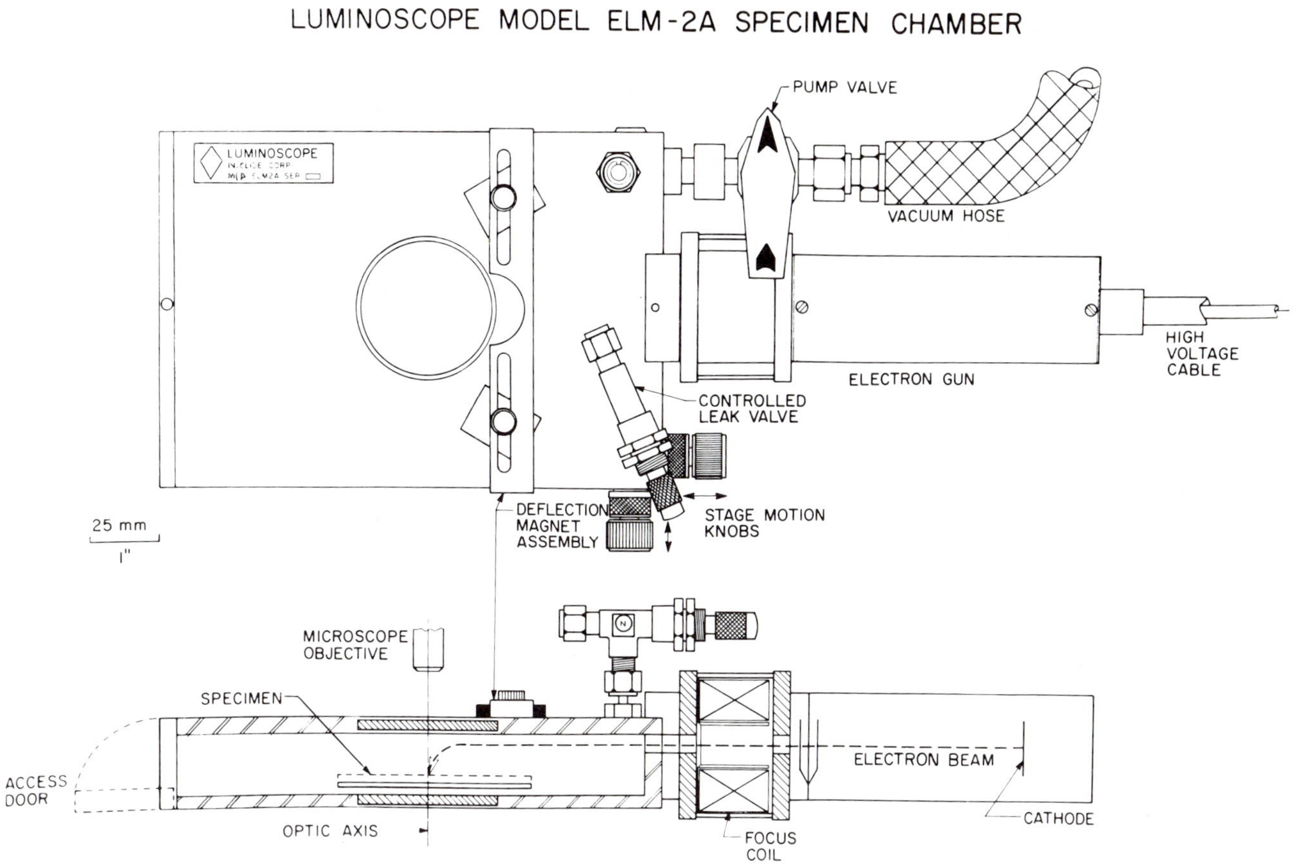

Figure 3.2 Top view and cross-sectional side view of the Luminoscope®. The electron beam is formed in a cold cathode discharge, passes through a focusing coil, and then comes into the influence of a deflection magnet assembly which bends the beam on to the sample. Moving the deflection magnet assembly allows for positioning the beam on the microscope optical axis for samples of varying thickness. The focus coil is used either to increase or to decrease the beam diameter at the point of incidence on the sample. Since the CL is observed from the same surface that is being bombarded, either thin or thick samples can be examined. (From Herzog *et al.* 1970.)

the sample surface. The position of the point of incidence can be changed by sliding the deflecting magnet assembly and this permits the operator to line up the point of impact exactly with the optical axis of the microscope. This arrangement also enables thick specimens (slabs) and loose grains to be studied.

The weak magnetic field provided in the deflected beam arrangement prevents either positive or negative ions from striking the sample.

3.3.4 *Cold cathode electron gun*

A cold cathode electron gun is the source of the electron beam on both the Sippel unit and the Luminoscope®. Since this type of electron gun is somewhat unusual and unfamiliar to many, its operation will be briefly reviewed.

The basis of operation is that an electrical discharge is established between two electrodes which have a potential difference between them (Fig. 3.3). In order for the discharge to be established, it is necessary for the gas pressure in the region between the electrodes to be in the appropriate range. Electrons emitted at random from the cold cathode (the processes of field emission or thermionic emission may initiate the process) ionize some of the gas molecules and the positive ions created are accelerated to the cathode. Positive ion bombardment of the cathode produces more electrons and the cascade process leads to the

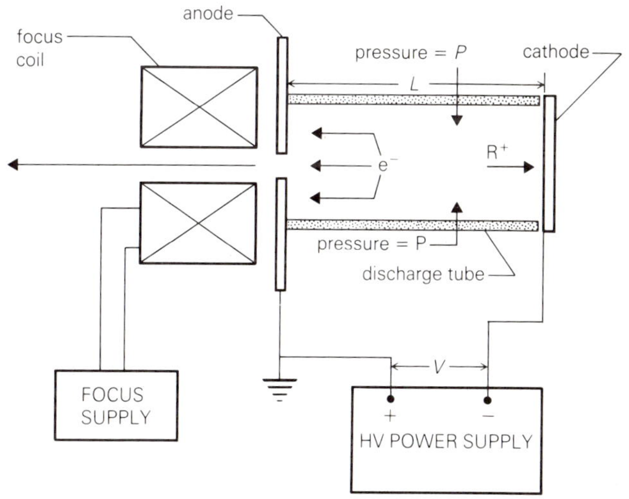

Figure 3.3 Cold cathode electron gun. The principal discharge takes place between the cathode and anode which are separated by a length, *L*, typically of the order of a few inches. At the proper set of operating conditions (*P*, *V*, *L*), a discharge is established between the anode and cathode. The principal components of the discharge are neutral gas atoms or molecules, positive ions (R^+), and electrons. Minor amounts of negative ions and metastable species may exist also. The gas can be air, He, or other. The ions are accelerated to the cathode; the electrons to the anode. Part of the electron beam strikes the anode and this represents unused beam current. The portion of the electron beam that passes through the anode aperture is the useful electron beam current which bombards the sample.

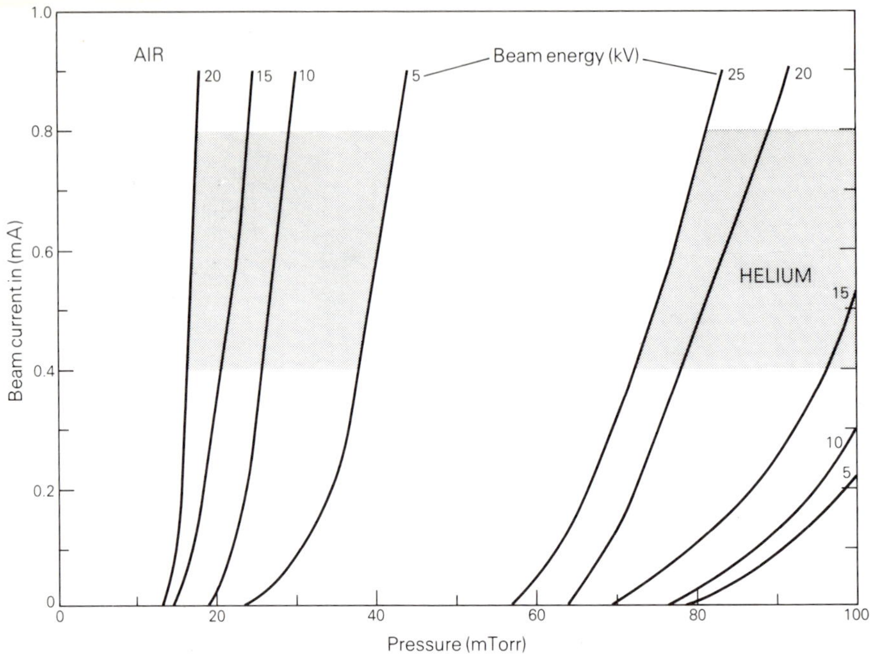

Figure 3.4 Characteristic curves for a cold cathode electron gun. These show the typical manner in which the electron current varies with pressure at a fixed value of voltage between the cathode and the anode. The curves on the left side are for the case when air is admitted to the discharge region, and those on the right are for the case when helium is admitted. The pressure scale values for air are read from a thermocouple vacuum gauge; for helium they are also read from such a gauge and then corrected for the different response of a typical thermocouple vacuum gauge to helium and air. (Actual gauge readings for helium would be off-scale to the right, i.e., in the 100–300 mTorr region.) There is sufficient variation in the calibration of different thermocouple gauges, and differences in their location with respect to pump lines and gas inlets, that the actual readings observed for air can vary from those shown on this figure by as much as 10–30 mTorr. They should be looked upon only as an indication of the relative pressure values. The voltage, *V*, is an upper limit to the beam energy. The electrons actually reaching the sample will have a distribution of energies, ranging from this maximum value to lower values, due to various energy losses incurred in passing through the discharge region. The usual range of operational values for voltage and current for geological samples is indicated by the shaded areas.

creation of a stable discharge. The discharge consists of electrons, traveling predominantly in the direction toward the anode, and positive ions, traveling toward the cathode. Small amounts of negative ions may be formed also. The electrons leave the discharge region through the anode aperture and strike the target.

The cold cathode electron gun will operate with either AC or DC voltage. If an AC power supply is used, the exit beam will be electrons on only one half cycle and positive ions on the other half cycle. If the arrangement is such that the positive ions can strike the sample, the positive ions may seriously modify (damage) the surface and the luminescence properties of the sample. When DC power supplies are used, as is nearly always the case, only electrons (and small amounts of negative ions in some arrangements) bombard the sample.

In order for the discharge to be created, the gas pressure in the gun region must be in the appropriate range. If the pressure is too low, not enough ions are produced to form the sustained discharge, and if it is too high, the mean free path of the electrons and ions is too short for the discharge to form. Between these two extremes, the discharge will form; the discharge current varies with the pressure. It is possible to prepare a set of 'characteristic curves' which describe the performance of a given design of cold cathode electron gun. These are similar to a plot of the characteristics of a vacuum tube or of a solid-state device. One of the principal differences is that the pressure is one of the parameters in the cold cathode electron gun characteristics.

A representative set of characteristic curves for a gun with an electrode spacing of several centimeters and applied voltages in the kV range is shown in Figure 3.4. For most CL observations on geological materials done with a CMA, the desired operating range is within the region bounded by the dotted lines. This range of voltages and currents is determined by the following considerations: the electron energies must be sufficiently high to develop the luminescence in the sample and the beam current must be high enough to enable the luminescence to be seen, but not so high that the specimen will be damaged. For the cold cathode gun described by these curves, one can choose an operating voltage and then estimate the pressure required within the gun to establish the desired level of beam current. On any one of these curves, i.e., at fixed beam energy, the beam current varies directly (but not linearly) with the pressure. In order to change the beam current, at fixed beam energy, the pressure must be varied.

The usual range of operating pressure is between about 20 and 100 mTorr*. The pumping system requirements in this pressure range become rather simple. A standard mechanical vacuum pump (often called a forepump), two-stage, in good operating condition on a leak-free system will pump to a pressure of less than 10 mTorr. From the curves of Figure 3.4 we observe that a pressure of 10 mTorr is so low that no

* The SI unit is the pascal. Conversion is 1 Torr = 133 Pa. Therefore 1 mTorr = 0.133 Pa. Older units still in use include the mm of Hg which very nearly equals 1 Torr and the micron of Hg which very nearly equals 1 mTorr.

beam current can be established for the range of voltages shown. It is actually necessary, in practice, to increase the pressure within the chamber and electron gun, either by introducing gas into the chamber or by reducing the pumping rate, to bring the pressure into the range where a beam can be established. The pressure increase is not done in a single burst but rather in a steady-state manner. The pressure equilibrium is reached when the chamber and sample outgassing, and any in-flow through a leak valve, are balanced against the pumping action of the pump to arrive at the desired operating pressure and hence the desired beam current.

One may logically ask why a cold cathode electron gun, with its unusual mode of operation, is used. There are several advantages. First, the vacuum system is very simple, and diffusion pumps, cold traps, ion gauges, and other complexities and hence costs of high vacuum systems are not needed. A second advantage is that conductive coatings are not required on most samples. Positive ions are created in the discharge and some of these find their way to the sample and neutralize any charge build-up. This is a definite advantage over the electron probe and other instruments which require a conductive coating on the sample. Such coatings represent an extra sample preparation step and they may also reduce the intensity of the luminescence appreciably. Finally, the cold cathode electron gun is quite rugged. Even as extreme a misoperation as venting the system when the electron gun is on at full power will not normally cause serious damage or require extensive cleaning of the chamber. Filament burnout, a common concern with heated filament electron guns, does not happen. The cathode does eventually become pitted from ion bombardment and requires replacement but this pitting takes place gradually, over a period of weeks or months.

Usually, air is admitted to support the cold cathode discharge. However, other gases can be used. Sippel (1965) noted that both hydrogen and helium are superior to air with respect to minimizing the stray light from the discharge. When helium is used, the gas inlet leak is opened more widely than when air is used, for the same voltage and discharge current conditions. The adjustment of the leak valve is easier to do with helium than with air. Some authors have commented that CL intensitics are greater when helium is used, and it also appears that the sample remains slightly cooler because the helium molecule is a more efficient heat conductor than the air molecule. In practice, the use of helium is worthwhile when samples that are difficult to observe, such as quartz, are studied, but not necessary with the usual carbonate and feldspar samples. Hydrogen, with the extra safety precautions required for handling, is rarely if ever used.

3.3.5 *Electron luminescence microscope design of Le Poole et al.*

Le Poole *et al.* (1968) described a more elaborate luminescence microscope design which utilizes a heated filament, up to 30 keV electron energy, and beam focusing and deflection. A high vacuum system is required because of the heated filament. They provided for TrL observations and also included a UV irradiation capability and provision for measuring the X-ray emission. Color photos of cassiterite and scheelite CL were given.

3.3.6 *CMA design of Harmer et al.*

Harmer *et al.* (1968) described an instrument, somewhat similar to that of Sippel (1965). However, they arranged for the electron beam to strike the same specimen surface that the optical microscope is focused on, thereby permitting both thin sections and opaque samples to be studied.

They provided an electron beam with energy up to 15 keV and included simple magnetic focusing and positioning.

Ceramic and glass microstructure samples were of interest in their work.

3.4 Other excitation sources

Many different excitation sources for luminescence have been described, some of which are suitable for use on a microscope attachment. We will briefly describe some of the other electron beam based units and comment on X-ray excitation. The reader interested in photoluminescence or ion-excited luminescence instrumentation should consult the appropriate references. A brief comparison of the various excitation techniques is given in Table 3.1 (Section 3.4.6).

3.4.1 *Tesla coil*

The Tesla coil discharge is a commonly used tool in laboratories with glass vacuum systems and it was a natural choice for evaluation as an electron source for CL studies. Gallup, in 1936, used such a source. Greer *et al.* (1967) used a Tesla coil on a CMA and Sommer (1972a) did some of his work with a Tesla coil attached to a Pyrex aspirator bottle.

The Tesla coil is inexpensive and easily set up, but it has the disadvantages of low intensity, and the spot location and size are not easily controllable. The power density is relatively low. Also it is an AC source, and thus it produces and accelerates both electrons and ions. The accelerated ions tend to be destructive to the sample.

3.4.2 *X-ray excitation*

A number of authors have reported on the similarity between CL and X-ray excited luminescence (XL). The XL process is related to CL although the instrumentation involved is clearly different. Medlin (1963a) used X-ray excitation (Mo tube, operated at 35 kV and 20 mA) as part of his study of the thermoluminescence of calcite, dolomite, and other carbonate minerals. Sommer (1972c) has reported on the use of a Siemens high intensity Cr X-ray tube (25–30 kV, 20–70 mA). With this X-ray milliprobe, the X-ray beam could be directed on areas as small as 100 μm in diameter. Sommer gave examples of spectral analysis results obtained from several $Ca_{1-x}Sr_xCO_3$ samples with this mode of excitation.

A quantitative comparison of incident beam and luminescence intensities for X-ray and electron beam excitation systems is difficult. It appears that the intensity of XL, at least with the intensity of X-ray beam provided by the systems presently used, is significantly less than CL and direct viewing of the luminescence with a microscope is not possible.

The X-ray excitation source has the advantage that a vacuum system is not required and therefore it would be possible to study samples at elevated pressures and/or elevated temperatures. Sample damage may be less.

The source of the luminescence excited by X-rays is different from that produced by CL because the penetration depths into the solid differ. For materials whose surface and bulk specimen luminescence differ one would expect to see corresponding differences between CL and XL.

3.4.3 *Scanning electron microscope*

CL can be observed also on the scanning electron microscope (SEM) (Muir *et al.* 1971, Steyn *et al.* 1976, Rémond 1977, Grant & White 1978, Tovey & Krinsley 1980). The SEM utilizes a scanned electron beam with a small spot size. A detector can be used to measure the secondary electrons produced or the CL can be measured with a photomultiplier (PM). By rastering the electron beam, synchronized with the *X, Y* scans of a display oscilloscope (CRT), and modulating the CRT intensity with the PM output, a magnified map of the CL of the specimen is obtained. Total PM output intensities can be used for the intensity function, or spectral analysis can precede the PM input.

The SEM offers the possibility of better spatial resolution than the CMA because of the smaller spot size. Grant (1978) discusses the resolution question and points out the two important limitations to resolution in the SEM. The first limitation is due to the size of the volume sampled. When the electron beam impinges on the crystal, it penetrates and excites CL over a volume (Fig. 1.2). This volume increases as the beam energy increases. Grant estimates the volume to be a sphere of 4 μm diameter for a 30 kV beam on quartz. The second limitation is due to phosphorescence phenomena. As the spot is scanned across the sample, one would like to measure only the CL from the instantaneous location, i.e., to measure only the fluorescent component. However, since the PM gathers total light from any emitting area of the sample, any phosphorescence from previous electron beam locations will be captured also. This will cause a smear. Grant shows that a mineral with a 0.1 ms, or shorter, rise and decay time will have a resolution limit set by beam energy (i.e., by beam penetration and the dimensions of the volume excited). For longer rise and decay times, phosphorescence will limit the resolution. Some minerals have long phosphorescence and the CL of important cases such as calcite cannot be effectively studied on the SEM because of this.

The SEM offers advantages over the CMA for some CL studies, advantages which Grant (1978), Tovey and Krinsley (1980), and others have discussed. These are higher magnification, the capability of combining CL and SEM observations on the same area, and sometimes higher power density over small sample areas. However, the sample usually requires a conductive coating. Another disadvantage is that viewing is usually done on a monochromatic reconstructed image.

It should be noted that the SEM shares the characteristic of other electron beam instruments with low values of beam current. One does not have the capability of direct viewing, in color, of the CL of the sample.

The SEM has been particularly useful with quartz and we will discuss some of the findings in Chapter 6.

3.4.4 *Higher-energy CMAs*

Smith and Stenstrom (1965) observed, with an electron probe, a quartz sample in which the authigenic component did not luminesce at lower excitation energies but did show a slight luminescence at higher energies. Since that time, other workers have confirmed the dramatic increase in luminescence of some quartz with higher beam energy (Zinkernagel 1978, Ramseyer 1982c).

Because studies of quartz are so important, some CMAs have been developed which provide higher energies (Richter & Zinkernagel 1975, Zinkernagel 1978, Ramseyer 1982a,b). The general concepts of these systems are shown in Figures 3.5a and b. Their development has been strongly motivated by the need to study those quartz samples that do not luminesce at the lower voltages usually provided. Typically the higher-energy units provide up to 25 or 30 kV output. Of course, most electron probes have this energy

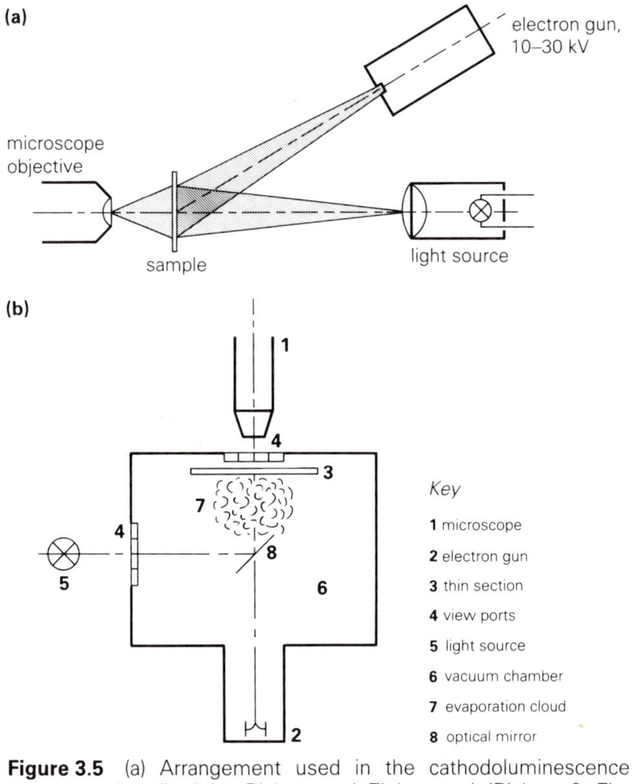

Figure 3.5 (a) Arrangement used in the cathodoluminescence apparatus described by Richter and Zinkernagel (Richter & Zinkernagel 1975, Zinkernagel 1978). The electron gun operates at voltages up to 30 kV in a high vacuum. The electron beam is incident on the sample at a steep angle and the transmitted light source is perpendicular to the thin section as in ordinary transmitted light observations. (b) An alternative design by Ramseyer (1982a) which also operates at voltages up to 30 kV in a high vacuum. The TrL source is introduced with a mirror, permitting the electron beam to be perpendicular to the specimen. In both designs, the CL emission from the sample is viewed with a microscope from the back side of the coated thin section.

capability but they frequently do not have the convenience of good microscope viewing optics or the current required to illuminate a conveniently large sample area (see Section 3.5). The use of instruments with higher electron energies introduces more technical complexity and cost, as compared to the 15–18 kV units. This tends to restrict their application to certain important applications such as those quartz overgrowths that do not show CL at lower electron energies.

Some authors have developed a jargon of 'hot' versus 'cold' electron sources to distinguish the HV (25–30 kV) from the LV (15 kV) sources. This is not really correct. It so happens that hot filaments were selected (e.g., by Zinkernagel and by Ramseyer) for their 30 kV design and that cold cathodes were selected by Sippel (1965) and by Herzog *et al.* (1970), but neither of these electron sources has a severely restricted energy range. Long and Agrell (1965) used a 'hot' filament at only 9 kV, for example. And cold cathodes have been operated at voltages greater than 20 kV. More appropriate than 'hot' and 'cold', it would seem, would be the use of a terminology such as 'low' (0–10 kV), 'moderate' (10–20 kV), and 'high' (20–30 kV) to describe the energy range used. If there is a difference in the CL excited by these two sources, it may be due to the differences in the ion content of the beam which is greater for the cold cathode source. The hot filament sources, as noted previously, require the sample to have a conductive coating.

3.4.5 *Other*

There are other electron beam instruments with cathodoluminescence attachments. There are attachments for the electron probe (Reid & Smith 1969, Lebiedzik *et al.* 1974) which use a phototube, with or without a monochromator. These usually provide an oscilloscope display which is generated by scanning the incident electron beam over the sample surface in synchronization with the spot scan on a CRT. The CRT spot intensity is controlled by the intensity of the PM output.

These more elaborate, and more expensive, instrumentation systems may offer advantages in terms of higher optical magnification of CL images, higher beam voltage, and a smaller spot size. But they do not, in general, possess the main advantage of the CMAs – that is *the capability of illuminating a sample area of significant size with sufficient power density for direct miscroscopic viewing in full color and immediate alternation between transmitted light and CL!*

3.4.6 *Comparison of excitation methods*

Some general comparisons can be made between CL (using either a hot or cold cathode) and other means of exciting luminescence in terms of practical matters such as intensity, convenience of use, sample preparation requirements, etc. These are presented in Table 3.1. The grading of luminescence intensities is based on the beam intensities readily available in instrumentation for producing X-rays, UV, positive ions, and electrons, though clearly one could engineer stronger sources of any one of these if desired.

3.5 Beam power and beam power density

An electron beam energy source can be partially characterized by its energy and beam current but it is also often necessary to think in terms of beam power and beam power density. The beam power is the product of the beam energy (volts) and the beam current (amperes), and can be expressed in watts (joules/second). Beam power density is just the beam power divided by the area of the beam. Both beam power and beam power density vary widely among the different types of electron beam instruments.

$$\text{beam power (watts)} = V\,\text{(volts)} \times I\,\text{(amperes)} \quad (3.1)$$

$$\text{beam power density (watts/cm}^2) = (V \times I)/(\pi d^2/4) = 1.27\, V \times I/d^2 \quad (3.2)$$

for a circular beam of diameter, d (cm).

The equations give the power delivered to the sample surface. The amount absorbed depends also upon how much is reflected. In the case of an electron beam, most of the reflected power is in secondary and scattered electrons and this is only of the order of a few percent of the incident power.

Table 3.2 is a compilation of the published operating characteristics of a number of electron probes and other electron gun devices, including the Luminoscope®. The parameter values given in the table are those published in various articles (not necessarily articles dealing with CL) and may not represent the best set of conditions available for CL studies on these same instruments.

Note that almost always the beam current measured is the total current flowing in the return line to the low end of the power supply. This is made up of the beam current to the sample plus any portion of the beam which strikes grounded parts of the apparatus (e.g., limiting apertures). The latter

Table 3.1 Comparison of various methods of exciting luminescence.

Excitation method	Lum. int.	Damage to sample	Variety of minerals	Depth sampled	Vacuum requirement	Sample preparation
electrons (cold cathode)	highest	medium	greatest	1–5 μm	yes (low vac.)	least
electrons (hot cathode)	highest	medium	greatest	1–5 μm	yes (high)	coating required
UV	least	least	intermediate	greatest	no	least
X-rays	least	least	intermediate	greatest	no	least
ions	highest	greatest	greatest	surface	yes (high vac.)	coating required

Table 3.2 Comparison of operating conditions for electron probes, Luminoscope®, CMAs, and other electron beam devices.

Instrument type	Max. beam energy (keV)	Max. beam current (μA)	Min. beam diam. (μm)	Max. beam power (W)	Beam power density at maximum power density (W/cm^2)	Beam power density for 1 mm diam. spot (W/cm^2)	Beam power density for 1 cm diam. spot (W/cm^2)	Reference
electron probe	30	0.01	10	0.0003	380	0.038	0.00038	Sommer (1972a)
electron probe	40	0.002	50	8×10^{-5}	4.1	0.01	0.0001	Sommer (1972a)
electron probe	15	0.015	2	0.00022	7,200	0.029	0.00029	Pingitore (1976)
electron probe	30	0.03	1	0.0009	115,000	0.115	0.0012	Greer *et al.* (1968)
cathode ray tube	10	200	360	2	2,000	255	2.55	Leverenz (1968, p.451)
^{90}Sr ß-source	2,300	—	—	—	3.7×10^{-5}	—	—	Greenman & Milton (1968)
CMA with hot filament	9	—	—	—	0.009	—	—	Long & Agrell (1965)
Luminoscope®	16	1,000	1,000	16	2,038	2,038	20.4	Herzog *et al.* (1970)
SEM (JEOL)	30	10^{-6}	0.01	3×10^{-8}	$3.8\times10^{+4}$	4×10^{-6}	4×10^{-8}	JEOL USA Inc. (1983)

component causes no excitation and is wasted. The effective beam current is always less than the total, but the true proportion is rarely known, though it is often assumed to be of the order of 90 percent. We make no attempt here to distinguish between total and effective beam current. The effective beam current is difficult to measure because of the scattering of primary electrons and the generation of secondary electrons. Unless these are completely eliminated by suitable suppressor electrodes, they will subtract from the target current, and a simple measurement, e.g., with a milliammeter in the return lead from the target to ground, will give a value lower than the actual target current.

In addition to providing the wattage and the maximum power densities available, Table 3.2 also lists the power density when the beam is defocused (if necessary) to a diameter of 1 mm, which might be typical of the beam size desired for many types of observation of CL on petrographic microscopes. Similarly the power density for a 1 cm diameter beam is listed. The CMA devices clearly show their advantages over the electron probe when moderate power densities are required over large areas. In order to develop a desired intensity of CL, one needs an electron beam with a certain level of beam energy and a certain power density. If the beam is sharply focused, as in the electron probes, then the power density will be high and the CL will be at its brightest, but over a small area. If the same power level is present in a beam of large area, then the power density will be low and no portion of the sample being bombarded may luminesce brightly. The advantages of the CMA are also apparent when the beam is defocused to a diameter of a few centimeters as required for macrophotography (Section 3.12).

One of the assumptions inherent in the table is that the beam has a uniform current density. More typically, the distribution of current density will be quasi-gaussian and the center of the beam will be more intense than the periphery.

A logical question is, what is the trade-off between the beam voltage and the beam current, if the total beam power and power density are kept constant? There is no simple answer to this question. Leverenz (1968, pp. 356, 439) has a discussion directed towards CRT phosphors. The factors he considers include electron range, saturation effects, scattering, and self-absorption losses within the sample.

Electron ranges are proportional to the square of the beam energy in the kilovolt energy range, and therefore the depth sampled increases rapidly with the energy of the beam. As an example, Leverenz gives a calculated range of 0.2 μm for a 6 keV electron in Zn_2SiO_4. This electron would have a calculated range of 2.2 μm at 20 keV and 5 μm at 30 keV. As the electron range increases, excitation within the volume of the sample becomes more important relative to excitation of the surface. Volume CL may be more efficient than surface CL, according to Leverenz, so this leads to an increased CL output.

In geological samples there appears to be a beam energy threshold in the keV range, below which CL is not visible to the eye. For a few minerals, this threshold is low, 1–3 keV; for many carbonates, it appears to be in the 6–8 keV range. Quartz has a high threshold, often 10 keV or more, and some authigenic quartz may not reveal subtle luminescence contrasts unless beam energies are 20 keV or higher.

3.6 X-ray, high voltage, and implosion safety considerations

3.6.1 *X-ray precautions*

When electron beams interact with matter, X-rays are generated. The energy and intensity of the X-rays produced increase with the electron beam energy but also depend on the nature of the material being bombarded. Most samples for CL studies are relatively inefficient in the conversion of a portion of the incident electron beam energy into energetic X-rays. However, some samples, e.g., those containing galena, will have appreciable cross sections for the generation of X-rays. Also, a portion of the beam may bombard metal parts of the apparatus which will similarly have appreciable cross sections. Unless proper precautions are taken, X-rays may be measured. These precautions include the use of lead-glass windows and sometimes supplementary shielding. The power supplies usually have a specific limit for the maximum high voltage and they should never be used over this voltage.

As a minimum, the user should have an initial radiation survey made and it should be repeated on a regular basis. Even better is a permanent monitor set-up for daily checks. This is especially valuable when several people are sharing an instrument. Radiation surveys are especially important whenever any changes or maintenance, e.g., new windows or gun cleaning, are made. When making the radiation survey, the usual sample thin-section target should be replaced with a heavy metal target such as a piece of lead sheet or some other metal with a large cross section for energetic X-ray production. The text by Shapiro (1972) is a guide to X-ray safety.

3.6.2 *High voltage safety*

The equipment for CL work typically includes a 10–20 kV DC power supply, capable of providing as much as 1–5 mA of current. These power supplies deserve respect! The basic design should be such that the user cannot come into contact with any portion of the HV output. Equally important is the maintenance of good grounding throughout the system, including the vacuum chamber. The electrons striking the chamber and the sample must eventually flow through the chamber ground and thence return to the positive end (grounded side) of the power supply. If this chamber ground is not present, the electrons cannot return and instead they will accumulate in the chamber. If this happens, the chamber charges up and its potential increases. Eventually, either the chamber potential will equal the high voltage, V, and the electron beam will be cut off, or the chamber potential will be so high that an arc-over will take place somewhere and discharge to some well grounded object, e.g., to the microscope or through the operator! Good grounding prevents this from happening.

3.6.3 *Implosion safety*

This is usually only a minor concern. Because the working distances required are so long, the microscope objective may be very close to the observation window, and thus it is possible to bring the objective into contact with the window and break it. With the vacuum inside, the glass will implode, not explode, but even this represents a possible safety problem. A realization of the possible problem and careful instrument operating techniques are clearly the best answers.

3.7 Sample temperatures

One of the questions which sometimes arises is, what is the actual temperature increase produced in the sample by the electron beam? The temperature increase of the specimen can have an important effect on the intensity of luminescence as well as contributing to more mundane problems such as slide damage.

The superficial indications of the sample temperature rise are that some cements such as Lakeside® and Canada Balsam may bubble up under the beam. The effect is localized at the beam impact, however, and the thin section as a whole will normally be barely warm to the touch when it is removed from the chamber. Nevertheless, these cements cannot be used and one of the epoxy resins, suitable for high temperatures, is required.

An estimate of the maximum temperature increase at the center of the beam can be obtained by considering the case of a point source on a semi-infinite solid, as illustrated in Figure 3.6. This type of problem is solved by Carslaw and Jaeger (1959), for example, and a solution by Castaign (1952), appropriate to electron probe conditions, is

$$T = (V \times I)/(2\pi JKr) \qquad (3.3)$$

where T is the temperature increase in degrees centigrade, V is the beam voltage, I is the beam current, J is Joule's constant, K is the thermal conductivity, and r is the radius of the beam. In cgs units,

$$T = (V \times I)/(26Kr) \qquad (3.4)$$

If this expression is used with the excitation conditions of a CMA, one can calculate for quartz (K = 0.01 cgs), for example, temperature rises ranging from 40–60 °C, for a

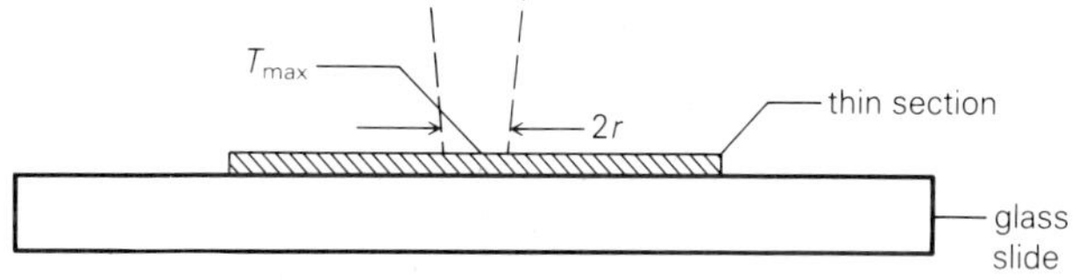

Figure 3.6 Parameters involved in the calculation of temperature increase. The incident beam has an energy, V, a current, I, and a radius, r. The beam power is $V \times I$ (in watts when V is in volts and I is in amperes). The temperature, T_{MAX}, calculated from equation 3.4 is the maximum temperature that exists at the center of the beam.

partially defocused beam, to temperatures as high as 500–600 °C for a tightly focused beam operating at full power. It should be stressed that these calculated temperatures are those directly at the center of the beam impact point and not averaged over the thin section. The thermal conductivity can be a highly variable quantity, especially within a thin section, and is a major unknown in making this calculation.

For samples of lower thermal conductivity, or beams of smaller area (higher power density), the temperature increase will be greater. It is easy to visualize that the heat remains concentrated near the source point in a good insulator and therefore the temperature at this point increases more than it would for a poor insulator. In practical situations, this effect may be seen in a pure quartz sandstone where individual grains may become hot enough to glow from incandescence, with little or no response by nearby grains. (See comments on Owen's work in Sec. 6.3 and Owen 1984b.)

Because thermal quenching is important in some minerals, a better knowledge of sample temperature could be important. A more thorough investigation, including better values for thermal conductivities in thin sections, would be a useful contribution.

3.8 Sample preparation and sample damage

3.8.1 Sample preparation

In principle, one can study nearly any kind of sample which will fit into the chamber and can be bombarded by the electron beam. This includes thin sections, loose grains, chips, and slabs.

Thin sections must not have cover slips. The electron beam penetration depth is only a few μm and this is not great enough for the electrons to pass through the cover slip and still retain sufficient energy to excite the specimen beneath.

Stained specimens can be studied but the stain can reduce the luminescence intensity by absorption of the emitted radiation. However, Ebers and Kopp (1979) reported that carbonate zones are more distinctly revealed with CL when the thin section is first stained with potassium ferricyanide solution.

Conductive coatings are not required when the cold cathode electron gun is used. The positive ions present in the cold cathode discharge are sufficient in number to neutralize the charge build-up on the sample. (Under some conditions, the bluish-purple plasma discharge can be seen extending outside the cold cathode electron gun, into the chamber and striking the sample surface.) Conductive coatings are required when heated filament type electron guns are used. Such coatings reduce luminescence intensity significantly if the emitted radiation must pass through the coating where some absorption will take place. If the CL can be viewed from the opposite side of a *thin* section as suggested by Long (1963) then only the absorption in the specimen itself is of concern. This approach was used in Sippel's original design (Fig. 3.1) and also by Zinkernagel (1978).

Thin sections should be mounted with a cement which is stable at the temperatures developed by the electron beam. These temperatures are not appreciable when averaged over the entire thin section. Usually, it will feel barely warm to the touch when it is removed from the chamber unless prolonged observation has been made. But the localized temperature at the point of beam impact can be much higher and cements such as Canada Balsam or Lakeside® may 'bubble up'. (These two materials start to melt at about 80 °C.) Many (but not all!) of the epoxies are suitable mounting media.

There are a variety of suitable epoxies available and there are also some which do not work well with CL samples. The epoxy and the curing procedure should be tried out carefully, repeating the steps on a few samples to be sure that the result is an epoxy which does not outgas excessively or otherwise prove to be unsuitable. Note that the epoxies themselves will characteristically exhibit luminescence, e.g., a dull green or blue color. In porous samples, areas of the epoxy exposed to the electron beam will undergo discoloration after extended observation.

The magnification to be used will dictate the surface finish required on the sample. For quick survey work, rough, unpolished chips may be studied and one could even use a simple hand lens for observing the CL. Unusually rough surfaces should be avoided, even in survey work, because of the possible shadowing effects on the electron beam. But for the highest magnification, a highly polished surface is used. This is especially important if TrL photography is to be done because immersion oils cannot be used inside the chamber under vacuum. The intensity of the CL emission appears to be enhanced also if the surface is highly polished (Mugridge & Young 1984), probably because there is less surface scattering.

A typical preparation procedure might involve the usual lapping with 600 grit, followed by polishing with ¼ micron grit. The same type of surface that would be prepared for reflected light microscopy is ideal.

During the sample preparation procedures, care should be taken to minimize the amount of compound remaining on the sample. Common lapping and polishing compounds (i.e., diamond, alumina, and silicon carbide) are brightly luminescent and show up as a galaxy of small bright specks if they remain on the sample. An ultrasonic cleaning, with solvents selected to remove fingerprints and any other organic materials present, is sometimes used.

Slabs of varying thickness can be examined in some CMA units. The most frequently encountered problem here is poor (slow) pumping due to absorbed water. A good approach is to heat the slab in an oven, and then store it in a desiccator until the CL observations are to be made.

Loose grains can be made into mounts (e.g., by dusting them on a slide coated with a suitable epoxy) or they can be looked at as they are. The latter approach is particularly useful if one wishes to monitor the progress of a mineral separation process. After the CL observations, the grains can be returned to the separation device for further separation if necessary.

Ryan and Szabo (1981) described a technique for mounting loose grains separated from a sandstone to permit point counting of the feldspar content. Their procedure consisted of sprinkling the loose grains on a slide coated with a suitable epoxy, and then curing it. Next the slides were soaked in HCl to remove carbonates and iron coatings. After rinsing, and an additional soaking with oxalic acid, they were dried and the CL observations were made. Their results are discussed in Chapter 5.

Sometimes problems are encountered with the charging-up of loose grains. When charged up, they are repelled from the area of the beam. Sometimes, by using the lowest possible beam voltage and the highest possible beam current, one can maximize the number of positive ions available for neutralization and succeed in looking at loose grains without any special preparation.

The usual glass slide will show a dull blue CL if the beam impinges directly on it in uncovered regions; this is not usually objectionable. Indeed it can be a useful way to focus the beam to the desired shape and diameter. However, the author has occasionally seen other types of glass which show a strong white CL which can interfere with the CL of the sample itself.

3.8.2 *Sample damage*

There are a number of possible ways in which the sample can be changed or damaged in CL work and even though they rarely occur the investigator should be aware of them.

Some samples may show a brown discoloration on the surface after prolonged bombardment. This is apparently due to the presence of traces of organic material in the sample, oils left from handling, or pump-oil deposits. The electron beam can interact with these organic materials causing pyrolysis and leaving a dark residue. The residue can be minimized by keeping the sample clean and using a trap in the pumping line. If the deposits do form, they can usually be removed by light polishing.

Some epoxies will discolor when exposed to the electron beam, and in some instances they may 'bubble' and outgas when the beam strikes them. Power densities must be kept low if there are large areas of epoxy exposed to the beam in unusually porous samples. The type of epoxy used and the curing procedure should be monitored carefully.

The most serious possible damage is the actual breaking of slides. This may occur when high power levels and densities or very long observation times are used. Most frequently this situation occurs on those samples with faint luminescence. Quartz sandstones are an important example here. Many quartz-containing samples show faint CL, at least in comparison with that of the carbonates and feldspars, and higher powers and power densities are required when studying them. If the sample is an unusually pure quartz sandstone, the problem is intensified because of the poor thermal conductivity of the quartz. The heat developed is localized instead of being distributed over a wide area of the sample. Occasionally one can see a single grain heat up to incandescence with little or no response from surrounding grains.

Some investigators will make 'thick sections' for those samples that they suspect will require more intense bombardment. These are typically 200 μm thick instead of the 30 μm thickness of normal thin sections. The thicker section or 'slab' distributes the heat better and is therefore less prone to thermal damage. The two sections can be made in pairs from immediately adjacent areas. Then conventional optical and CL observations can be performed on the appropriate section.

To a large extent, the discussion of slide breakage is now out of date because improvements in instruments, better microscope transmission, improved photographic techniques, and faster films, as well as a better understanding of the problems which lead to sample slide breakage, have all contributed to reducing the problem. Many investigators will never encounter it.

Physical damage such as slide breakage is a relatively straightforward problem to assess. There are some other forms of sample damage which are more subtle.

The electron beam can cause another type of sample damage by causing physical or chemical changes in the sample which lead to changed CL characteristics. Quartz is the most important example of a mineral whose CL properties may be changed noticeably under the influence of the beam. Its CL may change from a relatively bright blue to a dull red over a period of tens of seconds. But this is the case for only certain quartz samples. The changes associated with the electron bombardment may be due to heating effects which may in turn be causing annealing, crystal damage, mobilization of impurities, or other effects of a thermal nature. It is also possible that some of the transient luminescence observed when quartz samples are first bombarded may be due to thermoluminescence.

In an electron probe operating at 30 kV, 0.01 μA, and a 10 μm spot (a power density of 380 W/cm^2), Sommer (1972a) observed that $MgCO_3$ loses its CL if bombarded for less than a second. But when the probe was operated at 40 kV, 0.002 μA, and a 50 μm spot (power density of 4.1 W/cm^2), the CL properties of the magnesite did not change rapidly. Sommer (1972a, 1972c) also extended his observations to include XL and here he found no visible evidence of damage after 30 minutes exposure.

Some early work in luminescence was done with accelerated ion beams, protons in most cases (see the references to the work by Geake and coworkers, for example), and this is still a subject of interest. In contrast to electron beams, ion beams can cause appreciable damage to the sample luminescence, as Geake *et al.* recognized. In one of their examples, a proton beam of 60 keV, 1 μA/cm^2, caused a factor of three decrease in luminescence after one minute of operation on 'fresh' silicates. Note that the power density is only 0.06 W/cm^2 for this case.

The early CMA instruments used a cold cathode electron gun, operating on alternating current. In this case, positive ions are accelerated to the sample on alternate half-cycles and damage from ion bombardment by 'air ions' occurs. This is one of the reasons why AC power supplies did not come into wide use. With a DC power supply, positive ions are not accelerated to the sample*. Only electrons, and possibly negative ions, are accelerated. Negative ions could

* But they are accelerated to the cathode and represent an important reason for cathode deterioration in the cold cathode type of electron gun. Commonly, after weeks or months of use one will notice a 'hole' in the center of the luminescence pattern on all samples and this can be traced to a hole in the center of the cathode which is not emitting electrons. A doughnut-shaped beam is being produced. In some cases, these holes pass completely through the cathode discs and even into the cathode support!

cause sputtering damage, similar to that of positive ions, but the cross section for their formation is much smaller, at least from air components, and they are only a small component of the total beam*.

3.9 Observations of cathodoluminescence

3.9.1 Choice of microscope

The CMAs have been used on many different types of microscope, the choice often being dictated by the microscope already available when the CMA has been acquired. Recently, more attention seems to be devoted to selecting a microscope with good performance characteristics for CL observation, i.e., assembling an integrated package.

The microscope design and performance parameters which are important for CL include:

(a) adequate space for mounting the CMA vacuum stage and electron gun;
(b) objective lenses with adequate working distance (WD);
(c) transmission;
(d) polarizer and other accessory convenience;
(e) illuminator convenience.

MOUNTING SPACE

This first consideration is self-explanatory. Most of the 'older' petrographic microscopes had adequate mounting space. Some of the newer streamlined models have reduced this space, however, and therefore may not be suitable for CL work.

WORKING DISTANCE

The working distance of an objective lens is defined as the distance between the lens surface nearest to the sample and the sample, when the microscope is in focus. The distinction may also be made between working distance (measured from the lens element to the sample surface) and free working distance (which allows for space occupied by protective or decorative covers surrounding the lens elements; Fig. 3.7).

Some electron probes incorporate a system where the electron beam passes through a hole in the center of the light optics objective lens. This arrangement permits the objective lens to be quite near the sample although one must have concern for such things as charge build-up on insulators and evaporated coatings on the lens. The viewing optics of such systems are adequate, but rarely exceptional.

In other types of probes, and with some of the CMAs, the electron beam path and the microscope axis are not coincident except at the ultimate point of electron beam impact. This introduces the problem that the working distance of the microscope objective must be relatively long to provide sufficient space for the electron beam to strike the sample surface.

In the Luminoscope® chamber, shown in Figure 3.2, which allows for either thin or thick sections, a working distance of 20–22 mm is necessary. This may be reduced somewhat by elevating the sample within the chamber but the beam shape is poorer when this is done, since it hits the sample at more of a grazing angle. A re-entrant or recessed window can be used with the Luminoscope® and this reduces the working distance to about 10–12 mm.

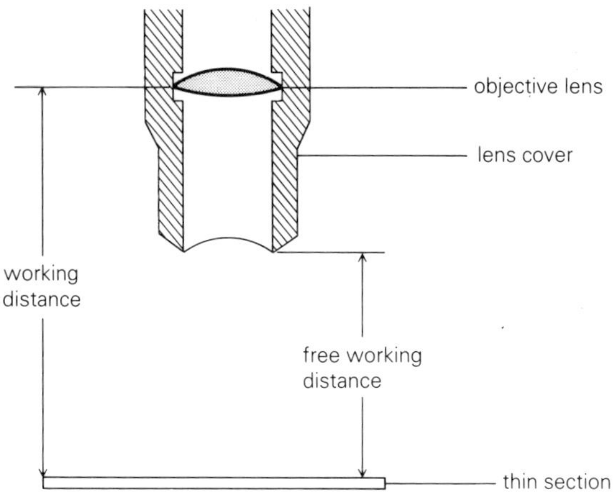

Figure 3.7 Definition of working distance and free working distance. The working distance is defined as the distance measured from the actual lens element to the specimen surface in focus. Part of this region, however, may be occupied by the protective or decorative cover so the usable working distance or free working distance may be less than the working distance.

Sippel's instrument design, which is suited to thin sections, has a short working distance requirement. The thin section is bombarded from one side and the viewing is done from the other, through the thin section glass. Even here one has, as a minimum, the thickness of the thin section slide, the thickness of the window, and the possibility of the mechanical hardware required for the sample motion. These add up to several millimeters also.

For the usual high-quality objectives furnished with most petrographic microscopes, the lower-power objectives may have adequate working distance but the WD decreases rapidly as the magnification is increased (see the first three entries in Table 3.3). The so-called universal stage objectives often offer the best choice for adequate working distance at higher magnifications.

It should be borne in mind that there is a practical limit to the useful total magnification for visual observation, because there is a limit to the available CL intensity. For a given CL intensity, as M increases, the intensity/unit detector area decreases as M^2. For example, if one has a photographic exposure time of 30 seconds at 50×, 120 seconds would be required at 100×. An analogous restriction applies to visual observations. The practical limits for routine observation depend on the sample type and the microscope transmission properties and are usually around 200×, with much work being done in the range of 25–50×.

TRANSMISSION

It is only recently that the importance of the last factor – microscope transmission – has been fully recognized, yet it may be the single most important factor in microscope selection for CL. In ordinary TrL and PoL work, transmission

* Note also that when a deflected beam is used, as in the Luminoscope®, the magnetic field is not sufficiently strong to deflect the negative ions of full energy on to the sample and this type of damage, therefore, is minimal.

Table 3.3 Working distance of some microscope objectives.

Objective	Magnification	Numerical aperture	Working distance (mm)	Comments
Leitz 559212	4×	0.12	24	strainfree for TrL
Leitz 559213	10×	0.25	6.8	
Leitz 559214	25×	0.50	0.48	
Leitz 559121	10×	0.22	15.8	U stage obj.
Leitz 559123	25×	0.22	14.7	
Nikon CF Brightfield	4×	0.10	13.8	Plan Achromat
Nikon CF ELWD	20×	0.4	10.5	
Nikon CF ELWD	40×	0.5	10.1	
Olympus	4×	0.10	18.23	strainfree Achromat
Olympus ULWDCDPL20X	20×	0.4	10	Plan Achromat
Unitron	3×	0.10	50	measuring objective
Unitron M10XLWD	10×	0.22	12.2	
Unitron M20XLWD	20×	0.30	12.2	
Zeiss 460100–9903	3.2×	0.07	30.0	Achromat
Zeiss 460320	6.3×	0.20	10.8	Neofluar
Zeiss 462044	16× (10)	0.17	10.0 (13.5)	UD Achrom
Zeiss 462042	6.3× (4)	0.12	13.5 (19)	UD Achrom

Sources of information: Leitz Brochure 550–061Engl. *Leitz Laborlux 12 Pol* 1982; Unitron Brochure *Universal measuring microscope*; Olympus 1985, personal communication; Zeiss, Microscope Optics Catalog, 41–9048/83; Nikon, CF Objectives, Brochure 1033–/c, May 1985.

This table is not intended to be a complete representation of the objectives available from these and other manufacturers but rather to give some information on the range of working distances and numerical apertures available and to furnish some indication of the manner in which the working distance varies with magnification. The universal stage (UD) objectives have different magnifications and working distances when they are used as a normal objective and the (reduced) magnifications and (increased) working distance, where available, are shown in parentheses after the standard values. The manufacturers should be contacted for further details.

is not an important consideration because one can usually ‘crank up’ the illuminator intensity as needed. There is a limit to CL intensity, however, and additional ‘cranking up’ of electron beam intensity leads only to incandescence or to slide breakage.

The author has asked several of the leading microscope manufacturers about transmission and it turns out that there are no generally accepted standards for microscope transmission within the industry at this time (1987). There certainly are differences, however, and some manufacturers’ literature notes that recent improvements in lens coatings, etc. have led to increased transmission. But the statements are qualitative.

As one illustration of the possible range of variations, Figure 3.8 shows the spectra obtained with a scanning monochromator on different combinations of microscopes and objectives. A difference of a factor of four exists among these four cases. This would represent the difference, in practice, between a 120 second and a 30 second photographic exposure.

Even though transmission standards are not available, we can note several factors which will influence the transmission:

(a) magnification of objective (M);
(b) numerical aperture of objective (NA);
(c) magnification of ocular;
(d) transmission of individual optical elements;
(e) details of optical path, splitters, polarizers, number of optical elements, etc.

The rôle of magnification and numerical aperture in determining brightness is expressed by the brightness being proportional to NA^2 and to the reciprocal of M^2 (Eastman Kodak Company 1980):

$$\text{brightness} \propto NA^2/M^2 \tag{3.5}$$

For example, the relative brightness of two objective lenses, one with an NA of 0.1 and M of 3.5 and the second with an NA of 0.12 and an M of 3.2 is:

$$\begin{aligned}&\text{brightness 'one'/brightness 'two'}\\ &= (0.12)^2 \times (3.5)^2/(3.2)^2(0.1)^2\\ &= 1.7\end{aligned} \tag{3.6}$$

i.e., a 70 percent increase in brightness.

The transmission of the individual optical elements, i.e., lens, prisms, and mirrors, is sometimes considered to be near to 100 percent although it can only approach this ideal value. This is a critical point because the number of elements and surfaces in the light path may be as many as 15 or 20. In the survey mentioned earlier, two manufacturers have stated that the improvements in lens coatings in ‘recent’ years (since the early 1960s?) have led to at least a 2× increase in transmission, so the transmission in older lenses must have been much less than 100 percent.

The influence of a splitter in the optical path is clear. In an ideal splitter, each individual tube after the splitter in a binocular microscope will receive only half the light that a monocular tube would receive. In a real splitter there will be other light losses so the number will be less than 0.5.

Polarizers have transmission of the order of 37 percent and their presence results in about a 3× loss of intensity (Fig. 3.9).

The lead-glass window used for X-ray protection has a transmission in the visible range of the order of 80–85 percent. The inside of the window on the vacuum chamber is

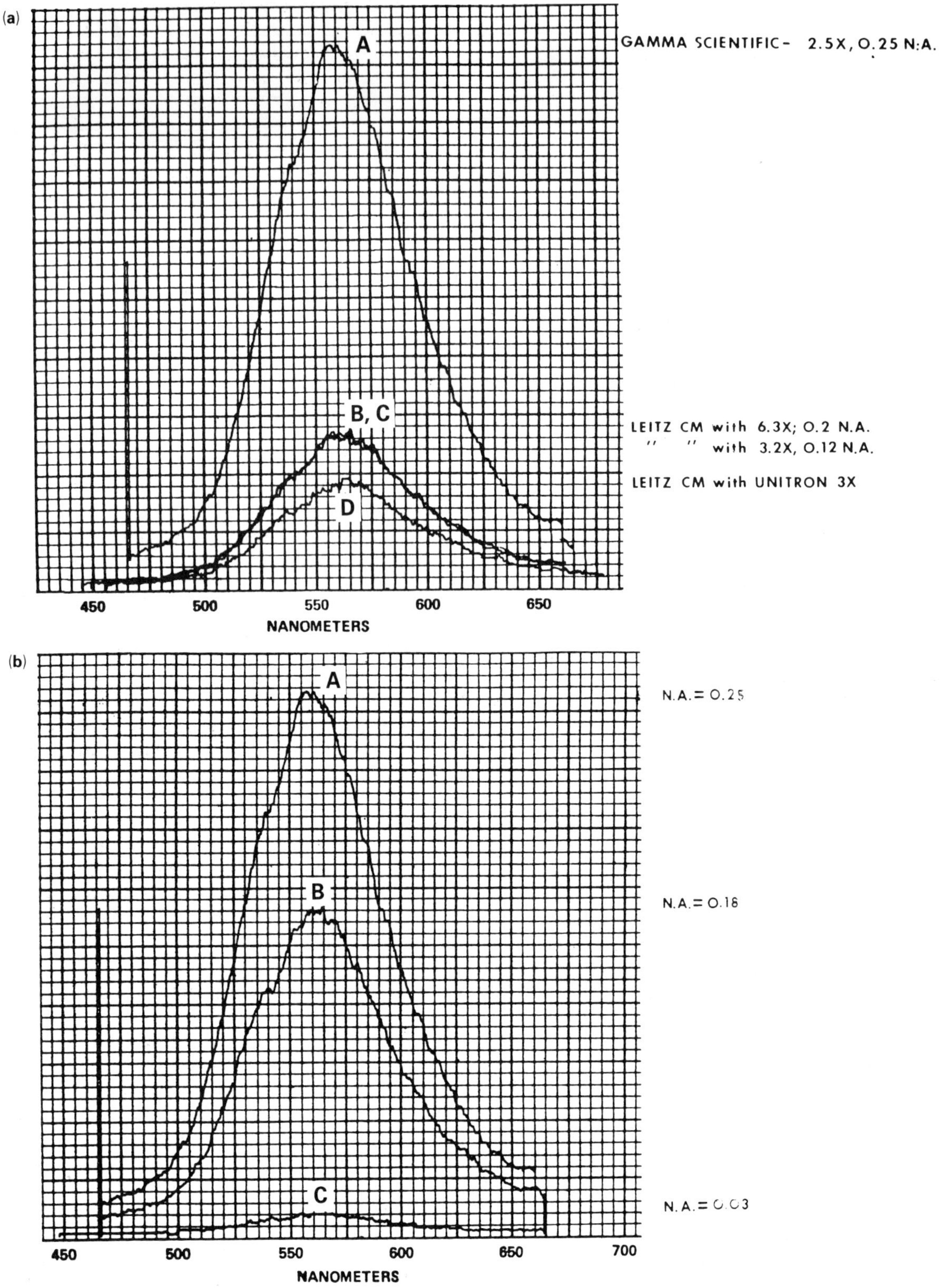

Figure 3.8 An illustration of the influence of the microscope system on observed CL intensities. (From Marshall 1980b.) (a) This is the CL spectrum from a feldspar grain in the Peggy's Cove granite, obtained with a spectrophotometer. Beam conditions and spectrophotometer sensitivity were maintained constant but the microscope and the objectives were changed. The same fiberoptic-containing ocular was used for all cases. A, Gamma Scientific Microscope, Model 700–10–1, with 2.5×, 0.25 NA objective; B, Leitz CM microscope with Leitz 6.3×, 0.2 NA objective; C, Leitz CM microscope with Leitz 3.2×, 0.12 NA objective; D, Leitz CM microscope with Unitron 3×, 0.10 NA objective. The extremes of the four cases shown represent about a factor of five difference in intensity as seen by the observer. (b) A spectrum from the same sample with the same microscope (Gamma Scientific) and same objective lens (2.5×) used for all cases. The objective lens has a variable numerical aperture and the effect of varying this is shown. A: NA = 0.25; B: NA = 0.18; C: NA = 0.03.

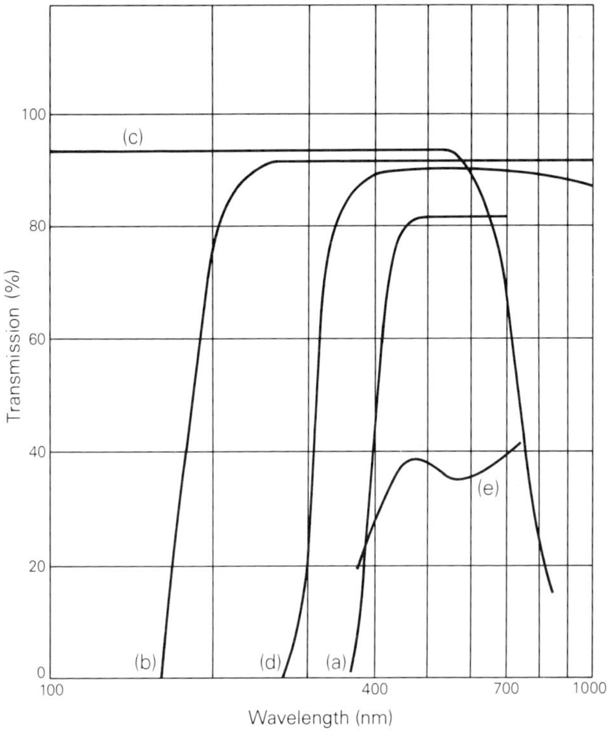

Figure 3.9 The transmittance of various materials as a function of the wavelength. (a) Lead-glass (8.5 mm thick) (Kaufman Glass Co. 1984). (b) Dynasil (fused silica, 0.5" thick) (Dynasil Corp. 1970). (c) MgF_2 (7 mm thick) (Harshaw Chemical Co.). (d) 7740 Pyrex (⅛" thick) (Kaufman Glass Co. 1984). (e) Polarizer sheet, Type HN38 (Polaroid Corp. 1978).

quite susceptible to coatings arising from the interaction of the beam and the sample. These coatings have been known to build up sufficiently to reduce window transmission, and hence the observed CL intensity, by a factor of ten or more.

SUMMARY

The summary we can make of the optical system is that one should use the simplest microscope available, at least for very low intensity samples. In addition, all elements of the system, including the viewing window itself, should be kept quite clean.

3.9.2 Polarizers for cathodoluminescence microscope attachments

Two general situations may be of interest. One is the capability for doing standard polarized light observations on samples mounted in a CMA. The other is the capability for studying the polarization of the emitted luminescence itself.

Standard polarized light studies may require rotation of the thin section, mounted on the microscope stage, between fixed crossed polarizers. When a CMA is mounted on the microscope stage, rotation over any appreciable angle is not possible because of the interference from vacuum system hoses, HV cables, etc.

Polarizer assemblies which straddle the CMA vacuum chamber have been designed to circumvent this problem (Sippel 1965, Herzog *et al.* 1970). These assemblies include two separate polarizer sheets, oriented at extinction, which can be rotated simultaneously about a fixed sample. This is equivalent to rotating a sample between fixed polarizers. Some commercial microscopes incorporate a similar arrangement, i.e., the top and bottom polarizers are connected by a 'C' shaped handle and can be rotated together. These arrangements work, but share at least two difficulties: first, full 360° rotation is not achievable; secondly, the transmission of polarizer sheets is only about 37 percent (Fig. 3.9) and it is important, therefore, that the top polarizer be removable for most CL observations – otherwise its presence would seriously hamper the observation of weakly luminescent materials such as quartz.

The CL emission from some minerals is polarized. This is usually revealed as a modest variation in CL intensity as the top polarizer is rotated. If this is the only information required it might be obtained simply by placing a piece of polarizer sheet over the observation window and rotating it.

3.9.3 Illuminator convenience

Frequently, one wishes to alternate quickly between CL and TrL on the same region of the specimen to compare features with both techniques. So the use of the standard microscope illuminator is still necessary when CL observations are being made.

Intensities of TrL are usually not critical for direct viewing. The viewing of CL is facilitated if the standard TrL illuminator is kept at as low a level as possible to ease the strain on the eye when switching back and forth between CL and TrL.

Rapid response of TrL is desirable. Some microscopes have very large illuminator lamps which may not switch 'on' and 'off' instantaneously. This can be aggravating.

Most investigators find that an automatic circuit which simultaneously switches off CL and switches on TrL, and vice versa, is the best technique. But some prefer to have a separate, variable intensity, control on the TrL so they can either turn it off completely or blend in a small amount of TrL with the CL to enhance certain features (Kopp 1980).

3.10 Spectral analysis of cathodoluminescence

The majority of investigators presently using cathodoluminescence make visual microscopic examinations of CL and may photograph selected samples. For many applications this is sufficient for presentation and discussion in publications. There are, however, important practical difficulties associated with the possible subjective definition and reporting of colors and also with the faithful reproduction of colors in those articles where color plates have been published.

Some investigators have supplemented visual observations with the addition of an accessory to display the CL spectrum, a very useful addition to a CL apparatus. This instrument can furnish a spectrum of CL intensity versus wavelength, information which is useful in the determination of activators, and even their charge states, in certain minerals (Mariano & Ring 1975). The spectra may also contribute information on the major constituents (Sommer 1972a) and on deformation effects (Sippel & Spencer

Table 3.4 Monochromator characteristics.

Author	Type	Resolution (nm)	Repeatability (nm)	Diameter of area sampled (μm)
Sippel & Spencer (1970b)	quartz prism	5	2.5	40 (by beam aperturing)
Sommer (1972c)	diffraction grating	—	1	—
Hanusiak (1975)	quartz prism	—	—	—
Mariano & Ring (1975)	grating	—	±5	50–100 (by fiber optics)
Hall & Ribbe (1971)	interference filter	—	—	300 (electron beam diameter)

1970b). It seems certain that progress towards better understanding of the physical phenomena of CL will require more widespread use of such systems.

Spectrophotometer is the generally accepted term which describes this type of instrument. A spectrophotometer includes a wavelength-selecting device (monochromator) and a suitable detector. The instrument is often referred to in discussions by the type of monochromator which is incorporated. We will here, for the most part, follow the terminology used by the authors cited. (See Table 3.4.)

Many different types of spectrophotometers have been used for CL studies. They usually represent a rather expensive addition to a CL instrument. No doubt the choice of instrument used by some investigators was dictated by the type of spectrophotometer already available in the laboratory, rather than by the optimum type for CL studies. We will briefly describe a few of the published systems here; there is not yet enough published information to permit us to make thorough comparisons of such performance factors as transmission, resolution, and accuracy of wavelength scale.

Sippel and Spencer (1970b) added a quartz prism monochromator to the CMA described earlier (Sippel 1965). They limited the area from which CL was gathered into the monochromator by limiting the electron beam size with apertures. The sensitivity was sufficiently high that they were able to use an aperture size as small as 40 μm in diameter, i.e., they could measure the spectrum of a 40 μm grain or they could sample a larger grain with 40 μm spatial resolution. The monochromator was calibrated with Hg, Ar, He, and Ne gas discharges. The spectral resolution varied from about 50 Å, based on the width of a peak at half intensity (FWHM), to a more typically used value of 200–300 Å (FWHM), depending upon the slit settings. Repeatability was stated as 25 Å. One example of their results is shown in Chapter 5.

Mariano and Ring (1975) described a very convenient system for use with a CMA. The monochromator was attached to the microscope ocular via a fiber optics light pipe which led the light into the monochromator. A rigid fiber optics probe was mounted in the eyepiece and terminated in the objective focal plane. The radiation received by the end of the fiber optics was led into the monochromator input. Separate probes provided for measurement areas of 1250, 700, and 400 μm in diameter – the areas in the objective focal plane. If the objective magnification was, for example, 10×, then they corresponded to areas of 125, 70, and 40 μm diameter on the actual sample surface. For an example of their results see Figures 4.4–4.7.

Sommer (1972a) used a wedge interference filter with 10 nm bandwidth and 35 percent transmission (in the green portion of the spectrum). This was fitted over the optical microscope ocular in the electron microprobe. He also used a grating monochromator in his work in X-ray excited luminescence (1972c). He gave a reproducibility of 5 nm for the wedge filter and 10 Å for the grating. One of the observations which he made was the systematic variation of the wavelength of the emission peak in the calcite–magnesite solid solution series with the content of $MgCO_3$ (see Ch. 7).

Hanusiak (1975), as part of his studies of quartz luminescence, attached a quartz prism monochromator to the electron probe and reported an accuracy of ±10 nm for the peak widths (FWHM) and ±5 nm for the peak position. An example of his spectra for quartz at −100 °C is shown in Figure 6.4.

Hall and Ribbe (1971) used an arrangement of interference filters on an electron probe and cited the fact that the transmission of these filters is 10–1000× greater than that of grating or prism monochromators. Transmission is always an important consideration in luminescence work because of the often low intensity of luminescence. They used ten filters to cover the 400–700 nm wavelength range. Figure 4.2 is an example from their work on cassiterite.

Bailiff *et al.* (1977) also used this type of system for their spectral analysis of low-level ThL emission. Sixteen filters, each with an approximately 10–20 nm wide transmission band, bracketed the range between 341 and 643 nm. The filters were disposed on a rotating disk which could cover the entire range in times as short as 125 msec. Intensity data were acquired and stored and then processed in the desired fashion after the run.

We could briefly describe the systems used by other authors for the spectral analysis of cathodoluminescence but the few already presented illustrate the range of approaches. So instead let us try to identify the important performance parameters to aid the selection of an optimum system.

One aspect of performance is the size of the area sampled. Some arrangements measure the entire radiation within the objective field of view. With these systems, the spectral analysis of luminescence from small areas requires aperturing of the electron beam (Sippel & Spencer 1970b). Other systems use the approach of locating one end of a small fiber optics light pipe in the objective focal plane (Gamma Scientific 1983, Mariano & Ring 1975). The position of the fiber optics probe can be seen through the ocular in the normal fashion. This arrangement facilitates looking at small grains and does not interfere seriously with the normal viewing.

A second important consideration is transmission. Cathodoluminescence intensities are frequently low and therefore this parameter is even more important than it might normally be. It is also related to the previous statement – the size of the area sampled – since one of the limits to this is the integrated intensity required to generate the desired signal strength and signal-to-noise ratio. If the transmission is too low, the area sampled will have to be relatively large.

Speed of operation, the time required to scan the entire spectrum, is sometimes important. For example, some quartz CL change as the sample is being bombarded, in times of only seconds. To obtain meaningful spectra, scans may need to be taken in a few seconds or less for these quartz samples.

Another important aspect of the system performance is the resolution. It is not always critically important, because the luminescence spectra of many geological samples are broad-band in character. But there are examples, and rare earth activators are the most important of these, where line spectra are produced and the ability to resolve and identify the line spectra permits the identification of the particular activator present. Mariano and Ring give examples of rare earth line spectra with spacings as small as 25 nm; e.g., Eu^{3+} has lines at 590 and 615 nm.

Finally, the usual parameters of the overall spectral range and the accuracy and reproducibility of the wavelength values are always important considerations.

Note that we have not stressed the importance of the 'transmission-wavelength' calibration. The magnitude of the variation of the transmission with the wavelength varies among the different types of instrument and a few authors (e.g., Sippel & Spencer 1970b) comment on this and provide data. At the present level of sophistication of use of cathodoluminescence spectra, i.e., as a qualitative tool, this variation is not important. This situation may change as more users recognize the value of spectral analysis. Also the possibility of eventually being able to make quantitative analysis for the activators in some minerals should not be overlooked. One of the factors which should be kept in mind is that the spectral information may be influenced not only by the passband of the monochromator used but also by the transmission of the lead-glass in the CMA and the optics of the microscope. Any overall calibration needs to be done on the completely assembled system, including the CMA window and the microscope, not just the monochromator.

Dynamic range is another parameter of importance. At the present time, most of the results reported do not cover a wide dynamic range, i.e., the intensity scale is not expanded to look at small peaks. Many published spectra are of relatively low-intensity CL and these spectra show an appreciable noise level (e.g., Figs 4.4b & 5.3). Any small peaks obtained by scale expansion in this instance would be lost in the noise. But there are many examples where noise level is not yet visible in the spectra (Figs 4.5b & 4.6c) and repeat spectral scans at expanded sensitivity might well yield additional data of significance.

Later chapters will contain examples of studies where spectral analysis has proved useful.

3.11 UV and IR investigations

Various investigators have noted that there may be useful CL information in the UV and IR regions of the spectrum, outside the visible range (Potosky & Kopp 1970, Geake & Walker 1975), and much work has been done on IR luminescence under photoexcitation (Barnes 1958).

Davey (1965) presented an example of the CL of calcite, studied in the UV. The work was done on an electron microprobe which was equipped with a fused silica vacuum window and a special UV microscope objective. The UV picture was a reconstructed image, obtained at the particular wavelength selected by a grating monochromator.

Investigations in the UV and IR become of even more interest when a wide range spectral analysis instrument is available. Limitations to measurements in the UV and IR can arise from the reduced transmission of the elements in the optical system including the lead-glass viewing window, the microscope optics, and the monochromator instrumentation. The importance of the first two items can be judged from the several transmittance curves gathered from the literature and shown in Figure 3.9.

Note that the transmission of a lead-glass window at 400 nm is only half its maximum value so this is a critical element in UV work. With alternative X-ray shielding, one might substitute a quartz or MgF_2 window and extend the range to below 20 nm, at least for the window.

Information on microscope lenses for IR and UV is less readily accessible but fluorite and quartz lenses of various types are available (Oriel Corp. 1979, Rolyn Optics Co. 1981) to extend the microscope performance into the IR and UV.

So many new optical materials are continually being developed that no doubt additional choices will be available for components with enhanced IR and UV transmission.

3.12 Photographic techniques

At first glance, it would appear that photomicrography of the CL of samples is not very different from photomicrography of other subjects – transmitted light and polarized light studies, for example. To some extent this is true, but there are a few important differences.

CL intensities are very low in comparison to those available for many other types of photomicrography subjects. And unlike TrL or PoL, CL intensities cannot be increased arbitrarily by increasing the intensity of excitation. What can be done is to use high-speed film and the best transmission optics available.

Another aspect of photomicrography of CL is that CL is emitted radiation, whereas many other subjects are light absorbers. The emitted CL is much brighter than absorption spectra and more difficult to get true rendition of on film.

Finally, the range of wavelengths and intensities present in many CL studies taxes the exposure range and wavelength range of film. For example, one might encounter both bright blue feldspars and also dull red or brown quartz in the same sample. Frequently, this leads to the requirement that the feldspar be over-exposed in order for the quartz to be recorded.

Even though it is difficult to write an exact formula for exposure time, it is possible to give a relationship which indicates how the exposure time will vary with several important parameters, if the sample, the exciting beam energy, and microscope–camera system are kept the same. This relationship is

$$t \propto (M^2 \times A)/(\mathrm{ASA} \times I \times \mathrm{NA}^2) \qquad (3.7)$$

where M is the magnification between the sample and the film, A is the area of the electron beam (assumed larger than the area actually being recorded on the film), NA is the numerical aperture of the objective lens, ASA is the film speed rating, and I is the beam current. The other important variables of electron energy, sample type (CL efficiency), and microscope–camera system cannot easily be incorporated into such an expression because there are simply not enough data available on relative CL efficiencies of different minerals, on the variation of CL efficiency with electron energy, or on microscope transmission.

For a typical case of a brightly luminescing carbonate, ASA 400 film, 0.5 mA beam, 15 kV energy, 5 mm diameter spot, and a monocular microscope–camera system producing an overall magnification of 20× on the film, the exposure time will be in the range of 10–20 seconds.

Some investigators record on black & white film, instead of color film, especially for zoned carbonates whose features are displayed quite clearly in this type of presentation. Faster speeds and high contrast are advantages of black & white film.

Exposure times reported in the literature range from a few seconds to many minutes, depending on sample, excitation, microscope system, film speed, etc. At this time there seems to be no easy way to categorize them. Some of the reasons for this have been presented in the discussion on microscopes in Section 3.9. We will be content to list the excitation and exposure conditions for as many as possible of the color plates presented elsewhere in the text.

In addition to photomicrography, macrophotography of CL has also been done. In this work, the beam is defocused to a large area (1–2 cm diameter or greater) and the microscope is replaced by a camera with a macro lens. The technique has been used to record the CL of chips of lunar samples (Geake *et al.* 1971), hand specimen size geological specimens (Mariano 1983a), gems (Ponahlo 1983), and archeological specimens (Sutton 1978).

3.13 Suggested standards for reporting CL results

It is useful to report a full set of experimental conditions when reporting CL results (Marshall 1978). It is especially important to note the conditions for samples which are found to be nonluminescent. Nonluminescent, or faintly luminescent, samples observed on one instrument and microscope combination, may be more (or less) readily observable on others. It appears that there may be as much as a factor of 10 to 50 variation in overall sensitivity among the hundreds of CL systems in use. For a bright sample, this difference might express itself, for example, as a difference in photographic exposure times of 5 seconds versus 50 seconds. In the case of a dull sample, the difference could be a 1–2 minute exposure versus a 30–40 minute exposure or longer, for the same sample!

A discussion of the various important parameters and equipment specifications has been prepared (Marshall 1978). From the discussion in the earlier sections of this chapter, we can state that the intensity of bombardment can be specified by reporting the beam voltage, the beam current, and the beam current density. The sample preparation procedure may be important. The measured (observed) intensity of the CL will be affected by the condition of the viewing window. The vacuum side of the viewing window will acquire a coating and this window coating may lead to as much as a factor of 50 to 100 reduction in transmission if it builds up over a period of time and the window is not cleaned. This coating arises from interactions between the electron beam and hydrocarbon vapors in the chamber produced by the pump or emanating from the sample. The components of the microscope and photographic system can have a large effect. The type of microscope, the magnification of the objective and ocular, and the photomicrograph set-up can differ widely among different investigators. And, finally, the film speed and development process (especially if forced development is used), and the exposure time should be furnished.

Fairchild (1983) suggests that detailed reporting of experimental conditions is not of practical use but the present author believes it is worthwhile, even with all of the uncertainties that presently exist, for the orderly development of the technique and its application.

Presently, the tendency is to report CL data in general terms of bright luminescence, nonluminescent, or dull luminescence. More quantitative statements of relative intensity information, along with the more complete statement of experimental conditions, would facilitate the comparison of results among different workers and different laboratories.

Quantification of the CL 'color' and intensity is more difficult unless one has access to a recording spectrophotometer such as described earlier in this chapter. Even with these instruments, a need exists for an intensity calibration standard.

Finally, we note again that some samples show time-dependent effects, i.e., a gradual change in the intensity and/or color of the CL with time, under bombardment. The more efficient the instrumental system, the less important this effect, because excitation power can be lower and exposure times shorter. Time-dependent effects should be reported but the best format for doing so remains to be determined until the effects are better understood.

Chapter 4

Cathodoluminescence properties of some common minerals

The published information on the CL properties of some of the more common rock-forming minerals is summarized in this chapter. The minerals are listed in the order used by Mason and Berry (1968). If there is no published information, the mineral is not listed. Expanded discussions on carbonates, feldspars, and quartz appear later in separate chapters; a few of the rarer minerals are included in the discussion in Chapter 8. These later chapters also indicate how CL information is being used to solve geological problems.

The amount of published information available for many minerals is quite limited, often only one or two references. This makes it difficult to make a critical review of the information and, for the most part, none is attempted. The reader should therefore bear in mind the many factors discussed in earlier chapters which can influence CL. Differences in CL reported by various authors may exist because of variability of samples, differences in strength of excitation, and different viewing techniques. For those minerals which exhibit polarization of CL, the results may be influenced also by the orientation of the sample. Of all of these factors, strength of excitation is particularly important. Excitation which is of too low intensity (or too low electron energy) may not produce emission which might otherwise be visible at higher excitation. On the other hand, if the excitation is too strong, one can easily approach incandescence conditions which push the observed spectra to the red end of the spectrum or make irreversible changes to the luminescence properties of the mineral.

As an example of the wide variability in the amount of data available on the CL properties of any given mineral we note that most of the information presented here on cassiterite is extracted from the 15-page paper by Hall and Ribbe (1971) concerning this one mineral. Rémond (1977), also, has done an extensive study of cassiterite. In contrast to this, comments on many of the other minerals are based on the useful survey by Görz *et al.* (1970) which reported on the color, intensity, polarization, rise time, and decay time of 28 different minerals. However, they may have been able to sample only one specimen or, at most, a very small number of specimens of each. As CL applications expand and further research is done, the information in this chapter will certainly be further developed.

It is important to keep in mind that, whereas a mineral may be a precisely defined quantity in terms of crystal structure, hardness, and many other physical properties, this is not the case for the CL properties. As outlined in Chapter 1, the CL properties may depend on subtle differences in amounts of different impurities and on crystal defects. In the nomenclature of the phosphor industry, a natural mineral is not a chemical compound, it is a material. Therefore each of the several observations reported for a given mineral will, in all likelihood, have been made on a different material, and until the possible activator, quencher, and defect structure contributions to the CL are thoroughly understood, it may be difficult to put all the observations into a consistent framework.

A large volume of luminescence literature is in Russian and the original papers are accessible only with difficulty. Most of the Russian literature through 1979 is, however, discussed in the text by Marfunin and usually we reference only this text. The interested reader will be able to pursue the original literature from there. It seems that much of the Russian literature is on photoluminescence, not CL, but often one cannot tell from the limited amount of information given.

Quite useful compilations of CL of minerals have been prepared previously by Nickel (1978) and Amieux (1982).

4.1 Class I: native elements

4.1.1 Semimetals and nonmetals

Diamond C Isometric Crookes (1879) had little difficulty observing the luminescence of diamonds in his apparatus. He states (p. 660):

> The only body I have yet met with which surpasses the luminous sulphides both in brilliancy and variety of colour is the diamond. Most of these gems, whether cut or in the rough, when coming from the South African fields, phosphoresce of a brilliant light blue colour. Diamonds from other localities shine with different colours, such as bright blue, pale blue, apricot, red, yellowish-green, orange, and bright green. One beautiful green diamond in

my collection when phosphorescing in a good vacuum gives almost as much light as a candle: the light is pale green – almost white. A beautiful collection of diamond crystals kindly lent me by Professor MASKELYNE phosphoresce with nearly all the colours of the rainbow, the different faces glowing with different shades of colour . . . .

Casual observations by the author of various samples of natural and synthetic diamonds, and diamond abrasive retained on incorrectly prepared thin sections, reveal a wide variety of different, intense CL colors, confirming Crookes' observations.

Leverenz (1968) discusses some observations on diamonds. He notes that thermal quenching is important and quotes an example where strong luminescence at −180 °C is no longer detectable at +350 °C. He also notes that some diamonds at room temperature show blue luminescence under UV excitation whereas others 'which have equal or superior gem quality do not give perceptible luminescence at room temperature.'

Gaal (1976–77) shows color plates of gem minerals under CL including several examples of diamonds. He states that blue CL is the most common, but red and green, and an example with a very weak blue-yellow-green CL color, were seen also. Crystal zoning and structural defects were better revealed. He notes that some examples which initially luminesced blue became pink at higher beam power densities.

Mendelssohn *et al.* (1979) have studied haloes in diamond with CL. Optical microscopy sometimes reveals haloes but many more are seen with CL. The authors note that the damage haloes are different for diamonds of different types and that they also depend on the subsequent thermal history of the diamond. They carried out various experimental attempts (irradiation, annealing) to distinguish internal sources for the haloes from external sources and used this to help to unravel some of their geological history. They mention the possible influence of quenching due to defects, analogous to concentration quenching in activated CL. The CL intensity reaches a maximum as the defect concentration increases and decreases at higher concentrations. They also noted the effect of beam energy on halo resolution. At 30 kV, the resolution is poor. At 10 kV the resolution is much better. Their publication contains a color plate with ten examples.

The properties of diamonds have attracted a great deal of attention from the solid-state physicists. Some of this work is directed to understanding of optical properties but CL studies have also been included. The references by Mendelssohn *et al.* (1979), by Davies (1974), by Hanley *et al.* (1977), by Lang (1979), by Collins and Woods (1982), and the book by Field (1979) will lead the interested reader into the extensive literature in this field.

Davies (1974, 1979) has presented an experimental and theoretical treatment for many aspects of CL and related properties of diamonds, including photoluminescence and absorption spectra. As one indication of the complexity of this mineral, Davies (1974) states: 'Over fifty absorption and luminescence bands have been reported in the optical spectra of diamonds.'

Some of the CL from diamonds is associated with specific localized crystalline imperfections. For example, a blue, linearly polarized, luminescence in natural and synthetic diamonds has been attributed to dislocations (Kiflawi & Lang 1974, Woods & Lang 1975). Both polarized CL in the IR (Lang 1980) and yellow CL (Collins & Woods 1982) have been associated with 'giant platelets' (planar defects with dimensions up to several tens of microns). An example of the yellow CL of 'giant platelets' is shown in Plate 1a and an example of the blue-luminescing diamond is shown in Plate 1b.

Marfunin (1979) states that: 'In diamonds all the luminescence centers are associated with a nitrogen impurity which substitutes carbon in the structure of the diamond.'

In most cases diamonds are examined as bulk specimens, not thin sections. As Davies (1974) points out, in such a case it is better to observe the CL from the excited face to minimize the effects of absorption. Spectra collected from the far side, a technique which is often used in CL studies of thin sections, might be appreciably modified by absorption of the CL by the diamond crystal.

4.2 Class II: sulfides

Most of the sulfides are relatively good conductors and do not exhibit luminescence. The most important exception to this is sphalerite.

Sphalerite ZnS Isometric Cubic ZnS is important in the phosphor industry and Leverenz (1968) has over 100 index entries to this one material. He notes that self-activation may be important in very pure ZnS, possibly produced by loss of small amounts of sulfur during high temperature crystallization of the phosphor. He also reports that ZnS with 0.01 percent Ag exhibits CL at 4,550 Å, ZnS with 0.01 percent Cu exhibits CL at 5,280 Å, and ZnS with Mn exhibits CL at 5,910 Å. Information is also provided on other activators in synthetic ZnS including Au, Ni, Sm, and also several cases of two activators being present simultaneously.

Goni and Rémond (1969) and Rémond (1977) associated the CL of sphalerite with gallium and copper (red CL) and germanium and copper (green CL). The former publication includes a color plate with numerous examples of sphalerite CL. In the comprehensive paper by Rémond in 1977, a combination of techniques including electron probe, SEM, secondary ion mass analysis, as well as CL were used. Among other interesting effects he noted a red CL superimposed on a green CL. The red CL intensity depended strongly on excitation conditions and decreased at higher excitation currents.

Gaal (1976–77) reported intense orange CL in sphalerite.

Marfunin (1979) comments also on the luminescence of ZnS (apparently photoluminescence). He states that Ag^+ produces blue luminescence, Cu produces green (520 nm), Cu–Ga and Ag–Ga produce luminescence in the range of 620–680 nm, and Cu^+–Tl^+ produces a dark red (825 nm) luminescence. He states that 0.1 percent or more of iron will quench the luminescence.

Ebers and Kopp (1979) used CL in their study of the

gangue dolomite associated with sphalerite mineralization. They noted that sphalerite 'does not normally luminesce because of its high Fe content, but locally it may luminesce bright yellow.'

Hagni (1984) observed intricate growth zoning in some sphalerites, brought out only by CL. He also noted many examples of nonluminescing sphalerite.

Cinnabar HgS Trigonal Tarashchan (1978; see Marfunin 1979) shows photoluminescence spectra and also excitation spectra of natural cinnabar with a peak in the luminescence spectrum at 630 nm.

Realgar AsS Monoclinic Tarashchan (1978; see Marfunin 1979) shows photoluminescence spectra and also excitation spectra of natural realgar with a peak in the luminescence spectrum at 660 nm.

4.3 Class III: oxides and hydroxides

Cuprite Cu_2O Isometric Leverenz (1968) reports that cubic Cu_2O:Cu is an infra-red emitting phosphor.

PERICLASE GROUP

Periclase MgO Isometric Saksena and Pant (1954) observed a strong CL emission in the 4,550–4,440 Å region in a fresh, pure, crystal of MgO but they noted that the CL intensity weakened with exposure.

Reid *et al.* (1964) observed red luminescence of synthetic periclase under X-ray excitation.

Greer *et al.* (1967) reported blue CL from MgO.

Leverenz (1968) reports that Sm can be an activator in MgO and he shows four lines in the spectrum at approximately 5,730, 5,830, 6,100, and 6,280 Å.

Mariano and Roeder (1983) noted that the periclase from the Kerimasi søvite contains 1.8 percent MnO but does not exhibit CL. (The iron content is 12.6 percent.)

SPINEL GROUP

Spinel $MgAl_2O_4$ Isometric Hummel and Sarver (1964) synthesized spinels using a controlled atmosphere system. They incorporated either Mn^{2+} or Fe^{3+} but not both elements together. In one spinel, the Mn^{2+} replaced 1 mole percent of Mg. In a separate sample, the Fe^{3+} replaced 0.2 mole percent of the Al. The Mn-activated spinel phosphor was excited by a 16 kV electron beam, at a current density of 1 μA/cm^2 (0.016 W/cm^2). The Fe-activated phosphor was excited by a 16 kV electron beam at a current density of 0.5 μA/cm^2 (0.008 W/cm^2). In both cases, the visual CL was a deep green. With a spectroradiometer, they observed a green peak from the Mn-doped sample. They observed the green peak and also a red peak from the spinel containing ferric ion (Fig. 4.1).

Reid *et al.* (1964) also observed green luminescence of a synthetic spinel sample under X-ray excitation.

Leverenz (1968) reports that the CL of synthetic cubic $MgAl_2O_4$ is weak green when Mn is the added impurity and a 'weak mottled green and red' when Cr is the added impurity.

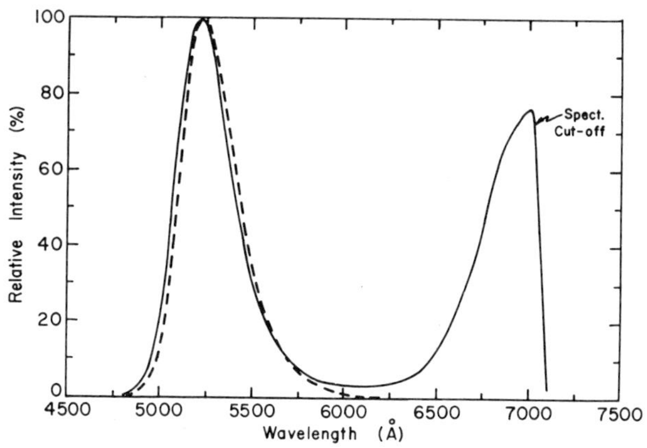

Figure 4.1 Normalized curves of cathodoluminescence emission spectra of $MgAl_2O_4$ activated with Mn or Fe. Solid line is for sample with 0.2 mole percent Fe replacing Al. Dotted line is for sample with 1 mole percent Mn replacing Mg. The CL of both samples appears deep green to the eye. (From Hummel & Sarver 1964.)

Hutcheon *et al.* (1978) studied inclusions in the Allende meteorite and noted yellow CL from some spinel, brilliant red and orange CL from spinel low in Fe, and blue CL from Fe-rich spinel.

Chrysoberyl $BeAl_2O_4$ Orthorhombic Crookes (1879) saw no CL in a sample of chrysoberyl. Gaal (1976–77) noted light yellow CL in a sample of gem quality chrysoberyl. He also reported that the alexandrite variety showed dull red or brown CL in those cases where it luminesced.

HEMATITE GROUP

Corundum Al_2O_3 Trigonal Leverenz (1968) states that the Cr activator in rhombohedral Al_2O_3 produces two sharp lines at 6,927 and 6,947 Å. Sm produces seven lines between 570 and 625 nm.

Chinner *et al.* (1969) saw red CL in natural corundum. Görz *et al.* (1970) reported on a sample of corundum with blue CL which was not polarized. Goebel and Patzelt (1976) and Dudley (1976) saw red CL in some samples. Weiblen (1965) saw faint red CL in a synthetic corundum.

Aluminum oxide is commonly used as an abrasive material for thin section preparation. If it is not thoroughly removed during the cleaning process, one will readily notice the residual grit because of its bright CL.

Plate 1f illustrates the CL of a synthetic corundum.

Ruby Al_2O_3 Trigonal Crookes (1879) made observations on the CL of ruby. To quote again (p. 661):

> Next to the diamond, alumina in the form of ruby is perhaps the most strikingly phosphorescent stone I have examined. It glows with a rich, full red; and a remarkable feature is that it is of little consequence what degree of colour the earth or stone possesses naturally, the colour of the phosphorescence is nearly the same in all cases; chemically precipitated amorphous alumina, rubies of pale reddish-yellow, and gems of the prized 'pigeon's blood' colour, glowing alike in the vacuum . . . .

Later he continues:

Ruby gives out a transcendentally fine crimson colour, exhibiting no marked distinction in the plane of its polarisation, though in one part of a stone the colour was extinguished by a Nicol prism with its long diagonal parallel to the axis of the crystal. Here, therefore, also the light was that of the extraordinary ray . . . .

Recent authors (Gaal 1976–77) have also noted the intense red CL and there is a great deal of information available in the laser literature because of the importance of ruby, doped with the activator Cr^{3+}, in this field.

Sapphire (Blue) Al_2O_3 Trigonal Crookes also investigated the CL of aluminum oxide in the form of sapphire, and reported: 'Sapphires gave out a bluish-grey light, distinctly polarised in a plane perpendicular to the axis. In this case, again, the ray developed corresponds to the extraordinary or quicker ray.'

McCauley (1975) has observed the CL of synthetic sapphires containing molybdenum wires and shows that the diffusion gradient of the Mo into the sapphire is nicely revealed by a progressive change in the color of the CL.

Hematite Fe_2O_3 Trigonal Weiblen (1965) did not observe CL in a hematite sample. Smith and Stenstrom (1965), however, showed an example of a red-luminescing hematite inclusion in an aventurine feldspar. We are unaware of other references to luminescent hematite.

RUTILE GROUP

Rutile TiO_2 Tetragonal The rutile samples studied by Weiblen (1965) did not show CL. Sprunt (1981), in her study of the reasons for CL in quartz, stated that 'Rutile is easily distinguished from quartz with CL because rutile does not luminesce.'

Leverenz (1968) lists Sm as an activator in TiO_2, producing several lines in the CL spectrum between 570 and 630 nm.

Cassiterite SnO_2 Tetragonal Tinstone was one of the minerals that Crookes (1879) investigated. He observed a 'fine yellow light, which was extinguished almost entirely when the long diagonal of the Nicol was perpendicular to the axis of the crystal.'

Cassiterite luminescence has been more thoroughly investigated than that of many of the more common minerals. Le Poole *et al.* (1968) included color photos of zoned, green-luminescing cassiterite in the paper describing their CL instrument. Hall and Ribbe (1971) and Rémond (1977) have made important contributions to this subject.

Hall and Ribbe (1971) used an electron probe, equipped with an interference filter for spectral analysis of the CL. They not only observed the luminescence but also made electron probe analyses for possible activators. The effects of temperature on CL were observed and they also measured decay times.

Forty samples were studied and Hall and Ribbe were able to place them in one of three categories.

They found a correlation between Ti and yellow luminescent zones and concluded that Ti is the most likely activator of yellow CL. The presence of both Si and Ti produced an orange CL.

Type	Characteristics	CL color
I	pleochroic, usually unzoned; they contain appreciable amounts of Ta, Nb, Fe, and Ti	no appreciable CL
II	contain appreciable Fe and Ti; usually no W, Ta, or Nb	thinly zoned yellow CL
III	moderate amounts of W, Fe, Ti; no Ta or Nb	blue CL

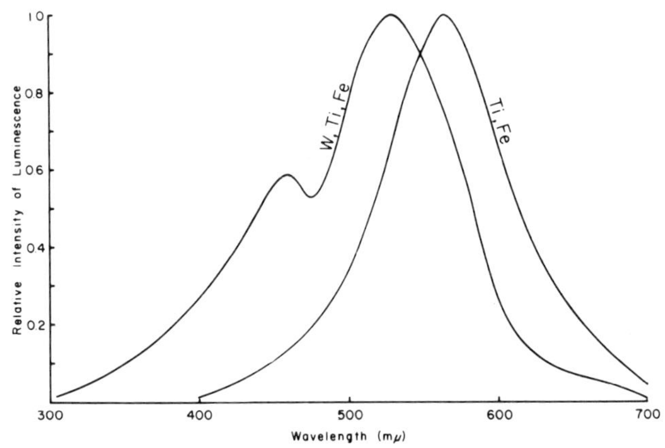

Figure 4.2 Typical CL spectra for (Ti, Fe)-activated and (W, Ti, Fe)-activated cassiterite. To the eye the samples appear yellow and blue respectively. (Hall & Ribbe 1971.) See also Plates 1c and d.

They found evidence that W is the activator for blue luminescence. Ta and Nb, when present, appeared to act as quenchers. Typical CL spectra from two of their sample types are shown in Figure 4.2. See also two of their examples in Plates 1c and d.

For one of the (Ti, Fe)-activated samples, they studied the effects of temperature on CL intensity. A rapid decrease of CL with increasing temperature was noted (Fig. 4.3).

They also reported the decay time for one of their Type II samples as 0.17 ms.

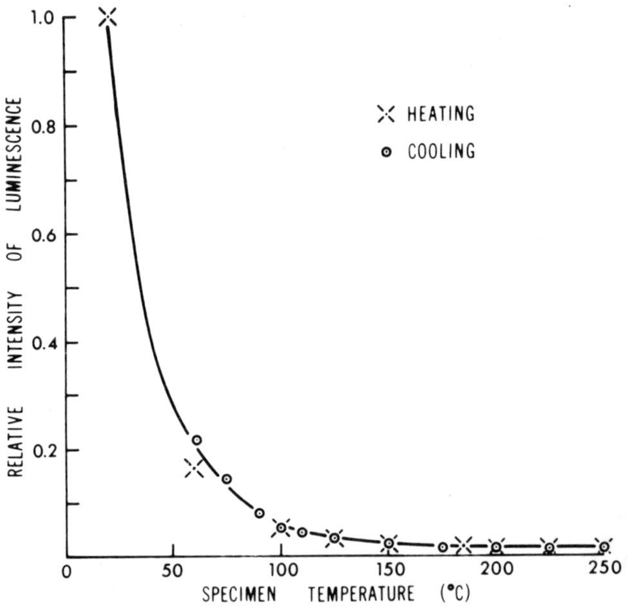

Figure 4.3 The variation of CL with temperature for a cassiterite activated with (Ti, Fe). (From Hall & Ribbe 1971.)

Hall and Ribbe reached the general conclusion that the natural specimens were too complex, in terms of impurities, to answer the question on activators and quenchers unequivocally. Synthetic cassiterites, carefully doped, will be required to answer these questions.

Rémond (1977) also presents results of a study of several samples of natural cassiterite using CL, ion mass analysis, SEM, and electron microprobe techniques. The association of yellow CL with Ti (and also Fe, Al, and V) was observed. He also noted the association of blue CL with W and the quenching effects of Nb and Ta.

URANINITE GROUP

Uraninite UO_2 Isometric Goebel and Patzelt (1976) reported orange CL from pitchblende.

Thorianite ThO_2 Isometric Leverenz (1968) lists Sm as an activator in synthetic ThO_2 producing four lines between 570 and 620 nm.

4.3.1. Hydroxides

LEPIDOCROCITE GROUP

Boehmite AlO(OH) Orthorhombic Dudley (1976) reported a deep blue CL in bauxite.

4.4 Class IV: halides

HALITE GROUP

Halite NaCl Isometric Pure halite shows weak or no luminescence at room temperature (Murata & Smith 1946, Teegarden 1966).

Murata and Smith (1946) studied the photoluminescence of several natural and synthetic halite samples. They found that neither manganese nor lead alone produced significant photoluminescence. 'But the two together in amounts of the order of a few thousandths of a percent will coactivate a vivid red fluorescence.'

Leverenz (1968) references Cu and Ag as activators in NaCl, producing CL peaks at 3,580 and 2,490 Å respectively. The 3,580 Å peak due to Cu exhibits a long phosphorescence at room temperature. Mn is also listed as an activator but the spectrum produced is not given.

Dickson (1980) observed strong blue luminescence 'which fades over a few hours' exposure.' The halite developed a yellow semipermanent coloration under the electron bombardment.

Sylvite KCl Isometric Pure sylvite at room temperature luminesces weakly, if at all (Teegarden 1966). However, it can accept many impurity activators. Leverenz (1968) references examples of several possible activators in KCl including Ag, Cu, Mn, Sb, and Tl. Cu produces a spectral peak at 3,960 Å, Ag in the UV at 2,720 Å, Sb at 5,700 Å, and Tl produces several peaks, the most intense ones being in the UV.

OTHER HALIDES

Fluorite CaF_2 Isometric Leverenz (1968) lists Sm as an activator in CaF_2, producing three lines at approximately 5,670, 6,060, and 6,130 Å.

Mariano and Ring (1975) presented spectra of both synthetic and natural Eu^{2+}-activated fluorites. The spectra showed a maximum between 410 and 435 nm and the fluorite exhibited 'a strong blue luminescence' (Fig. 4.4).

Marfunin (1979) also discusses REE as activators in fluorite. He states that examples of all REE^{3+} and REE^{2+}, except for Tb and Gd, can be found as activators of luminescence in natural fluorites. A spectrum with two peaks in the 4,100–4,300 Å range is shown for the case of Eu^{2+}. His remarks are apparently based on photoluminescence results.

Dickson (1980) also noted a strong blue CL in fluorite. He found that the fluorite developed a semipermanent purple color when subjected to prolonged electron bombardment (30 kV, 800 μA, 1 hour) and indicated how this technique might be used to identify fluorite, especially when present as small crystals. Internal features, e.g., growth zones, are also displayed by this induced coloration.

CL spectra of fluorite activated by other REE are reproduced in Chapter 8. This later chapter also includes an expanded discussion of the causes of CL in natural fluorites and the application of this information to the interpretation of the geochemistry of the crystallizing solutions.

4.5 Class V: carbonates, nitrates, borates

4.5.1 Carbonates

The carbonate minerals frequently exhibit luminescence. It is generally accepted that the most important activator is Mn^{2+}; Fe^{2+}, Co^{2+}, and Ni^{2+} can act as quenchers.

CALCITE GROUP

Calcite $CaCO_3$ Trigonal Calcite typically shows a bright luminescence if it contains Mn^{2+} in the amount of a few hundred to a few thousand ppm. The Fe^{2+} ion can serve as a quencher. We note again, for emphasis, that the Fe^{2+} (as well as the Mn^{2+}) must be present as a substitutional, or possibly interstitial, impurity to be effective, not simply as a surface coating or staining.

The color of calcite CL is usually described by some combination of the colors yellow, orange, and red.

Occasionally, large rhombohedral crystals of Iceland spar are found to exhibit a faint blue luminescence. The cause of this is not known to the author. It may be intrinsic. A similar dull blue luminescence is sometimes seen in limestone samples. Sippel and Glover (1965) have suggested that the dull blue luminescence is due to a defect center. These same authors also report on some calcite showing a green luminescence which they attribute to an unidentified activator. Barsanov and Sarsembaeva (1962, referenced in Sommer 1972a) suggested that the green luminescence in Iceland spar is due to the presence of interstitial Mn.

Schulman *et al.* (1947) found that calcite exhibited CL if it contained Mn but that in order for calcite to exhibit photoluminescence it needed to contain both Mn and a sensitizer which might be Pb, Tl, or Ce. Thus the combination of cathodoluminescence and photoluminescence observations of the same sample can serve as a test for the presence of these sensitizing elements.

Medlin (1963a) studied both the ThL and the XL of synthetic calcite containing various added impurities. He

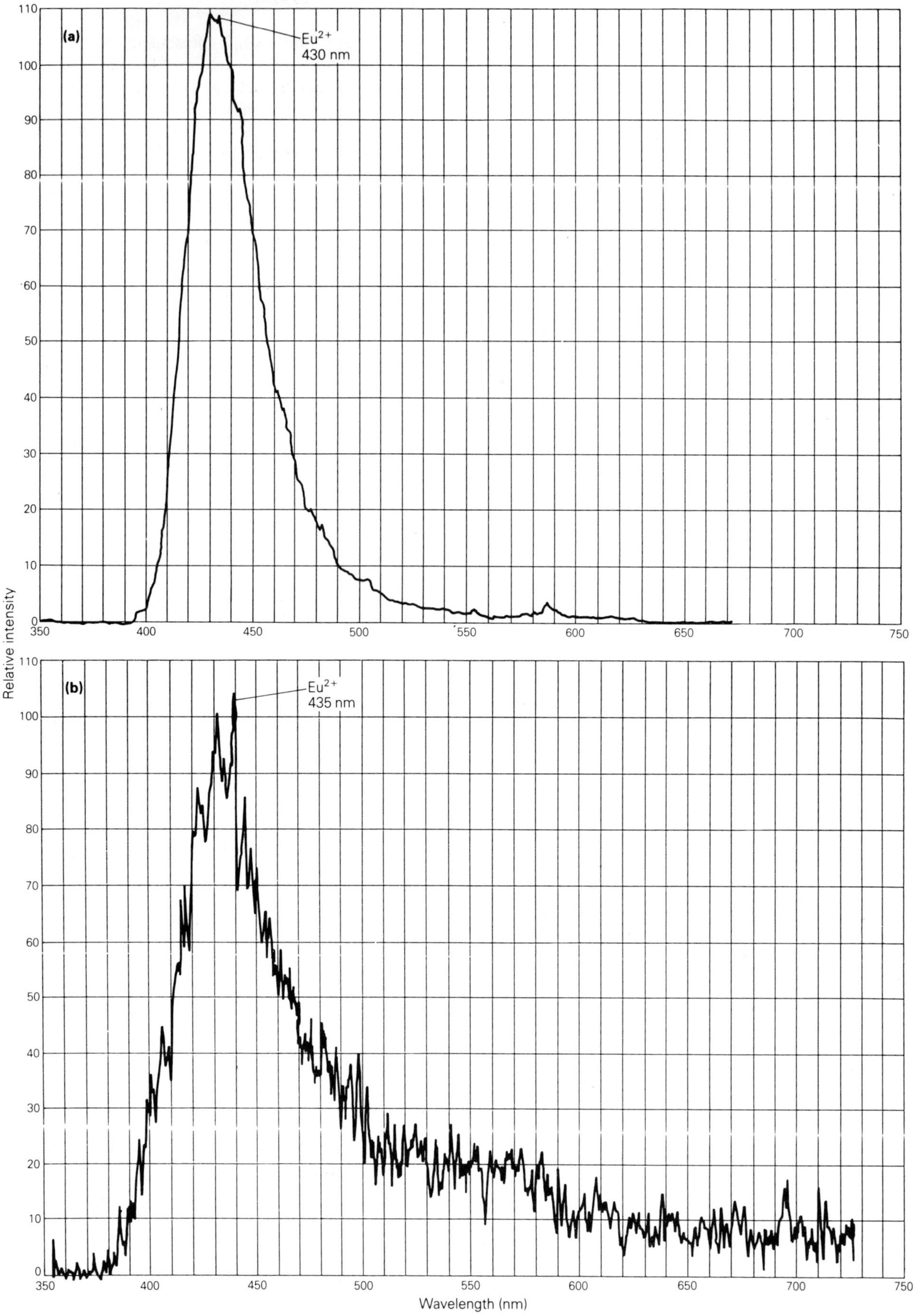

Figure 4.4 A comparison of (a) the CL spectrum of a synthetic, Eu^{2+}-activated, fluorite (CaF_2) to that (b) of a natural fluorite from a hydrothermal ore vein in Safford, Arizona. The color to the eye of both is a strong blue. (From Mariano & Ring 1975.)

found that Mn^{2+} and Pb^{2+} were the important activators, and Fe, Co, and Ni can serve as quenchers. The emission spectrum of the XL had a maximum at 6,600 and 6,400 Å for sample temperatures of 77 and 300°K, respectively.

Despite the fact that a great deal of use has been made already of the CL properties of calcites in textural studies, there is only a moderate amount of quantitative information available on the relationship between CL intensity and the concentrations of Mn and Fe. The information that is available is reviewed in Chapter 7. In general terms it appears that as little as 10–50 ppm of Mn may be sufficient for activation in Fe-free samples, whereas, for more typically encountered natural samples, about 100 ppm or more may be required. Some manganese-rich samples containing as much as 1 percent iron still exhibit bright luminescence. Little data have been published on calcites containing more than 1 percent iron.

In addition to the very important rôle of Mn as an activator in calcite, there is also demonstration that REE can serve as activators for UV-excited luminescence (Gies 1975, Machel 1983, 1985) as well as for CL. Chapter 8 has examples of Dy^{3+}- and Sm^{3+}-activated CL in calcite which, to the eye, looks identical to that produced by Mn^{2+}. Without access to accurate spectral analysis, the REE-activated calcite might pass for an Mn-activated calcite. Machel (1983) has also stressed the possible importance of the REE as activators in calcite and dolomite, especially for photoluminescence. Rare earth element activation is obviously important in environments such as carbonatites and this subject is treated in more detail in Chapter 8. Machel (1983) has emphasized that there are possible sedimentary situations which contain a source for REE, and the carbonates formed may contain them as activators.

Coy-Yll (1969/70) has measured decay times of calcite CL and found a value of approximately 12 ms. Occasionally samples of calcite show phosphorescence to the unaided eye, i.e., with decay times measured in seconds. It is generally considered that the decay is too long to permit ready use of the SEM for studying CL of calcite (see Section 3.4.3).

To the author's knowledge, no one has reported on polarization of the CL from calcite.

Chapter 7 contains an expanded discussion of the CL properties of calcite.

Magnesite $MgCO_3$ Rhombohedral Medlin (1963a) prepared synthetic magnesite samples and studied their ThL and XL. He found that Mn^{2+} is an activator and Co^{2+} and Ni^{2+} are quenchers. The emission spectrum of the ThL peaked between 6,600 and 6,800 Å. The emission spectrum of the XL in his samples was too weak to measure.

Sommer (1972a) reported that the CL spectrum of synthetic Mn-activated magnesite, containing 1,000 ppm Mn and 250 ppm Fe, has a peak at 675 nm. He also observed that magnesite CL is lost very quickly under bombardment in the electron probe at high-power densities (see Section 3.8.2).

Rhodochrosite $MnCO_3$ Trigonal Gaal (1976–77) noted dull orange CL. Goebel and Patzelt (1976) noted violet CL in some samples.

Smithsonite $ZnCO_3$ Trigonal Hagni (1984) reported bright blue CL in a sample of smithsonite.

A CL emission spectrum from a smithsonite sample is included in Figure 7.7; in this sample the peak is at approximately 650 nm, and at a longer wavelength than that of calcite, dolomite, and witherite.

Otavite $CdCO_3$ Trigonal Sommer (1972a) reported that the CL spectrum of Mn-activated otavite showed a maximum at 595 nm, similar to calcite with 15 percent $MgCO_3$.

Vaterite $CaCO_3$ Hexagonal Sommer (1972a) reported that vaterite has a CL spectrum similar to that of calcite with a peak at 595 nm.

DOLOMITE GROUP

Dolomite $CaMg(CO_3)_2$ Trigonal Dolomite luminescence is usually attributed to the action of Mn^{2+} as an activator and Fe^{2+} as a quencher. The CL of dolomite is usually more to the red end of the spectrum than that of calcite and, accordingly, one can often quickly distinguish between these two minerals in the same thin section by the CL color contrast, without resort to staining. Examples of this contrast can be seen on Plates 7c, d, and 10a.

Medlin (1961b) found that the ThL of synthetic samples of dolomite was due only to Mn^{2+}. Fe^{2+}, Ni^{2+}, and Co^{2+} served as quenchers. The maximum intensity of the ThL, as a function of the Mn concentration, occurred at about 0.2 mole percent. In a later paper (Medlin 1963a), he reported that the spectra from XL were the same as those of the ThL peaks, showing a broad band with a maximum at 6,700 Å. He checked the spectra for four synthetic samples, with Mn contents ranging from 160 to 15,000 ppm, and they were all the same. For a sample containing 500 ppm Mn and 500 ppm Fe, the spectrum was the same, with the addition of a small peak at 3,900 Å (Fig. 7.2).

Much additional data on the intensity of CL in natural dolomites as a function of their Mn and Fe contents have been accumulated in recent years and are summarized in Chapter 7. As is the case for calcite, there is still a need for more good analytical data to define the relationship between CL intensity and iron and manganese contents in dolomite. It appears that in natural samples with low iron contents, as little as 100 ppm molar of $MnCO_3$ may be sufficient to produce bright CL. Samples containing 0.1–1 percent of iron, and similar amounts of manganese, will also show bright CL. Samples containing above about 3 percent molar of $FeCO_3$ are usually quenched.

A more extensive discussion of CL in dolomite appears in Chapter 7.

ARAGONITE GROUP

Manganese is also the most important identified activator in aragonite. However, Mn does not substitute as readily in minerals of the aragonite group as it does in minerals in the calcite and dolomite groups, and the Mn content of natural samples of the aragonite group minerals is usually low.

Aragonite $CaCO_3$ Orthorhombic Medlin (1963a) found that 500 ppm of Mn^{2+} in synthetic aragonite produced a broad ThL emission spectrum with a maximum at 6,300 Å. The ThL emission is described as yellow.

Sommer (1972a) found that Mn^{2+} in synthetic aragonite

produced yellow-green CL with a maximum at 540 nm. His results for the intensity of XL as a function of the Mn concentration are shown in Figure 7.4.

Richter and Zinkernagel (1981) also reported yellowish-green CL in aragonite in both nonmarine and marine skeletons. Their paper includes color plates of luminescent aragonite.

Amieux (1982) noted green CL of medium to strong intensity in aragonite which he attributed to Mn^{2+} in interstitial positions.

Witherite $BaCO_3$ Orthorhombic Sommer (1972a) reported that Mn^{2+}-activated synthetic witherite, with about 300 ppm Mn and 100 ppm Fe, showed a CL emission spectrum maximum at 675–677 nm.

Figure 7.7 includes a CL spectrum from a witherite sample which peaks between 525 and 550 nm, in the green region of the spectrum.

Dudley (1976) reported bright green CL in witherite.

Strontianite $SrCO_3$ Orthorhombic Sommer (1969,1972a) observed a CL emission spectrum peak at 590 nm, orange, for a Mn^{2+}-activated synthetic strontianite. He also reported on the CL (1972a) of the aragonite–strontianite solid solution series and the XL (1972c) of $(Ca_{1-x}Sr_x)CO_3$ and noted in both cases a regular increase in the emission wavelength from 540 nm to about 590 nm as the $SrCO_3$ content increased from 0 to 100 percent (Fig. 7.5). The Mn contents for this suite of samples were 400–500 ppm and the iron contents were 100–200 ppm.

Mariano and Ring (1975) reported blue CL in a natural strontianite due to activation by Eu^{2+}. The spectrum of this luminescence is shown in Figure 4.5a. They noted the selective partitioning of Eu^{2+} to the strontianite when, for example, apatite and strontianite are found together. Mariano (1978) also presented the spectrum of a multiple rare earth- (Sm^{3+}, Dy^{3+}, and Tb^{3+}) activated strontianite which gave a light yellow CL (Fig. 4.5b). The significance of strontianite CL spectra as a reflection of the geochemistry of the crystallization environment is covered in Chapter 8.

Cerussite $PbCO_3$ Orthorhombic Hagni (1984) observed white CL in cerussite. The cerussite became readily incandescent under the beam and then became nonluminescent.

4.6 Class VI: sulfates, chromates, molybdates, tungstates

4.6.1 Anhydrous sulfates

Barite $BaSO_4$ Orthorhombic Leverenz (1968) has several references to synthetic $BaSO_4$ with Pb impurity exhibiting CL in the UV which extends into the blue end of the visible spectrum.

Goebel and Patzelt (1976) saw red and brown CL in some barite samples.

Harwood (1983) used CL in a study of relative dating of barite mineralization in limestone. The CL of the carbonates helped to 'unravel paragenetic sequences' but the barite itself, in these particular samples, showed no CL.

Hagni (1984) reports a distinct green CL in a barite sample at high voltage.

Celestite $SrSO_4$ Orthorhombic Leverenz lists Bi as an activator in synthetic $SrSO_4$.

Anglesite $PbSO_4$ Orthorhombic Hagni (1984) reports the CL of an anglesite sample as white and readily incandescent. It rapidly became nonluminescent under the action of the beam.

Anhydrite $CaSO_4$ Orthorhombic (crystals rare) Medlin (1963a, 1968) studied both the X-ray excited luminescence and the ThL of synthetic anhydrite. For ThL he concluded that Mn^{2+} and Zn^{2+} are the only important activators in most natural samples. Quenching is by Fe^{2+} and/or by Fe^{3+}. For Mn^{2+} he observed green (5,000 Å) ThL emission and a similar emission for XL.

Leverenz (1968) lists Bi as an activator in synthetic $CaSO_4$. He also notes that $CaSO_4$ with added Mn is important as a dosimeter material.

Mariano and Ring (1975) found that most natural anhydrites show REE activation. They reported that Sm^{3+} and Dy^{3+} are CL activators in a particular anhydrite sample from a porphyry copper hydrothermal mineral assemblage. The color to the eye of the CL of this sample is light tan. Chapter 8 contains an extensive discussion of CL, and zonation of CL, in anhydrites.

Tarashchan (1978) observed a variety of peaks in the photoluminescence spectra of anhydrite including peaks due to Sm^{2+} [*sic*; ?Sm^{3+}], Mn^{2+}, S_2^-, Tb^{3+}, Eu^{3+}, and Gd^{3+}.

4.6.2 Hydrated sulfates

Gypsum $CaSO_4{\cdot}2H_2O$ Monoclinic Leverenz (1968, p. 102) comments on photoluminescence of some natural gypsum with unidentified activator(s). Dudley (1976) reported no CL in a gypsum sample.

4.6.3 Molybdates, tungstates

Scheelite $CaWO_4$ Tetragonal Randall (1939) commented on the blue (435 nm) emission observed from carefully purified $CaWO_4$. He said it might originate in the WO_4^{2-} ion.

Leverenz (1968) has many references to tetragonal $CaWO_4$. He lists self-activated [$CaWO_4$:W] as producing a peak at 4,300 Å, self-activated with added Cr [$CaWO_4$:W:Cr] in the 4,200–4,700 Å range, Pb activator producing peaks in the UV and violet, and Sm producing an orange-red peak.

Le Poole *et al.* (1968) included a color plate of a bright blue-luminescing scheelite.

Mariano and Ring (1975) noted the blue cathodoluminescence from self-activation in scheelite. From analysis of the luminescence spectra, they also established that Dy^{3+}, Sm^{3+}, and Eu^{3+} can serve as activators in some natural scheelites. The spectra of three REE-activated samples are shown in Figure 4.6.

Goebel and Patzelt (1976) reported yellow CL in some scheelite samples.

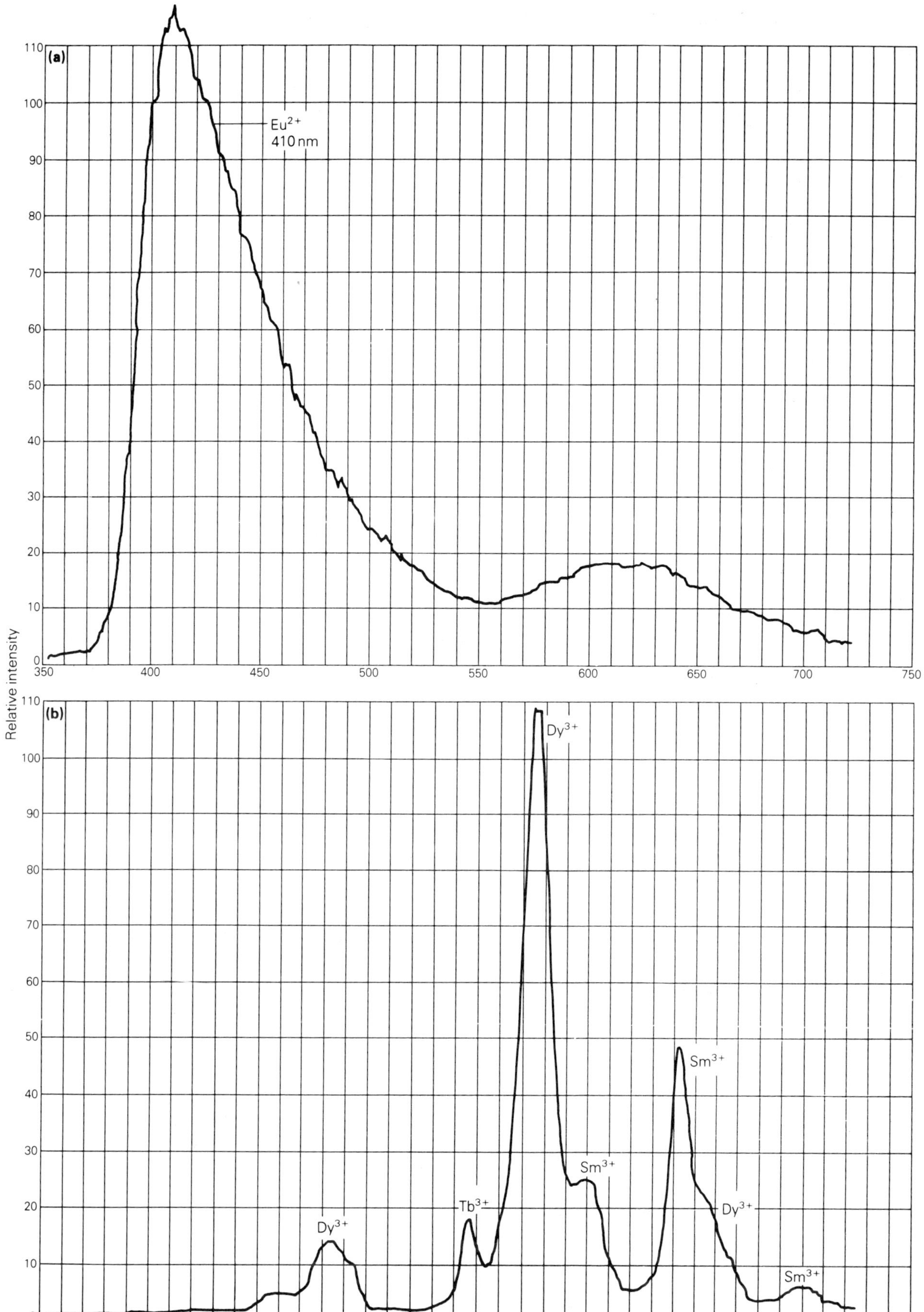

Figure 4.5 (a) Cathodoluminescence spectrum from strontianite from Mountain Pass, California. The broad-band emission at 410 nm is typical of carbonatite strontianites and has been interpreted as being due to Eu^{2+} activation. The color to the eye is deep blue. The weak broad band at approximately 625 nm is due to Mn^{2+}-activated calcite associated with the strontianite. (From Mariano & Ring 1975.) (b) Cathodoluminescence spectrum from strontianite from Argyllshire (Strathclyde), UK (Smithsonian # 48,904). The CL spectrum of this strontianite is due to heavy lanthanide activation. To the eye it appears to be light yellow. (From Mariano 1978.)

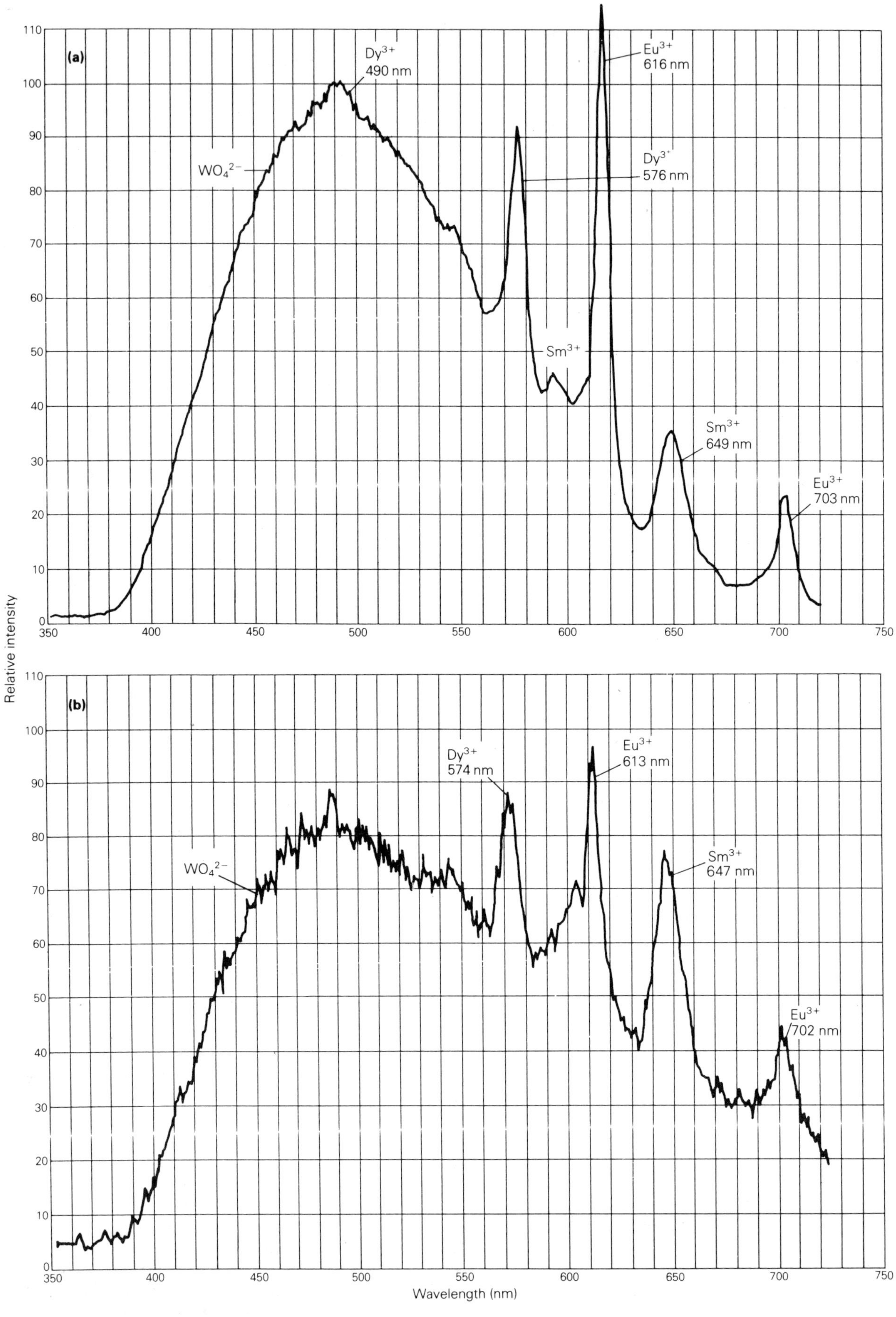

(a)
Dy3+ 490 nm
WO4 2−
Dy3+ 576 nm
Eu3+ 616 nm
Sm3+
Sm3+ 649 nm
Eu3+ 703 nm
(b)
WO4 2−
Dy3+ 574 nm
Eu3+ 613 nm
Sm3+ 647 nm
Eu3+ 702 nm
Relative intensity
Wavelength (nm)

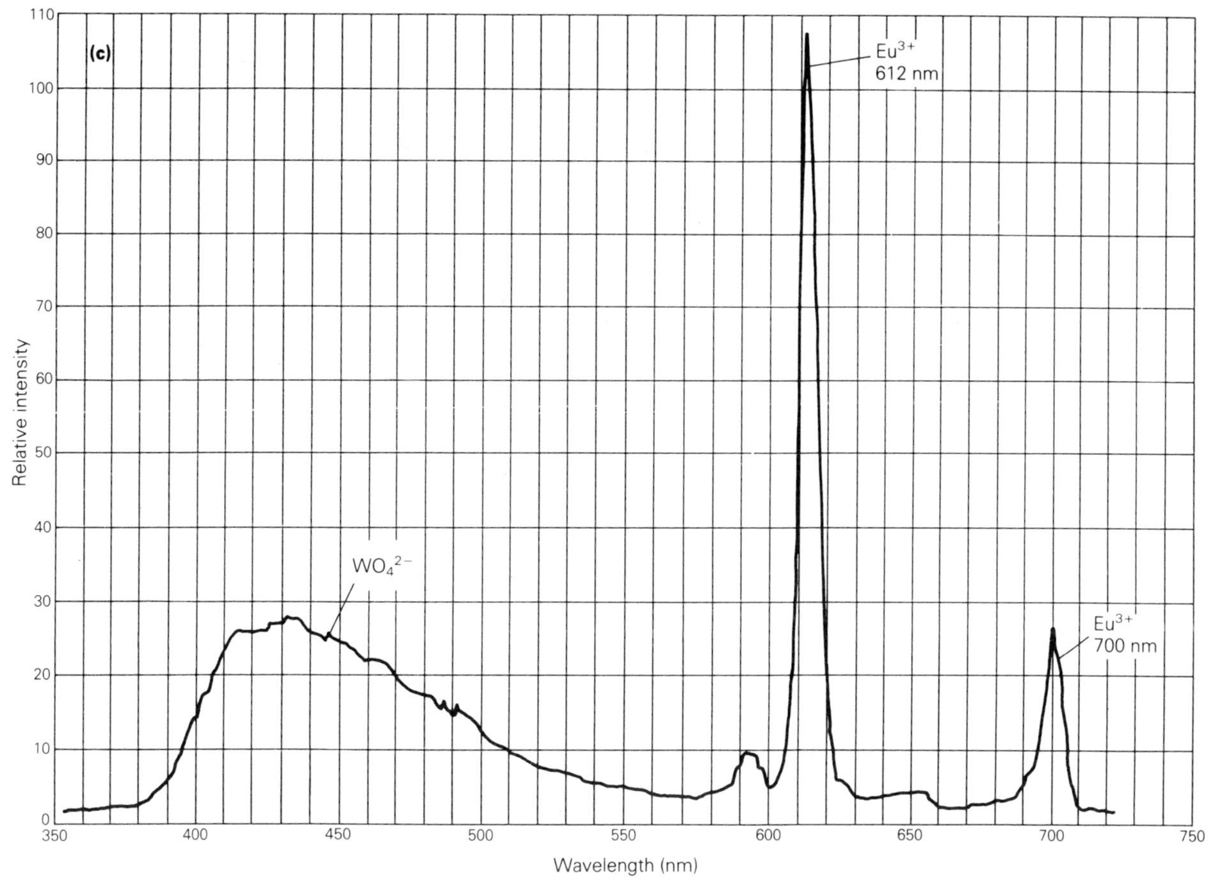

Figure 4.6 CL spectra of representative scheelite samples (from Mariano & Ring 1975). (a) Scheelite from Porcupine, Ontario, exhibiting intrinsic CL, broad-band, in the blue-green region. The scheelite CL appears bluish-white to the eye. Spectral peaks due to three rare earth activators, Eu^{3+}, Sm^{3+}, and Dy^{3+}, are superimposed on the broad band. (b) Scheelite from Safford, Arizona, occurring in the mineralized vein of a quartz diorite. Because of smaller grain size, the signal level is lower, and the signal-to-noise ratio higher, than the spectrum in (a) but otherwise they are almost identical. The CL of this sample is also bluish-white to the eye. (c) Synthetic $CaWO_4$, doped with Eu^{3+}, exhibiting the intrinsic CL of $CaWO_4$ and the narrow-band emission from europium. The color to the eye is light red.

4.7 Class VII: phosphates, arsenates, vanadates

APATITE GROUP

Apatite $Ca_5(PO_4)_3$(F,Cl,OH) Hexagonal Crookes (1879) looked for, but did not see, luminescence in apatite.

Smith and Stenstrom (1965) noted 'a wide variety of luminescence colors, including yellow, brown, lavender, and green . . . .' They noted that submicroscopic apatite crystals in feldspars could be detected because of the color contrast under CL, i.e., the yellow apatite and the blue feldspar. Weiblen (1965) also noted yellow-luminescing apatite in a basalt sample.

Sippel and Glover (1965) reported a light violet apatite in a carbonatite from Magnet Cove, Arkansas. A photomicrograph of this sample appeared in a black and white photo on the cover of *Science*. Sippel (1971) reported yellow or orange CL in apatite from lunar igneous rocks. The CL was polarized.

Leverenz (1968) reports on the CL of many $Ca(PO_4)$ materials containing various impurities. In his Table 63 (p. 72), he shows the spectrum of $3Ca_3(PO_4)_2{\cdot}CaCl_2$:Mn peaking at 6,000 Å.

Portnov and Gorobets (1969) reported on the photoluminescence of apatite from different rock types. They showed results for several activators including:

Activator	Color to the eye	Peak emission (nm)
Ce^{3+}	violet	365
Eu^{2+}	blue	450
Sm^{3+}	reddish-orange	3 lines at 560, 600, and 648
Mn^{2+}	yellow	576
Dy^{3+}		480, 580

Many natural apatites in granites and other samples show a bright yellow CL which is due to Mn^{2+} (Mariano & Ring 1975). In other samples there may be a variety of shades of yellow and brown, and occasionally zoned apatites in this color range are present. Fine-grained apatite, or apatite which is present in opaques, is easily located with the CL technique as shown in Plates 2a, c, d, e, and f.

Hagni (1984) investigated the problems in the beneficiation of some phosphatic iron ores. The phosphorus was contained in the carbonate–fluorapatite (collophane). This apatite was not easily detectable with reflected light or with conventional TrL or PoL techniques because there was fine-grained hematite disseminated throughout the crystals.

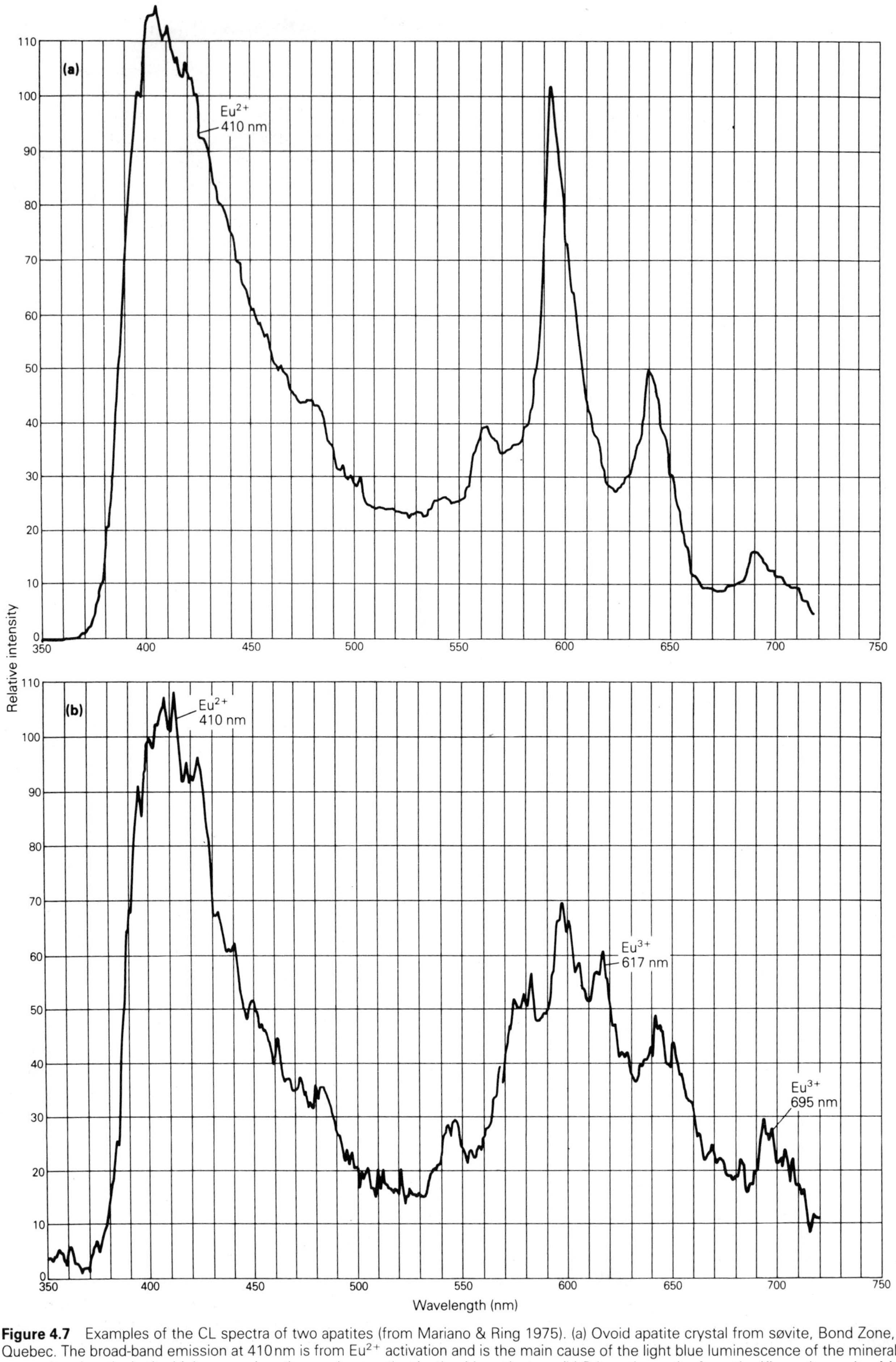

Figure 4.7 Examples of the CL spectra of two apatites (from Mariano & Ring 1975). (a) Ovoid apatite crystal from søvite, Bond Zone, Oka, Quebec. The broad-band emission at 410 nm is from Eu^{2+} activation and is the main cause of the light blue luminescence of the mineral. The narrow-band peaks in the higher wavelengths are due to other lanthanide activators. (b) Prismatic apatite from the Kimzeyite søvite, Magnet Cove, Arkansas. The luminescence spectrum consists of a broad-band emission (410 nm) from Eu^{2+} and two narrow bands at 617 and 695 nm from Eu^{3+}. All other lines are accountable to other lanthanide activators. The luminescence color of this apatite is a blue-violet.

However, they were readily observable with CL and mineralogical and textural data on the distribution of the apatite were readily obtained. Plate 10d is an example of the appearance of this type of sample with CL.

Apatite is an excellent host for REE activators and Mariano (1978) and Mariano and Ring (1975) have noted that the CL of apatite is a very sensitive test for the presence of rare earths. They present many examples including a light blue luminescence due to activation by Eu^{2+} (peak at 410 nm – see Fig. 4.7). They noted also blue-violet CL in a natural apatite where Eu^{3+} and Eu^{2+} peaks both exist, and violet from a synthetic Eu-doped apatite. The significance of apatite CL is further discussed in Chapter 8 and several additional emission spectra are shown.

4.8 Class VIII: silicates

4.8.1 Subclass tektosilicates

SILICA GROUP

Quartz SiO_2 Trigonal Not surprisingly, quartz was one of the many minerals that Crookes investigated for CL. The results were inconclusive. Since then many authors have looked at the CL of quartz, and various CL colors have been reported including red, blue, violet, gray, rust, and brown. The intensity of quartz CL is weak compared to that of feldspars and carbonates and many other minerals. The CL of authigenic quartz is dull or nonexistent.

Quartz CL is apparently often controlled by defects and crystal damage effects, rather than by activators. Alpha particle decay haloes, for example, have different CL characteristics from the bulk quartz (Morton 1978) but the state of knowledge of this matter is still far from complete.

Several investigators have observed that the CL of some quartz changes with time under electron bombardment. The change is such that a grain which shows initial blue CL changes to a dull red CL, and often the change is permanent.

Even though many observations have been made, there is still a great deal to be learned about the reasons for CL in quartz. Chapter 6 has an expanded discussion on what is presently known. Even with the present state of the knowledge, however, quartz CL has proved to be an extremely useful tool for the sedimentary petrographer. Frequently it provides otherwise unavailable information on the distinction between detrital grains and secondary quartz cement, evidence of fracturing, and the possibility of multiple sources of detrital material.

Chalcedony SiO_2 Cryptocrystalline Schluger (1976) saw deep purple CL in chalcedony.

Tridymite SiO_2 Hexagonal Sippel and Spencer (1970a, b) observed small pink-luminescing crystals of tridymite in lunar rock samples. They later corrected this observation, reporting that it was actually quartz. In a later suite of samples, Sippel (1971) observed CL in plagioclase, tridymite, cristobalite, quartz, and apatite and stated that tridymite exhibited the brightest luminescence, 'a brilliant blue emission, about 10 times more intense than the plagioclase.' It was feebly polarized. He found that with increasing excitation intensity, a point was reached where the blue CL was abruptly extinguished and was not recoverable. It exhibited a dull blue CL thereafter. He suggested that this change might be due to the beam heating causing a passage through the alpha–beta transition temperature. Further he stated that this phenomenon is useful in petrography to identify scattered tridymite and to distinguish it from the strong CL of alkali feldspars. One looks initially at the tridymite and then increases the intensity to extinguish it and emphasize the feldspar grains.

Hanusiak (1975) observed the CL of synthetic tridymite at room temperature and at −100 °C. The CL intensity at room temperature was about twice that of an orthoclase sample and increased by a factor of 80 when the temperature was decreased to −100 °C. The emission maximum for this tridymite sample was at 454 nm.

Cristobalite SiO_2 Isometric Cristobalite in lunar samples (Sippel 1971) had an identical spectrum to that of tridymite but was a much duller blue. The CL is relatively stable compared to tridymite. Under prolonged electron bombardment it developed red luminescence in the same manner as quartz. The CL was feebly polarized.

Hanusiak (1975) also looked at synthetic cristobalite and observed the CL emission maximum at 445 nm. At −100 °C, the intensity increased by a factor of 90 over its room temperature value.

FELDSPAR GROUP

Smith and Stenstrom (1965) and Long and Agrell (1965) noted several examples of luminescing feldspars. Red and blue luminescence and the polarization of the CL were some of the features noted. Further details from their work are presented in Chapter 5.

Görz *et al.* (1970) noted the CL from several feldspars which was, for the most part, blue or reddish-blue. All, with the exception of adularia, showed polarization of CL. Sippel (1971) also noted polarization of CL from all types of terrestrial feldspars, with the exception of sanidine and adularia.

The reasons for CL in feldspars have been investigated by Mariano *et al.* (1973). They introduced known activators into synthetic plagioclases (synthesized by J. Ito) and observed the luminescence properties. Four ions were shown to be possibly important in the production of CL in natural samples: Mn^{2+}, Ti^{4+}, Fe^{3+}, and Fe^{2+}. The first three produce luminescence colors of yellowish-green, blue, and red respectively. Fe^{2+}, also found to be an activator, produces a luminescence which to the unaided eye is similar to that produced by manganese.

Feldspar luminescence was of great interest in the lunar sample studies. Geake *et al.* (1971, 1973, 1977) observed green, blue, and infra-red luminescence and concluded that Mn^{2+} is the activator for the green peak and Fe^{3+} causes the red CL. They thought it likely that the blue peak in natural feldspars is due to lattice defects.

Kastner (1971) reported that authigenic feldspars do not luminesce. This provides a useful tool for distinguishing these from their detrital neighbors in sedimentary samples. Her results were obtained on the basis of studying some 34 specimens at electron probe voltages up to about 20kV. Sippel (1971) and others have also confirmed the lack of luminescence in authigenic feldspars.

Chapter 5 is devoted to the feldspar minerals and expands on the material in this short summary. Section 5.3 in particular has an expanded discussion on the CL of authigenic feldspars.

Scapolite series (Na, Ca)$_4$[(Al, Si)$_4$O$_8$]$_3$(Cl, CO$_3$) Tetragonal Burgner *et al.* (1978) studied wernerite, a variety of scapolite. They investigated natural samples from Grenville, Quebec, using an argon-ion laser for excitation. The samples exhibited a strong yellow photoluminescence under this excitation.

The luminescence they observed was enhanced by heating the sample to 900°C for 24 hours, but the samples were actually studied at 4.2°K. Spectral analysis revealed a series of bands from 500 to 700nm. These are considered to be due to the S_2^- luminescent center, substituting in the anion sites.

FELDSPATHOID GROUP

Leucite KAlSi$_2$O$_6$ Isometric Dudley (1976) reports pale blue CL in leucite.

Sodalite Na$_8$(AlSiO$_4$)$_6$Cl$_2$ Isometric Van Doorn *et al.* (1972) reported on optical and CL properties of doped synthetic sodalites. They observed a green CL from Mn^{2+}-doped and a red CL from Fe^{3+}-doped sodalite.

Gaal (1976–77) reported mottled violet, blue, and orange as CL colors in lapis, a semiprecious stone form of sodalite.

Tarashchan (1978) states that S_2^- can be an activator for photoluminescence in sodalite and in lazurite, producing a complicated multipeak spectrum for each.

Mariano (1978) observed CL in a number of natural sodalites and reported on variations in their spectra. All sodalites from nepheline syenites in carbonatites showed red CL from Fe^{3+} activation.

4.8.2 *Subclass phyllosilicates*

One of the difficulties associated with CL observations in this mineral group is the need for high purity, especially with respect to feldspar contamination. Even very small amounts of feldspar could significantly modify (increase) the apparent CL of the clay minerals.

Kaolinite Al$_4$Si$_4$O$_{10}$(OH)$_8$ Triclinic Hanson *et al.* (1981) noted that kaolinite produced a subdued blue background under CL. Hagni (1984) noted that kaolinite from alteration zones exhibits a light blue CL. Dudley (1976) reported 'bright deep blue' CL in kaolinite. Plates 4c and d have an example of kaolinite luminescence.

Dickite Hanson *et al.* (1981) noted a 'bright, light-blue color' under CL.

Nacrite Hanson *et al.* (1981) noted that nacrite produced a subdued blue background under CL.

Serpentine Mg$_6$Si$_4$O$_{10}$(OH)$_8$ Monoclinic Weber *et al.* (1967) examined 33 specimens with an electron probe, operated at 30kV. Blue fluorescence was the most intense but individual grains exhibited colors ranging from blue to red. They found 'some evidence that manganese and possibly chromium and titanium are serving as activator ions. . . .' Also they noted that 'Fe, and to a lesser extent Ni, may suppress fluorescence.'

Talc Mg$_3$Si$_4$O$_{10}$(OH)$_2$ Monoclinic Reid *et al.* (1964) looked at two samples of talc with X-ray excitation. One was nonluminescent and one luminesced white.

MICA GROUP

Muscovite KAl$_2$(AlSi$_3$O$_{10}$)(OH)$_2$ Monoclinic Blair and Edgington (1970) looked at muscovite from a granite under proton bombardment. They noted a peak in the red at 6,800Å.

Görz *et al.* (1970) observed the CL of a muscovite sample to be wine-red and unpolarized.

Schluger (1976) noted greenish purple CL in sericite–chlorite intergrowths.

Biotite K(Mg,Fe)$_3$(AlSi$_3$O$_{10}$)(OH)$_2$ Monoclinic Görz *et al.* (1970) observed blue, unpolarized CL from a biotite sample. Most biotites, however, are nonluminescent.

Lepidolite KLi$_2$Al(Si$_4$O$_{10}$)(OH)$_2$ Monoclinic Görz *et al.* (1970) noted that the CL of a lepidolite sample was weak blue and unpolarized.

Chlorite (Mg,Fe,Al)$_6$(Al,Si)$_4$O$_{10}$(OH)$_8$ Monoclinic Schluger (1976) noted 'greenish purple' CL in sericite–chlorite intergrowths.

Prehnite Ca$_2$Al(AlSi$_3$O$_{10}$)(OH)$_2$ Orthorhombic Dudley (1976) reported yellowish-white CL in prehnite.

4.8.3 *Subclass inosilicates*

AMPHIBOLE GROUP

The minerals of the amphibole group are typically iron-rich and therefore, for the most part, nonluminescent as Weiblen (1965) noted in the several samples he studied. But there are some exceptions.

Cummingtonite series

Grunerite (Fe,Mg)$_7$Si$_8$O$_{22}$(OH)$_2$ Monoclinic A sample of grunerite showed reddish-blue, polarized, CL according to Görz *et al.* (1970).

Tremolite–actinolite Ca$_2$(Mg,Fe)$_5$Si$_8$O$_{22}$(OH)$_2$ Monoclinic Ross *et al.* (1968) looked at amphibole specimens from several sources which specimens included exsolution lamellae. Under CL they observed golden yellow luminescence of tremolite, a dark red CL, and also a grayish-yellow. The red CL seemed to be associated with a Mn-rich, Ca-poor phase. Figure 4.8 shows the results of a traverse with the electron probe across a single grain which shows the three CL colors and illustrates the associations: yellow

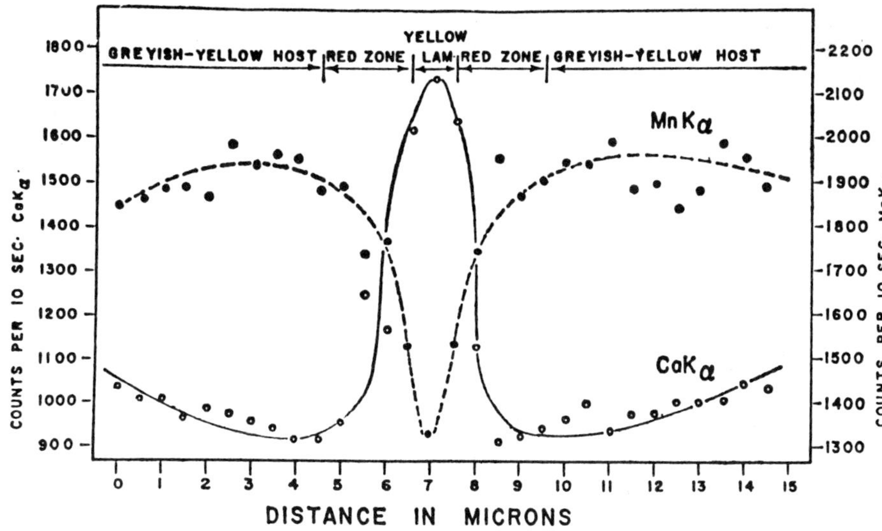

Figure 4.8 Electron microprobe scan for calcium and manganese across adjoining areas, luminescing grayish-yellow, dark red, and golden yellow, in a grain from a manganese-rich amphibole from International Talc Mine No. 3, Talcville, New York. The yellow-luminescing areas marked 'LAM' are exsolution lamellae which have tremolitic composition. The red and grayish-yellow areas are Ca-poor (and Mn-rich). (From Ross *et al.* 1968.)

CL with Mn-poor and Ca-rich; red CL with Mn-rich and Ca-poor. The grayish-yellow is an intermediate phase. The iron content of the samples was low, apparently 1,000 to 2,000 ppm. These data do not identify the cause of the yellow luminescence in tremolite.

Gaal (1976–77) noted yellow and orange CL in tremolite.

Hagni (1984) observed a bright red CL in tremolite from Gouverneur tremolite–talc deposits in New York State.

Hornblende $NaCa_2(Mg,Fe,Al)_5(Si,Al)_8O_{22}(OH)_2$ Monoclinic

PYROXENE GROUP

Enstatite	$MgSiO_3$	Orthorhombic
Hypersthene	$(Mg,Fe)SiO_3$	
Clinoenstatite	$MgSiO_3$	Monoclinic
Clinohypersthene	Mg,Fe,SiO_3	
Diopside	$CaMgSi_2O_6$	
Hedenbergite	$CaFeSi_2O_6$	
Augite	$Ca(Mg,Fe,Al)(Al, Si)_2O_6$	
Pigeonite		
Aegirine	$NaFeSi_2O_6$	
Jadeite	$NaAlSi_2O_6$	
Spodumene	$LiAlSi_2O_6$	
Johannsenite	$CaMnSi_2O_6$	

Crookes (1879) looked at enstatite and 'minerals of the augite class' but saw no luminescence.

Derham *et al.* (1964) reported on the proton-excited luminescence of enstatite from meteorite samples. They found, characteristically, both a red and a blue luminescence (Fig. 4.9). Individual grains might luminesce red (6,700 Å) or blue (4,000 Å), with the red-luminescing grains much more abundant than the blue-luminescing grains. While most of their experiments were done with proton excitation, they did some work with electron and X-ray sources also and saw similar results.

Reid *et al.* (1964) observed the X-ray excited (tungsten tube, 50 kV, 35 mA) luminescence of orthopyroxenes. One of their interests was to see if any correlation existed between luminescence color and disorder in $MgSiO_3$. Terrestrial samples, iron-bearing, had no detectable XL. A synthetic sample (disordered) had strong violet luminescence and the blue component was more intense than the red. Examples of ordered and disordered enstatite from meteorites gave moderate red, violet, or blue luminescence but they found 'no simple correlation between disorder and the intensity and colour of the light emitted.' They found a suggestion that 'some correlation does exist between the character of the luminescence and either disorder in the structure or some effect closely related to and perhaps cogenetic with disorder.'

They noted no X-ray excited luminescence in natural samples of hypersthene or bronzite.

In a later paper (Reid & Cohen 1967), they reported that the

> character of the luminescence is very sensitive to the amount of manganese in the enstatite .... Less than

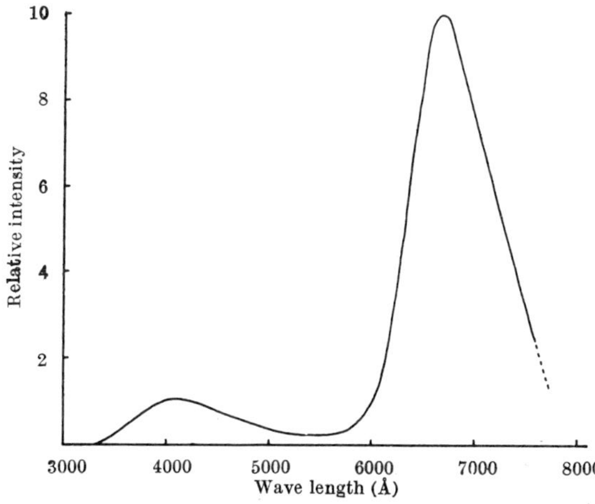

Figure 4.9 The luminescence spectrum obtained under proton bombardment of a dust sample from the Bustee meteorite. The mineral believed to be principally responsible for the two peaks shown is enstatite. This spectrum was obtained under 40 keV proton bombardment but similar spectra were obtained under electron or X-ray bombardment. (From Derham *et al.* 1964.)

350 ppm gives a bright blue luminescence, 350–620 ppm gives a bluish purple, 610–720 ppm gives reddish purple and 900–1200 ppm manganese gives a red luminescence. The blue color may be present in all samples but completely disguised by the red at higher manganese contents. Thus the presence and intensity of the red color can be directly attributed to the presence of manganese and the amount to which it is present.

The samples studied were from meteorites with little iron (<500 ppm) whereas terrestrial samples of enstatite apparently nearly always contain sufficient iron to quench the CL. This latter paper was based on results from electron (probe) excitation.

Weiblen (1965) noted bright blue CL from the enstatite in the Abee meterorite. He also reported that clinopyroxene with 8–34 percent FeO gave a faint blue CL and with 2–3 percent the CL is yellow-orange and polarized.

Greer *et al.* (1967) reported lavender CL from enstatite and Greer and Weber (1969) presented a comprehensive discussion of the luminescence of enstatite from meteorites. Greer (1970) also reported on the thermoluminescence of enstatite grains separated from meteorites. The effect of increasing amounts of Fe reducing the ThL output was noted.

Leverenz (1968, p. 230) stated that a sample of $MgSiO_3$ (monoclinic) with mole fraction 0.006 MnO produced CL with a peak wavelength of 6,740 Å.

Sippel and Spencer (1970a) saw nonluminescing pyroxene in their lunar sample work.

Finnerty (1977) prepared synthetic clinopyroxenes and orthopyroxenes, with Mn added. The samples were iron-free. He observed, under the electron probe, a deep red CL for orthopyroxene and a gray CL for clinopyroxene.

Leitch and Smith (1980) saw both red- and blue-luminescing enstatite in the Indarch chondrite. They found a correlation between minor elements and the CL color. The red clinoenstatite had higher concentrations of Ti, Al, Cr, Mn, and Ca and the blue-luminescing material had higher amounts of Na.

Diopside $CaMgSi_2O_6$ Monoclinic Smith (1949) prepared synthetic phosphors of diopside composition with added titanium. They showed an efficient blue CL with a broad curve, peaking at 4,150 Å. He found that this Ti-doped diopside was more efficient than Ti-doped willemite under CL, but it showed little or no emission with 2,537 Å photoexcitation. Cerium was also found to be an effective activator. Manganese also activated synthetic diopside but it was found to be much less effective than Ti. The Mn produced an emission spectrum with two peaks, one at 5,850 Å and one at 6,700 Å. Leverenz (1968) also states that monoclinic $CaMg(SiO_3)_2$:$0.02TiO_2$ exhibits a blue peak at 4,150 Å.

Derham *et al.* (1964), using proton bombardment, noted little or no luminescence from the diopside separated from the Bustee meteorite. However, Grögler and Liener (1968) noted that diopside in the Khor Temiki aubrite emitted a green-blue light.

Görz *et al.* (1970) reported the CL of diopside as 'bluish red patchy' and polarized, with a decay time of 0.3 ms.

Mariano (1973) lists Mn^{2+} as an activator in diopside.

Walker (1983) used luminescence excitation spectroscopy to predict site preferences in diopside. He found that Mn could be located on two different types of cation sites in diopside but only on one site in enstatite. The combination of excitation and luminescence spectra emerges as a powerful tool for studies of this sort.

Hagni (1984) noted dark green CL in a diopside.

Augite $Ca(Mg,Fe,Al)(Al,Si)_2O_6$ Monoclinic Schluger (1976) reported greenish-yellow CL from augite.

Spodumene $LiAlSi_2O_6$ Monoclinic Kunzite (gem quality spodumene) luminesces bright pink under CL and the luminescence is polarized according to Görz *et al.* (1970). They also measured relatively long rise and decay times, 6 and 8 ms respectively.

Gaal (1976–77) noted an intense orange CL from gem quality spodumene.

Jadeite $NaAlSi_2O_6$ Monoclinic Görz *et al.* (1970) looked at a sample of jadeite and noted that the CL was wine-red and decayed in intensity rapidly under their excitation ($8 W/cm^2$).

PYROXENOID GROUP

Wollastonite $CaSiO_3$ Triclinic Long and Agrell (1965) observed bright green-luminescing wollastonite in a sample that contained 0.11 percent Mn and 0.07 percent Fe.

Lange and Kressin (1955) and Geake *et al.* (1971) also reported on the luminescence of wollastonite. The emission maximum for β-$CaSiO_3$:Mn (β-wollastonite) is about 620 nm and for α-$CaSiO_3$:Mn (pseudo or α-wollastonite) it is about 560 nm (green).

Mariano (1973, 1978) lists Mn^{2+} as an activator for CL in wollastonite. It produces a 'very intense yellow' at 565 nm.

Gaal (1976–77) noted intense orange CL in a sample of white wollastonite.

Hagni (1984) reported a bright greenish-yellow CL from wollastonite 'which contrasts markedly with the dark green cathodoluminescent color of an associated diopside and the dull green of garnet.'

Rhodonite $MnSiO_3$ Triclinic Leverenz (1968) states that triclinic $MnSiO_3$ has been observed to be nonluminescent at room temperature but shows red luminescence at low (90 °K) temperatures.

4.8.4 Subclass cyclosilicates

Benitoite $BaTiSi_3O_9$ Hexagonal Benitoite is often used for beam current and spot size adjustments in the electron probe because of its bright blue CL which Heidel (1971b), Gaal (1976–77), and others have noted.

Axinite $(Ca,Mn,Fe)_3Al_2(BO_3)Si_4O_{12}(OH)$ Triclinic Gaal (1976–77) reports the CL of axinite as very dark red and orange.

Beryl $BeAl_2Si_6O_{18}$ Hexagonal

Emerald Crookes (1879, p. 661) reported for emerald that 'A small hexagonal prism gave out a fine crimson-red colour. The light was polarized, apparently completely, in a plane perpendicular to the axis . . . .'

Burns *et al.* (1965) prepared synthetic emeralds with Cr^{3+} impurities and observed CL emission at about 6,800 Å, the same as they observed for natural emerald crystals.

Gaal (1976–77) reported no CL for the emeralds he studied.

Escobar and Mariano (1976) have used CL as an exploration tool for emerald deposits.

Aquamarine Gaal (1976–77) noted no CL for the aquamarine variety of beryl.

Cordierite $(Mg,Fe)_2Al_4Si_5O_{18}$ Orthorhombic Hummel (1961) synthesized the pure Mg form with $2MgO \cdot 2Al_2O_3 \cdot 5SiO_2$. It exhibited red CL but did not luminesce under excitation by 2,537 or 3,650 Å UV.

Weiblen (1965) observed the CL in cordierite from a gneiss. The CL was 'bright blue and is readily distinguishable from quartz (red) and feldspar (light blue-green) in the same rock.' He further notes that without the use of CL, this distinction is difficult to make.

Gaal (1976–77) noted very dull red CL from the iolite variety of cordierite.

Tourmaline $Na(Mg,Fe)_3Al_6(BO_3)_3(Si_6O_{18})(OH)_4$ Trigonal Crookes (1879) studied tourmaline but saw no luminescence. Schluger (1976), however, reported that the tourmaline CL was bright green or dull yellow in samples from the red beds of the Perry Formation.

4.8.5 Subclass sorosilicates

Hemimorphite $Zn_4Si_2O_7(OH)_2 \cdot H_2O$ Orthorhombic Hagni (1984, p. 52) describes the CL of hemimorphite as a 'dull green to greenish blue which readily serves to distinguish it from adjacent sphalerite, which usually is nonluminescent, and from smithsonite . . . .'

Epidote $Ca_2(Al,Fe)_3Si_3O_{12}(OH)$ Monoclinic Smith and Stenstrom (1965) observed green-luminescing epidote in the Rhomb Porphyry basalts from the Oslo region. Weiblen (1965) also reported the CL of epidote as light green.

Renfrew and Peacey (1968) identified 'blood-red' CL of epidote in a marble from Syros.

4.8.6 Subclass nesosilicates

Olivine series

Cathodoluminescence in the olivine series is associated with the members that are iron-free or at least low in iron.

Forsterite	**Mg_2SiO_4**	**Orthorhombic**
Fayalite	**Fe_2SiO_4**	
Olivine	**Intermediate**	

One of the many minerals that Crookes investigated was peridot, a name that is used (at least nowadays) for transparent olivine of gemstone quality. He saw no luminescence. Gaal (1976–77) also noted no CL in peridot.

Reid *et al.* (1964) found that two samples of synthetic forsterite luminesced violet under X-ray excitation.

Weiblen (1965) looked at an olivine containing 8 percent FeO and saw no CL.

Geake and Walker (1966) noted orange–red luminescence of some terrestrial forsterite under proton excitation.

Greer *et al.* (1967) reported brick red CL from forsterite.

Görz *et al.* (1970) observed the CL of a sample of forsterite. The luminescence was reddish, with a patchy distribution, and was not polarized.

Sippel and Spencer (1970a) reported that the olivine in a lunar sample showed no CL.

Berkley *et al.* (1976), in their study of the Kenna Ureilite, an ultramafic meteorite, found that forsteritic olivine 'fluoresces dark red–orange.' Fayalite-rich olivine showed no cathodoluminescence.

Finnerty (1977) studied exchange equilibria at high pressure between olivine and clinopyroxene, among others. He found that CL was a useful tool in his iron-free system when Mn was added. Several of the minerals luminesced which 'greatly facilitated the analysis.' Under the electron probe, the CL of the olivine was 'orange–red.'

The work of Steele (1986a, b) on the CL of forsterites from extraterrestrial samples is discussed in Chapter 9.

Monticellite $CaMgSiO_4$ Orthorhombic Smith (1949) prepared synthetic monticellite phosphors and found that manganese activation produced CL in the red region. The CL was relatively inefficient. Leverenz (1968) saw green CL from monticellite as did Mariano (1978) in a natural sample from Vesuvius (brilliant green – 525 nm).

Mariano and Roeder (1983) noted that the monticellite from Oka and Magnet Cove carbonatites does not show CL but that from Kerimasi 'gives a rich green CL at 550 nm indicating that Fe^{++} is not sufficient in the structure for quenching.' The contents (electron probe) of FeO and MnO in the Kerimasi monticellite are 1.9 percent and 1.4 percent respectively.

Willemite Zn_2SiO_4 Trigonal Schulman (1946) prepared synthetic willemites with varying Zn:Be ratios and varying amounts of Mn. He studied their photoluminescence and found that Mn produced a peak at 5,250 Å in the pure Zn form. As the amount of Be increased, the 5,250 peak was suppressed and the 6,100 Å peak increased. He also determined that the optimum Mn concentration, for maximum photoluminescence intensity, was 1 percent by weight of Mn. His paper contains a detailed discussion of Mn as an activator in four-fold and six-fold coordination.

Leverenz (1968, p. 68) comments on a pale violet luminescence under X-ray or cathode-ray excitation. Altogether he has about 100 page references to willemite luminescence in the index to his text.

Bhalla and White (1970, 1972) studied natural willemite containing Mn^{2+} substituting for Zn^{2+}. The samples were iron-free (electron probe analysis). The CL spectra showed a peak at 5,250 Å. CL emitted from surfaces perpendicular to the *c* axis did not show polarization, but the CL from surfaces which contained the *c* axis showed strong polarization. They noted a small shift (100Å) in the emission maximum, depending upon the polarizer orientation. Later (1971) they reported on the CL of some single crystal natural willemite, 'free' (electron microprobe analysis) of Mn and Fe. Here they observed a 'weak pale violet-blue' CL which they considered to be intrinsic. If the samples were heated, the luminescence efficiency increased, possibly because Zn evaporated, resulting in an excess of silica which increased the efficiency of the intrinsic CL. Polarization effects, similar to those seen in the Mn-activated samples, were observed. They used molecular orbital theory and symmetry arguments to explain the anisotropic emission and concluded that the intrinsic CL was associated with the $(SiO_4)^{4-}$ tetrahedral complex and, in particular, to its reduced symmetry in willemite.

One of the experiments they did (1972) was to diffuse Mn into a natural willemite single crystal. They observed the maximum CL intensity at 2.9 weight percent of Mn. They noted that the CL emission spectrum maximum varied from 516 nm, for trace amounts of Mn, to 530 nm, when the concentration reached 12 weight percent. This shift of wavelength with concentration was explained with the help of crystal field theory.

ALUMINUM SILICATE GROUP

Chinner *et al.* (1969) used CL as one of three methods to study a group of natural samples of Al_2SiO_5 and, in particular, to detect alteration. Their work was done on the electron probe (25 kV). They noted 'The cathodoluminescence was quite variable in color and in intensity from one polymorph to another, from one rock specimen to another, and sometimes from place to place in a single grain.' They also suggested that examination of metamorphic minerals by CL may produce some examples of 'earlier fabric relicts' similar to those seen in some carbonates.

Mariano (pers. comm.) states that the CL from natural samples in this group is normally red. Sillimanite is the brightest, kyanite is the next brightest, and andalusite the least intense. Chiastolite is quenched.

Andalusite Al_2SiO_5 Orthorhombic Crookes (1879) looked at andalusite but saw no luminescence. Görz *et al.* (1970) also reported that a sample of andalusite did not luminesce.

Chinner *et al.* (1969) noted CL in andalusite and commented on 'abrupt spatial variation in cathodoluminescence from light to dark blue.'

Hagni (1984) reported the CL of an andalusite sample as a 'distinct blue when excited at the high voltage.'

Sillimanite Al_2SiO_5 Orthorhombic Chinner *et al.* (1969) noted dull blue CL in sillimanite.

Hagni (1984) also observed 'dull bluish white' CL in sillimanite associated with the Benson iron deposits in New York. The sillimanite he studied from elsewhere in the Adirondacks luminesced bright red.

Sillimanite in gneissic rocks often luminesces red (see Plates 6e & f).

Kyanite Al_2SiO_5 Triclinic White and White (1967) examined seven natural samples of kyanite with an interest in explaining the color under natural lighting. They noted also that five samples displayed 'bright red to violet' CL. Their electron probe analyses showed the luminescent samples contained Cr in the 50–150 ppm range and no detectable Mn, whereas nonluminescing samples contained 100 ppm Mn and no detectable Cr. They attributed the CL in kyanite to the presence of Cr^{3+} impurities.

Chinner *et al.* (1969) observed bright red CL in kyanite and noted that 'Lamellar variations of luminescence are quite common.' They also noted that zircon inclusions within bright red-luminescing kyanite 'are surrounded by blue-luminescing haloes, an effect presumably due to radiation damage.'

The CL of a sample of kyanite was observed by Görz *et al.* (1970) to be deep red and possibly polarized, with a 0.3 ms decay time. Dull red CL in kyanite was noted by Gaal (1976–77). Hagni (1984) observed a dark violet red CL in a kyanite sample from the Graves Mountain mine in Georgia.

Tarashchan (1978, in Marfunin, 1979) shows a spectrum of Cr^{3+} in kyanite with three peaks near 700 nm.

Topaz $Al_2SiO_4(OH,F)_2$ Orthorhombic Neither Crookes (1879) nor Görz *et al.* (1970) saw CL in the samples of topaz they investigated.

Mariano (1984 pers. comm.) reports often seeing CL in topaz, usually with a broad-band spectrum in the blue. Dudley (1976) also saw deep blue CL in topaz.

GARNET GROUP

One might suspect that natural garnet CL would be limited because of the iron content, and this has been so reported by Weiblen (1965) for an iron-rich garnet. But there are some examples (Plate 10c) which show with CL a spectacular, extremely complex, pattern.

Garnet Isometric Crookes (1879) looked at several garnets but saw no CL.

Schluger (1976) noted dull green and yellow CL in garnets from the red beds of the Perry formation. Hagni (1984) also noted dull green CL in garnets.

Finnerty (1977) found CL to be a useful tool in the study of equilibria at high pressures. He added Mn to the iron-free system which produced several minerals that exhibited bright CL. Some of his synthesized garnet grains, which luminesced a bright yellow, contained minute inclusions of clinopyroxenes which showed up well with CL.

Almandite $Fe_3Al_2(SiO_4)_3$ Isometric Görz *et al.* (1970) saw no CL in a sample of almandite. Dudley (1976) reported bright yellow CL in almandite.

Pyrope $Mg_3Al_2(SiO_4)_3$ Isometric Görz *et al.* (1970) saw either no CL or very weak blue CL in a pyrope sample.

Spessartine $Mn_3Al_2(SiO_4)_3$ Isometric Gaal (1976–77) noted dull red CL in spessartine.

Grossular $Ca_3Al_2(SiO_4)_3$ Isometric Gaal (1976–77) noted intense orange CL in a grossular sample. Hutcheon *et* *al.* (1978) noted the CL of grossular as yellow-green to yellow.

Zircon $ZrSiO_4$ Tetragonal Crookes (1879, p. 62) commented on the CL of zircon as follows:

> A crystal of hyacinth, however, introduced a new phenomenon. In this optically positive crystal the ordinary ray was of a pale pink hue, the extraordinary of a very beautiful lavender-blue colour. In another crystal, like the former from Expailly, the ordinary ray was of a pale blue, the extraordinary of a deep violet. A large crystal from Ceylon [Sri Lanka] gave the ordinary ray of a yellow colour, the extraordinary ray of a deep violet-blue.

Weiblen (1965) looked at zircon in glass shards from a volcanic tuff. He noted intense blue CL which exhibited zoning.

Leverenz (1968) has many references to the CL of tetragonal $ZrSiO_4$ with either Si, Mn, or V as activators. He states that the unactivated (or self-activated) material shows the usual UV and a pale blue CL. With Mn, the CL is a light yellow. $ZrSiO_4$:V has a broad spectrum with a peak near 3,800 Å.

Sippel (1968) observed yellow-luminescing zircons in sandstones.

Görz *et al.* (1970) reported that the CL of a sample of zircon showed blue and yellow zones and was strongly polarized.

Schluger (1976) noted yellow or green CL in zircons from the red beds of the Perry formation.

Gaal (1976–77) observed 'dull yellow and purple bands' in gem quality green zircon.

Dudley (1976) reported bright yellow CL from zircon.

Ono (1976) observed either blue CL or no CL in zircons from a schistose granodiorite. Chemical correlations were also reported for these zircons and their HfO_2 and YPO_4 contents, measured with an electron probe.

	HfO_2	YPO_4	
blue CL	1.5–2.5	0.0–0.5	(all weight
no CL	1.0–1.5	0.0–1.5	percent)

Some of the samples showed very fine oscillatory zoning with some zones less than 1 μm in width.

According to Marfunin (1979) the photoluminescence of zircon is 'basically related to Dy^{3+}, the Gd^{3+}, Sm^{3+}, Eu^{3+} lines being much less intensive.'

Mariano (1978) describes a zircon CL spectrum as a broad-band intrinsic at 590 nm (yellow-orange) with superimposed Dy^{3+} line spectra at approximately 480 and 575 nm (see Fig. 4.10). He points out that zircon has a very restricted crystal chemistry compared to other host minerals, that

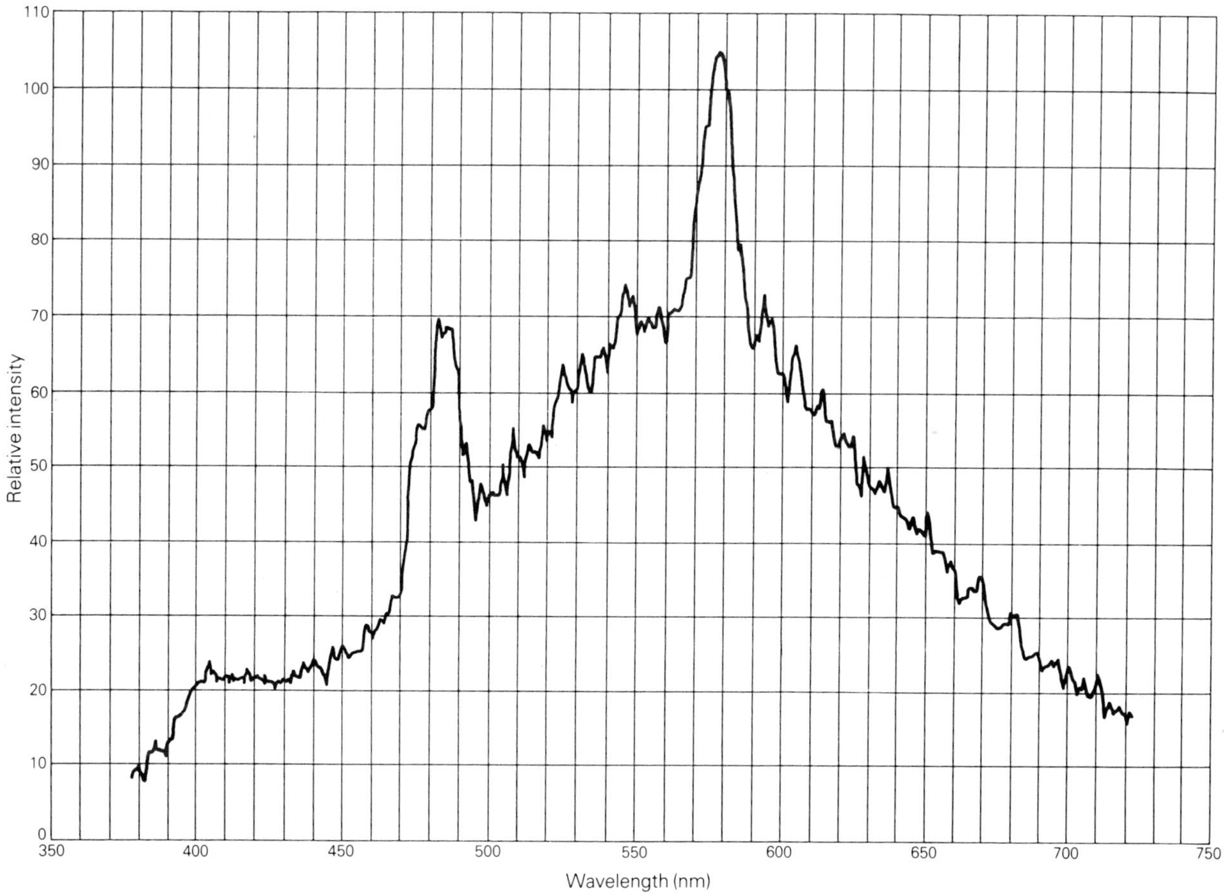

Figure 4.10 The CL emission spectrum from a zircon from country-rock gneiss from the Fen, Norway. This example shows the broad-band intrinsic, peaking in the 570–580 nm region (yellow-orange), and the two lines from dysprosium at approximately 480 and 575 nm are superimposed on the intrinsic CL spectrum. The color to the eye is yellow-white. Polished thin section, 15 kV, 0.5 mA. (Courtesy of A. N. Mariano.)

which strongly favors the incorporation of the HREE including, especially, Dy. As a result it will often show Dy^{3+} line spectra, normally indicative of an HREE environment, regardless of whether it has crystallized in a HREE or an LREE dominant environment.

Plates 5e and f are from two different zircon samples. These zircons were separated from sedimentary samples and mounted as grains for CL study. There is a clear visual difference between these two samples, and the implication is that one may be able to use zircon CL as another tool to assist in provenance studies.

Grauert and Wagner (1975) and Vocke and Hanson (1981) have used CL to assist in the interpretation of U–Pb ages from zircon. The information furnished helped in the evaluation of formation *in situ* versus inherited detrital components and other aspects of zircon paragenesis. Grauert and Wagner show examples of multiple zoning, euhedral crystal growth, and grain corrosion. They state (p. 685):

> A study of the cathode luminescence patterns on polished grain mounts (pl. 1) revealed a remarkable internal heterogeneity with respect to the distribution of heavy elements (uranium, thorium, lead). These inhomogeneities are not apparent under the microscope by means of either transmitted or reflected light. The patterns may be interpreted as reflecting a multistage zircon growth; they are indirect indicators for a presumably complex history of the rocks . . . .

Vocke and Hanson (1981) noted that zircons separated from a gneiss often exhibited a core with finely banded CL and a strongly fluorescent rim. Zircons separated from a granite often exhibited weak CL in the core but the layer surrounding the core showed a strongly luminescent zone and the rim exhibited finely banded CL. For both cases they stated that: 'the luminescing regions are generally correlated with regions of higher U abundance' (p. 251).

Owen (1984a) looked at zircon CL, as part of his study on the use of quartz CL color for provenance work (Ch. 6). He saw examples of zoned CL in zircon but did not actually use the CL characteristic as one of his tools in this particular study. One of his samples (Plate 4d) shows a zircon inclusion in a quartz grain with a radiation damage halo surrounding it.

Sphene $CaTiSiO_5$ Monoclinic Gaal (1976–77) reported dull brown CL in a sphene sample.

Datolite $Ca(OH)BSiO_4$ Monoclinic Gaal (1976–77) observed intense light blue CL in datolite.

Parisite Crookes (1879) investigated parisite but saw no CL.

4.9 Summary

This has been a rather succinct presentation of the published CL data for the common minerals: as mentioned at the outset, a critical review is difficult to do at this early stage of development. The complexity of the CL process, even in well characterized synthetic samples, and the difficulty of relating these synthetic samples to natural samples, present an enormous task. Indeed it would seem that the fundamental work required to understand activation, quenching, intrinsic CL, and other aspects of CL in a *single* mineral, or at most a *single mineral group*, would represent the work equivalent to at least a PhD thesis. We hope that some of the questions generated within this text will inspire others to take up these tasks.

Chapter 5

Cathodoluminescence properties of the feldspar minerals and some geological applications

Feldspars are the most abundant of all mineral groups. They are divided into two major groups. The potassium and barium feldspars are monoclinic or nearly so. The sodium and calcium plagioclases are triclinic.

K-FELDSPARS

Sanidine		
Orthoclase		
Microcline	**$KAlSi_3O_8$**	**Monoclinic**
Adularia		

Na–Ca PLAGIOCLASES (Na,Ca) (Al,Si)$AlSi_2O_8$ TRICLINIC

	Percent $NaAlSi_3O_8$	Percent $CaAl_2Si_2O_8$
albite	90–100	0–10
oligoclase	70–90	10–30
andesine	50–70	30–50
labradorite	30–50	50–70
bytownite	10–30	70–90
anorthite	0–10	90–100

Virtually all of the feldspars in samples from igneous and metamorphic rocks exhibit luminescence. The CL emission colors include bright blue, greenish-yellow, and red.

A discussion of feldspar luminescence in various samples formed a major part of the discussion in the paper by Smith and Stenstrom (1965). They discussed several examples of luminescing feldspars including microcline, perthite, albite, orthoclase, aventurine, and anorthoclase. They reported red and blue luminescence from feldspars and noted that the CL emission of microcline is polarized, i.e., the intensity of the CL is a function of the orientation of the mineral. They also noted that twinning showed up well under CL because of the variation of the intensity of emission as a function of orientation. Long and Agrell (1965) also observed CL in perthite. The two feldspars gave 'distinctly different intensities of a bluish-purple fluorescence.'

5.1 Fundamental studies

There has been a limited amount of work done on the reasons for luminescence in the feldspar minerals. A great deal remains, some of which will be identified at the end of this chapter.

An important piece of work dealing with the general question of the identity of the activators in feldspars is that of Mariano *et al.* (1973). They synthesized doped plagioclases and showed that the plagioclase luminescence can be caused by:

(a) Ti^{4+} activator produces blue CL (460 ± 10 nm);
(b) Mn^{2+} activator produces greenish-yellow CL (570 ± 5 nm);
(c) Fe^{2+} activator produces greenish-yellow CL (550 ± 5 nm);
(d) Fe^{3+} activator produces red CL (700 ± 10 nm);
(e) Cu^{2+} activator produces blue CL (420 ± 5 nm).

The manner by which Ti influences the CL is still not known. It is introduced in the doping process as Ti^{4+} but the effective valence state in the mineral may differ. It is possible that it is serving as an 'intensifier activator' (Ch. 1), i.e., its presence intensifies the intrinsic CL.

Their observation of the effectiveness of Fe^{2+} is interesting because of the usual rôle of Fe^{2+} as a quencher of CL, not an activator. The CL color produced by Fe^{2+} in synthetic plagioclases is impossible to distinguish, by eye, from that produced by the Mn^{2+} activator and therefore, in natural feldspars, one cannot be certain from just the CL evidence whether the greenish-yellow CL so commonly seen is due to Mn^{2+} or to Fe^{2+}.

Examples of CL emission spectra are shown in Figures 5.1–5.4.

Figure 5.1 The CL spectrum of a Ti^{4+}-containing feldspar. This particular example is a synthetic labradorite with 0.2 weight percent of TiO_2. The CL color is light blue. (Sample synthesized by Jun Ito.) (From Mariano 1978.)

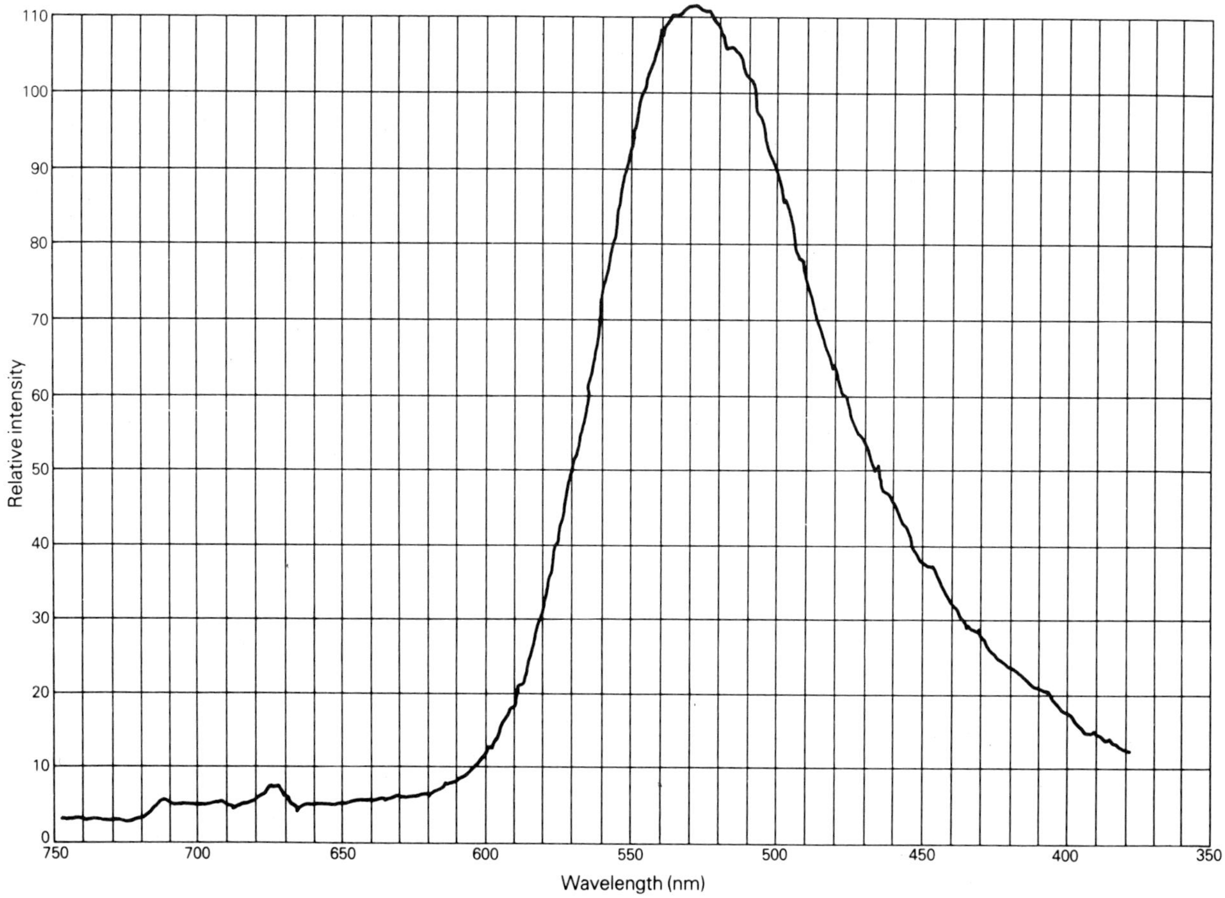

Figure 5.2 The CL spectrum of a Mn^{2+}-activated feldspar. This particular example is a synthetic labradorite with 430 ppmw of MnO. (Sample synthesized by Jun Ito.) The CL color is yellow. (From Mariano 1978.)

Figure 5.3 The CL spectrum of a Fe^{2+}-activated feldspar. This example is a synthetic plagioclase with 2 weight percent of FeO. (Sample synthesized by Jun Ito.) The CL color is green. (From Mariano 1978.)

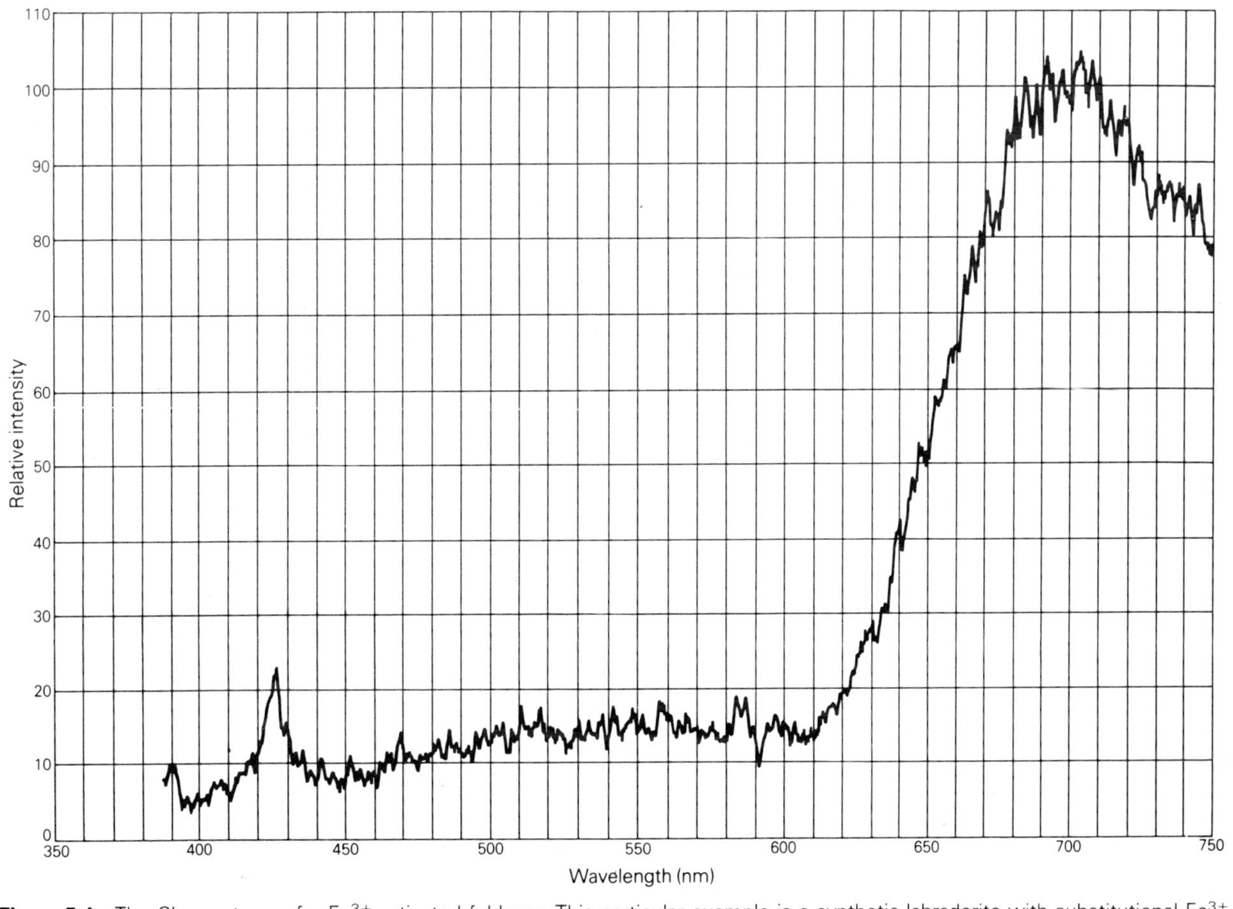

Figure 5.4 The CL spectrum of a Fe^{3+}-activated feldspar. This particular example is a synthetic labradorite with substitutional Fe^{3+} (0.18 weight percent of Fe_2O_3 added). (Sample synthesized by Jun Ito.) The CL is red. The broad band with a peak wavelength at approximately 700 nm is from Fe^{3+} activation. CL conditions: 11.5 kV, 0.3 mA. (From Mariano 1978.)

Table 5.1 Wavelength (nm) of the CL emission spectrum maximum for doped calcic-plagioclases (Mariano *et al.* 1973, Mariano & Ring 1975) and for terrestrial and lunar samples (Sippel & Spencer 1970b, Geake *et al.* 1973).

Activator	Mariano *et al.* (1973, 1975)	Sippel and Spencer (1970b)*	Geake *et al.* (1973)
Eu^{2+}	420	—	—
Cu^{2+}	420 ± 5	—	—
Ti^{4+}‡	460 ± 10	450	450†
Fe^{2+}	550 ± 5	—	—
Mn^{2+}	570 ± 5	559	560
Fe^{3+}	700 ± 10	705–730	700–780

* Sippel and Spencer did not assign the wavelengths they measured to specific activators. We have listed them beside the most likely activator and changed wavelengths reported in ångstroms to nanometers for uniformity.

† Geake *et al.* thought it likely that the blue peak is due to lattice defects.

‡ There is still uncertainty whether Ti serves as an activator or enhances intrinsic CL.

Mariano *et al.* (1973) listed the lower limits of activator concentrations which are effective in synthetic feldspars as

(a) titanium (0.05 to 0.1 weight percent; see ‡, Table 5.1);
(b) manganese (0.0*x* weight percent);
(c) ferrous iron (0.*x* to 1.0 weight percent);
(d) ferric iron (greater than 0.1 weight percent).

Geake *et al.* (1971, 1973) looked at the luminescence of feldspars in terrestrial and lunar samples using both proton and electron excitation. They noted that plagioclase showed a 'strong green' (5,600 Å), a 'weaker blue' (4,500 Å), and a weak IR (7,700 Å) peak. The position of the green peak varied between 5,550 and 5,650 Å with a FWHM of 1,000 Å. They examined the crystal field situation in some detail and concluded that the emission due to Mn^{2+} would occur somewhere between 5,500 and 6,000 Å. They also doped a sample of labradorite with Mn, by heating in an atmosphere of Ar and $MnSO_4$, and observed that the green peak intensity increased. Thus they concluded that it was due to Mn as an activator. They thought it likely that the blue peak, which they found common to most silicates, was due to a lattice defect. The cause for the IR peak was thought to be an activator, not yet identified. In 1973, they reported the successful doping of synthetic anorthite with Fe^{3+} and reported that this was probably the activator for the IR peak (Fig. 5.5). 100 ppm of Fe^{3+} was sufficient to produce CL in synthetic plagioclase.

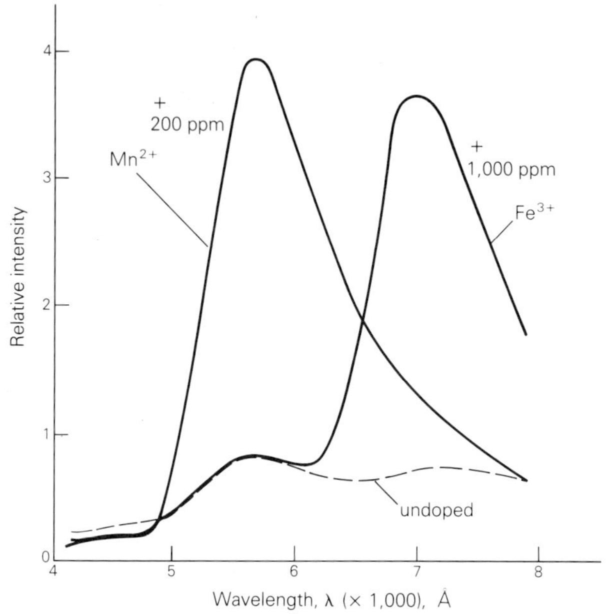

Figure 5.5 The luminescence emission spectra for undoped synthetic anorthite and anorthite doped with either Mn^{2+} or Fe^{3+}. (From Geake *et al.* 1973.)

Telfer and Walker (1975, 1978) and Geake *et al.* (1977) introduced results from the use of luminescence excitation spectra to confirm their earlier results on Mn and Fe. This work supported the conclusion that the Mn^{2+} was on the Ca^{2+} sites but might occasionally be on the Al^{3+} tetrahedral sites. It also supported the conclusion that Fe^{3+} occupied Al^{3+} sites. Telfer and Walker (1978) prepared doped synthetic anorthites and found that about 1 mole percent Mn^{2+} produced the maximum CL intensity (Fig. 5.6a). In separate samples, doped with Fe^{3+}, the maximum was found at 1.5 percent and was broader (Fig. 5.6b).

Tarashchan (1978) shows a photoluminescence spectrum of adularia containing Fe^{3+}. The emission peak is at about 700 nm. Tarashchan *et al.* (1973) also show the photoluminescence spectrum of Pb^{2+} in amazonite with a peak in the UV at 300 nm.

The rare earth element ions, Eu^{2+}, Sm^{3+}, and Dy^{3+}, were also found to be effective as activators in synthetic samples

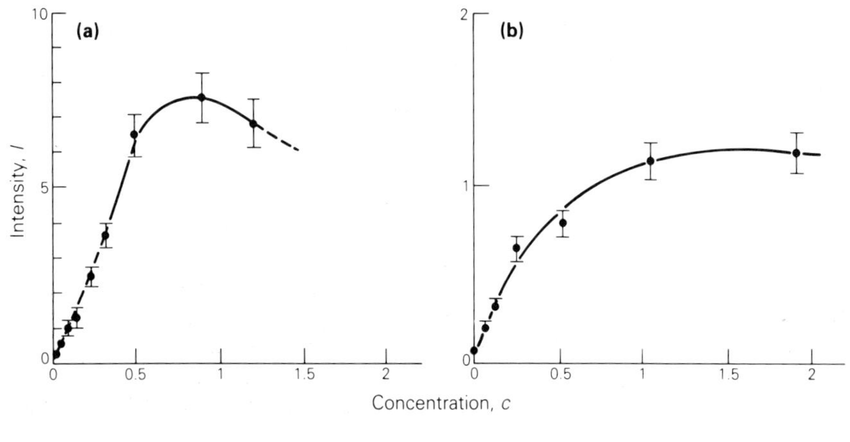

Figure 5.6 (a) The intensity of green emission as a function of Mn^{2+} content (mole percent) in synthetic anorthite. (b) The intensity of the IR peak as a function of Fe^{3+} content (mole percent) in synthetic anorthite. (From Telfer & Walker 1978.)

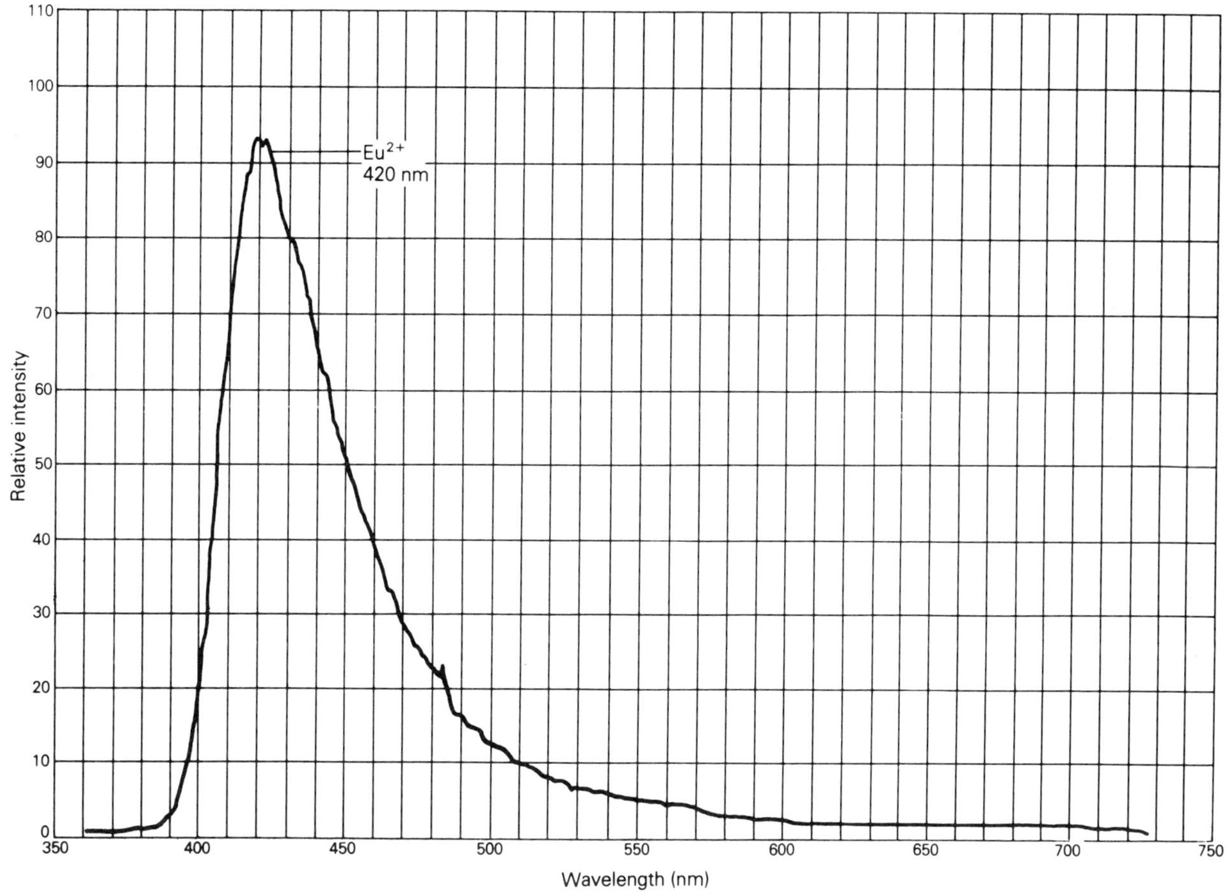

Figure 5.7 The CL spectrum of a Eu^{2+}-activated feldspar. This particular sample is a synthetic single crystal of anorthite with approximately 200 ppmw of Eu^{2+}. (Sample synthesized by Jun Ito.) The luminescence color is a deep blue. (From Mariano & Ring 1975.)

although these elements rarely, if ever, occur in sufficient amounts to be activators in natural feldspars (Mariano *et al.* 1973). In a subsequent paper (Mariano & Ring 1975), the CL spectrum of a synthetic anorthite crystal activated by 200 ppmw Eu^{2+} was shown, with a peak wavelength at 420 nm, a deep blue (Fig. 5.7). Laud *et al.* (1971) prepared synthetic anorthite with 1 mole percent Ce^{3+} and the CL at 3 kV was blue–green (490 nm). They also prepared synthetic anorthite with 2 mole percent Eu^{2+} and observed a blue peak at 470 nm. Drake, reported in Smith (1974b), also observed rare earth activation in feldspars, Eu producing an intense blue CL and Sm producing a deep red CL. Tarashchan (1978) reported that Eu^{2+} in oligoclase produces a photoluminescence spectrum peak at about 400 nm.

The exact wavelengths reported in three publications are given in Table 5.1.

To the author's knowledge, no one has reported on the quenchers in feldspars.

A number of investigators (Smith & Stenstrom 1965, Görz *et al.* 1970, Sippel 1971) have noted that the CL emission from some feldspars is polarized. A variation in intensity is found when a sheet of polarizer material, placed between the luminescing sample and the microscope objective, is then rotated. The change in intensity is modest, not complete extinction. More specifically, Smith and Stenstrom noted that the CL emission from microcline is polarized. Görz *et al.* (1970) reported that all their feldspar samples, with the exception of adularia, showed polarization of CL. Sippel (1971) noted that the CL emission from all feldspars, except sanidine and adularia, is polarized. He noted, further, that all three peaks commonly found in terrestrial plagioclase, i.e., blue, green, and red, showed polarization, but the green peak is more strongly polarized than the red or blue.

5.2 Observations in natural samples

Although the potential uses for cathodoluminescence in the studies of feldspars and other minerals in igneous and metamorphic rocks are numerous, the actual published information in this area is still quite sparse. Some of the information on granites will be discussed here, and further discussion will appear in Chapter 8.

Feldspars in igneous and metamorphic rocks often exhibit either the yellowish-green CL, due to Mn^{2+}- or Fe^{2+}-activated plagioclase, or a bright blue-luminescing, Ti-activated, K-spar. Examples are shown in Plates 2a–f, 6e and f, and 11a–f.

The calcic-plagioclase feldspars can accept any of the listed activators and natural samples may display colors ranging from the blue to the red ends of the spectrum. The alkali feldspars, however, apparently do not accept sufficient Mn^{2+} (or Fe^{2+}) to show the characteristic yellowish-green CL. Therefore, when they luminesce, the CL is either blue or red (Mariano 1983, Sippel & Spencer 1970a). These observations are consistent with the crystal chemistry substitutional rules for a divalent ion in the $KAlSi_3O_8$ or $NaAlSi_3O_8$ structures.

The principal colors of feldspar CL are as given above. But through the microscope with the unaided eye, or on photomicrographs, one will often observe small variations in color and intensity which presumably represent subtle variations in the activator concentrations and/or in the crystallographic orientation. This may then be a tool for distinguishing different generations of feldspar. In Plates 2e and f, for example, there is a clear difference in intensity between the phenocrysts and the fine-grained feldspar in the ground mass. Under microscopic examination, to the eye, there appear to be four different types of feldspar in the ground mass, all of which might be described as a shade of blue or grayish-blue, but they contrast well with each other when viewed simultaneously.

Within a single feldspar grain, alteration, exsolution, zonation, and related phenomena are often revealed by CL (Plates 2a and d).

Smith and Stenstrom (1965) show an example of an albite with 'eight rings of alternating dark and light blue luminescence.' These same authors commented that epidote impurities (green CL) in feldspar showed up well, as did inclusions of calcite and apatite. They pointed out that much of the Ca determination in microcline perthite may be calcite or apatite impurities, impurities which are readily distinguished by CL observations prior to analysis.

The red, Fe^{3+}-activated, luminescence of some feldspars is of particular interest in the study of alkaline rocks and particularly in carbonatite localities. Mariano (1978) has shown numerous examples of feldspars introduced in the fenitization process. The fenitizing solutions contain Fe^{3+}. The introduced feldspars incorporate Fe^{3+} and their striking red CL stands out in sharp contrast to the CL of the feldspars in the invaded rock. This subject is treated more fully in Chapter 8 and several examples are shown in Plates 11a–f.

The presence of small amounts of barium in feldspars is a common occurrence and a frequently asked question is whether the CL properties are influenced by the barium concentration. The author is unaware of any published observations which would indicate that barium plays a rôle in determining CL properties, i.e., serves as an activator. Conceivably, there might be some co-substitutional process in which the concentration of barium varied similarly or inversely with one of the activator elements but, to the author's knowledge, no study of this subject has been published.

5.2.1 *Granites*

Granites can provide striking CL images in most cases and there are several immediate uses for CL in the study of granitic rocks.

Feldspars in granites often exhibit both the yellowish-green CL, due to Mn^{2+}- or Fe^{2+}-activated plagioclase, and a bright blue-luminescing, Ti^{4+}-activated, K-spar. It is quite easy to map the distribution of the two feldspars using CL. Plate 2a is an example of a granite, showing the contrast between the two feldspars.

When two or more feldspars are present, it is usual that they will show different CL. Sometimes this may be a subtle variation in shade but in many samples it will be a clear contrast like the blue K-spar, yellow-green plagioclase example shown in Plate 2a. In either case it is usually quite easy to map the distribution of the two (or more) feldspars without resorting to staining or more time-consuming optical petrographic techniques. It is still necessary, of course, to establish by independent methods, the actual composition(s) of the feldspar(s) revealed by CL since the color of the CL alone is not an unequivocal guide.

When other brightly luminescing minerals, commonly apatite and calcite, are intimately associated with the feldspars, they are usually revealed clearly under CL. Knowledge of the location and amounts of such minerals can be critical if, for example, one is trying to make representative electron probe analyses of the feldspars. See Plates 2c and d for an example.

Lyons and Krueger (1976) studied alkalic granite plutons of southern New England; cathodoluminescence was one of the techniques used. Some of their comments regarding CL may be of interest. First they noted that some of their samples were examined (by A. N. Mariano) using CL and 'this was particularly helpful in detecting apatite, calcite, fluorite, and zircon.' In the coarse biotite granite they noted that:

> Traces of calcite were observed under cathodoluminescence around the margins of biotite, indicating alteration associated with chloritization of biotite. Cathodoluminescence studies show fluorite in all the major alkalic granites of southern New England; it appears as a secondary mineral.

In the riebeckite granites they also noted the use of CL to detect traces of secondary minerals – fluorite, apatite, and calcite. They also noted:

> Near the contact zone with the coarse biotite granite, the fine riebeckite granite has microperthite with light-bluish cores surrounded by bright-red halos, probably reflecting Ti^{4+} and Fe^{3+} activation, respectively, under cathodoluminescence. Similar zoning was found in the feldspar phenocrysts. . . .

I mentioned earlier the red luminescence of some feldspars in carbonatites. There are also examples of red-luminescing feldspars in granites (Lyons & Krueger 1976). Presumably those granites whose feldspars have incorporated Fe^{3+} and exhibit red CL, were formed under conditions which are different from those present during the formation of granites whose feldspars exhibit the blue or greenish-yellow luminescence, due to Ti^{4+}, Mn^{2+}, and Fe^{2+}. The possibility exists of developing a procedure to determine $Fe^{2+}:Fe^{3+}$ ratios in feldspars by CL spectral measurements (Mariano *et al.* 1973).

5.2.2 *Metamorphic rocks*

High-grade metamorphic rocks often contain luminescing feldspars. Plates 6e and f are examples of a biotite gneiss where the feldspar shows a uniform blue luminescence.

5.2.3 *Activator evidence from natural samples*

Several investigators have made correlations between variations in luminescence and the variation in the amounts of the major elements, i.e., Ca, Na, and K. The present state of knowledge, i.e., that Mn, Fe, and possibly Ti, are the responsible activators, implies that such correlations are indirect. The primary variable is the concentration of the transition elements. If the CL intensity and wavelength correlate with the major elements, then one may suspect that the transition element concentration is also varying simultaneously with one or more of the major elements.

One of the reasons that better correlations between CL and the activator concentrations in natural samples have not been made is the difficulty of measuring the transition element concentrations at the low levels involved. Mariano *et al.* (1973) found that concentrations of only 500–1,000 ppm were sufficient for activation. Further support to the effectiveness of this range of concentrations is given by the observation that feldspars in igneous rock samples virtually always show some CL. Smith (1974a, b) tabulated the results of many analyses of feldspars, primarily from igneous rocks. Some of these include data on the Mn and Ti content. The range of Mn concentration in his tables is 10–400 ppm in plagioclases and most values are in the 50–100 ppm range. In K-spars, Mn is usually 50 ppm or less. Titanium concentrations averaged 30–300 ppm in the alkali feldspars reported. CL for these samples was not reported but the experience mentioned earlier, that virtually all feldspars from igneous rocks show CL, leads us to infer that these low concentrations may well represent the minimum effective concentrations needed to produce it. They are at or below the detection limit for many electron probes (Heidel 1971a).

Sippel and Spencer (1970b) made particular note that the wavelength of the green peak in terrestrial feldspars is not sensitive to the composition of the plagioclase. They found that the blue peak showed a variability in position and shape which did not correlate with composition. They speculated that there might be more than one cause for the blue CL.

Hutcheon *et al.* (1978) made a detailed combined ion microprobe, electron microprobe, and CL study of the plagioclase inclusions in the Allende meteorite. The Ti concentration, averaging about 200 ppm, did not correlate with the CL color. They observed a strong negative correlation between CL (blue) intensity and Na and Mg content. The significance of these correlations is not yet clear.

Thus, despite the clear evidence from synthetic samples on the identities of the activators, the corroboration of this by measurements and analysis of natural samples has not yet been achieved.

5.3 Feldspars in sedimentary rocks

When sedimentary rock samples are observed with CL, one frequently observes detrital feldspar grains luminescing brightly, apparently with the same luminescent properties as their counterparts in crystalline rocks. The feldspar grains may stand out vividly in contrast to the quartz since they luminesce much more brightly than quartz for the same excitation conditions. Feldspar twins also show up nicely because of the anisotropic nature of the CL emission from most feldspars (Thomas 1974).

Table 5.2 Statistics comparing the mineralogy determined by cathodoluminescence to that determined by staining (from Ryan & Szabo 1981).

	Cathodoluminescence	Staining
number of samples	20	20
mean quartz content	60.2%	60.7%
mean feldspar content	39.8%	39.3%
variance	70.4	55.3

Because of the intensity difference between feldspars and quartz, it is often quite easy to establish the percentage amounts of the feldspars by a combination of CL and point counting. Ryan and Szabo (1981; see also Section 3.8.1) described a technique for mounting sand grains with epoxy on a glass slide and then identifying and counting the feldspars, which were distinguished from the quartz by their CL color and intensity. They compared this method with the more common method which involves staining, a less predictable and more time-consuming process. They presented the comparison of the results of the two approaches which is shown in Table 5.2. The comparison is quite good.

CL may be thought of as a reversible staining – the color differences appear and disappear as the beam is turned on and off, with no permanent change to the thin section.

Plates 3e and f contain a good example of a weathered feldspar being altered to clay. The unweathered grains still stand out brightly under CL. Also, some very small fragments of larger grains which remain incompletely weathered are still easily visible with CL.

In sharp contrast to the CL of detrital feldspar grains, authigenic feldspars do not luminesce. This subject was investigated extensively by Kastner (1971). She studied many examples of feldspars in carbonates and low-grade metamorphic rocks and she found in every case that the authigenic feldspar or feldspar from low-grade metamorphic rock did *not* luminesce. She used both the Luminoscope® and an electron probe. Electron energies were apparently as high as 20 keV in the electron probe. Similar observations on authigenic feldspars in sandstones have been reported by Sippel (1971), and by Escobar and Mariano (1976) in connection with their studies on the origins of Colombian emeralds. Plates 3a, b, c, and d contain examples of authigenic feldspars.

Kastner noted that neither feldspars from low-grade metamorphic rocks nor authigenic feldspars show CL so the lack of luminescence is a necessary but not sufficient condition for authigenic feldspars. She concluded, however, that 'a combination of cathodo-luminescence and twinning can be used unmistakably to distinguish between authigenic and other feldspars.'

There are at least three possible reasons for the non-luminescence of authigenic feldspars:

(a) absence of the necessary activator ions in the environment of their formation; or
(b) the incorporation of quenching elements; or
(c) at the temperature of formation of authigenic feldspars,

the crystal structure cannot take up the activators in the necessary amounts.

Kastner considers that (c) is the probable answer, i.e., the temperature of formation is (to quote her directly):

> too low for a sufficient amount of activators to be taken up by the feldspar structure. It is generally true that the concentration of impurity ions or atoms increases with the temperature. The present study supports the suggestion of Smith and Stenstrom (1965) that cathodoluminescence properties of minerals depend on the temperature of formation. . . .

As Kastner has pointed out, the answers to the above would help to establish the upper limit to the temperature of formation of authigenic and low-grade metamorphic feldspars.

Although choice (c) is the most likely explanation, conclusive demonstration will require analyses of authigenic and low-grade metamorphic feldspars for the possible activators and quenchers. These analyses must be valid at the lowest concentration levels of the element which are sufficient for activation. This, in turn, assumes that the identity and minimum required amounts of activators for the feldspar structures have been established. As noted earlier, Mn^{2+}, Fe^{2+}, Fe^{3+}, and possibly Ti^{4+} are believed to be the most important activator ions in natural feldspars which exhibit CL. Concentrations in the range of roughly 50–1,000 ppm are effective. These concentrations probably represent the detection capabilities required to establish the purity of authigenic feldspars.

The author is aware of only one published analysis of authigenic feldspars which included a careful search for Ti, Mn, and Fe at low levels. Escobar and Mariano (1976) observed nonluminescing authigenic oligoclase in emerald deposits. This nonluminescent feldspar replaced bivalve remains, ammonites, and other fossils. The state of preservation of the fossils pointed to a low temperature of formation for the oligoclase and also, therefore, for the emeralds which were formed simultaneously. A wet chemistry analysis (by K. Thomas) revealed less than 100 ppm of Mn, Ti, or Fe in these authigenic feldspars.

The electron probe might seem to be a logical choice for analysis of authigenic feldspars. However, the detection limit of the probe, in a very favorable situation (Heidel 1971a), is only of the order of 700 ppm (Mn) and 180 ppm (Ti), which may not be low enough. The advent of ion and laser probes, with spot diameters of 1–2 μm and detection limits in the 1–10 ppm range or lower (Conzemius & Svec 1978, Hutcheon *et al.* 1978, Shankai *et al.* 1984) offers the promise of providing more data on the subject. Proton-induced X-ray emission (PIXE) (Blank *et al.* 1984) also has the required performance for this type of analysis. A careful program of analyses of authigenic feldspars by one or more of these techniques, coupled with CL studies, would be an important contribution.

Despite the lack of complete knowledge of the *raison d'être* for nonluminescing authigenic feldspars, the empirical facts are still put to good use in detecting their presence. Buyce and Friedman (1975), for example, have studied the Cambrian–Ordovician carbonate rocks of the proto-Atlantic shelf. It was important in their study, to know not only the abundance of K-feldspar, but also the relative amounts of detrital and authigenic K-spar. Their photographic study was facilitated by their development of a combined CL and TrL technique. The thin sections are stained with sodium cobaltinitrite stain which stains both the detrital and authigenic feldspars a light yellow color. Under CL the authigenic portions are black but the detrital cores are bright blue. By superimposing a CL exposure on a very short TrL exposure, a picture of the total feldspar outlined in yellow, and the detrital portion in blue, is obtained (see Plates 3c & d). Other authors have also noted the usefulness of CL for distinguishing authigenic from detrital feldspars. Mou and Brenner (1982), for example, commented on this in their study of the Tensleep sandstone.

Ali and Turner (1982) and Turner *et al.* (1982) made a comprehensive study of authigenic K-feldspar from the Triassic Bromsgrove sandstone formation (central England). They reported, apparently for the first time, optically zoned overgrowths which correlated with a variation in the concentration of BaO and Na_2O. However, consistent with the observations of other authors already cited, the authigenic overgrowths did not luminesce, so there was no correlation of CL with the compositional variations in the overgrowths. All their detrital feldspar cores displayed bright blue CL. Ali and Turner did extensive electron probe analyses of the cores and the overgrowths on several samples, reporting K_2O, Na_2O, BaO, CaO, SiO_2, and Al_2O_3. No systematic differences in composition between core and overgrowth were noted. Minor elements, which might be activators, were not detected in the overgrowths by electron probe analysis.

Hearn and Sutter (1985) also studied the occurrence and possible mode of formation of authigenic potassium feldspar in Cambrian carbonates. They observed the backscattered electron images and the cathodoluminescence images of the feldspars produced with an SEM. The authigenic feldspars occurred both as overgrowths on detrital grains and as a fine-grained matrix. In either case, they showed a general lack of cathodoluminescence.

Even though authigenic feldspars show negligible CL at electron energies up to about 20 keV, it is conceivable that they might show weak luminescence at higher energies. To my knowledge, no one has reported on this subject yet.

Some investigators (Geake *et al.* 1973) have suggested that the blue CL so commonly seen in feldspars may be due to lattice defects instead of an activator. If this is so, then the nonluminescence of authigenic feldspars implies that they may have a significantly different level of lattice defects than that found in other feldspars.

5.4 Lunar and meteoritic feldspars

Sippel and Spencer (1970b) studied the CL properties of lunar samples. They reported calcic plagioclase to be the 'dominant luminescent mineral in crystalline rocks and breccias. . . .'

Their CL apparatus was equipped with a means for making spectral measurements between 3,700 and 8,200 Å.

For the lunar feldspars they observed broad bands centered at 4,500 and 5,590 Å, values which agree with their observations on terrestrial feldspars.

Their observations on terrestrial feldspars are best summarized by a direct quotation:

> Of the 47 feldspar specimens, microprobe analysis shows 20 of them to be alkali feldspar, 20 plagioclase, 6 approximately albite, and one of mixed calcic–alkali composition. The green peak (~5590 Å) does not appear in the spectra of alkali feldspar, but is present in each of the sixteen plagioclase spectra where mole per cent anorthite exceeds 11 per cent. It probably results from a divalent activator ion substituting for calcium in the plagioclase structure. The position of the green peak is not sensitive to composition within the plagioclase series and spectrum corrections do not result in any appreciable change in the peak position. The blue peak (~4500 Å) shows some variability from specimen to specimen in peak position and in shape. This variability does not correlate with composition, and its cause is unknown. In some cases, more than one luminescence center may be involved in the blue emission. In any case a definite blue emission band is found in 45 of the 47 specimens. The two exceptions are both alkali feldspar. The red–infrared peak (~6850 Å) in the uncorrected spectrum shown in Fig. 3A falls in the region where correction is important. The peak actually occurs at about 7050 Å when corrected. The red–infrared is modestly sensitive to composition. It shows a range of about 400 Å and its position correlates roughly with mole percent albite, without regard to whether the feldspar is calcic or potassic. The red–infrared peak of alkali feldspar is usually less intense than that of plagioclase. Potassium-rich specimens usually exhibit a broad blue band and a small, but definitely observable, red–infrared peak. In three cases, alkali feldspar shows no red–infrared peak at all. In the other 44 specimens, the red–infrared peak is clearly present. The mean position is about 7300 Å corrected.

They also noted a 'red–infrared' peak in their terrestrial plagioclases which fell between 7,050 and 7,300 Å depending on the sample. No similar peak was noted on over 100 grains of lunar plagioclase. They speculated that an activator was lacking in the crystallization environment of the lunar samples. From the work of Mariano *et al.* (1973) and Geake *et al.* (1973) we know that Fe^{3+} is an activator which produces red feldspar CL and therefore is presumed to be the missing activator in lunar feldspars.

For the plagioclase from the lunar crystalline rocks, Sippel and Spencer continued:

> The blue and green peaks are present, but the red–infrared peak is entirely missing. The spectra of more than one hundred grains of plagioclase have been measured in lunar crystalline rocks and breccias and no evidence of a red–infrared peak was found in any of them. This is anomalous because in our suite of terrestrial rocks all 20 plagioclase specimens show a red–infrared peak. . . .

Sippel and Spencer (1970b) made interesting observations also on maskelynite in some of their lunar samples. Maskelynite is a shock-formed, noncrystalline phase resulting from vitrification of plagioclase by shock waves. It retains the external features of crystalline plagioclase. They observed dull reddish CL in such samples and also described an example of one grain, which had undergone intermediate damage, with orange-appearing CL. They suggested the observed CL was due to a 'shifting and broadening of the green plagioclase peak.' Further they observed that the material had become isotropized and the CL was no longer polarized. The CL spectra of some maskelynite changed with time under continued bombardment. Figure 5.8 is from this work.

Sippel and Spencer (1970b) also noted a lack of correlation between the shape and position of the blue peak with

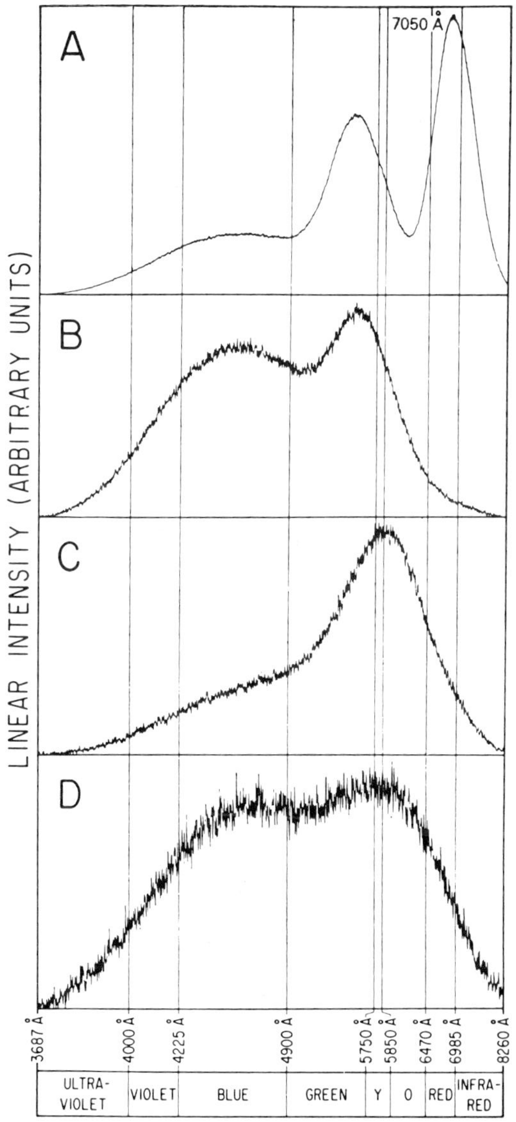

Figure 5.8 Feldspar CL spectra: (a) Terrestrial plagioclase, AN_{85}. The IR peak falls at 7,050 Å when corrected for spectrophotometer response. (b) Plagioclase from lunar crystalline rock. The red–infrared peak is absent. (c) Plagioclase grain from lunar breccia showing intermediate degree of shock damage. Note shift and broadening of the green plagioclase peak. The luminescence of this specimen appears orange. (d) Maskelynite grain. Note that the abscissa is nonlinear and the spectra are not corrected for overall spectrophotometer response. (From Sippel & Spencer 1970b.)

composition and suggested that more than one luminescent center may be involved in producing the blue luminescence.

Geake and coworkers (Geake *et al.* 1970, 1971, 1973, 1977, Geake & Walker 1966, 1975) carried out extensive investigations of luminescence in lunar materials using proton and electron excitation. Their work has been referred to several times in this chapter. They found plagioclase to be the main luminescent component of lunar material, and the peak at 5,600 Å was the strongest. They noted the lack of the near IR peak in lunar samples whereas they stated its dominance for most terrestrial plagioclase. Based on synthetic doped samples and theoretical work they concluded that Fe^{3+} and Mn^{2+} are the important activators and that the blue peak is due to lattice defects. They concluded that Fe^{3+} is the activator lacking in the lunar environment.

Hutcheon *et al.* (1978) used CL as one of the tools in their study of inclusions in the Allende meteorite. They found it to be an important tool for the study of the plagioclase content and noted that:

> Four types of luminescence patterns in plagioclase are due to: (1) twinning; (2) primary sector zoning; (3) alteration of boundaries and fractures; (4) shock effects. . . .

The CL was highly variable and regions of uniform CL were less than 50 μm in size. They did detailed analyses of two inclusions and found that only Na, Mg, and Ti were greater than 50 ppmw. In some cases, the bright CL correlated with low Na and low Mg content. The Ti content was highly variable and there was no correlation of the Ti with the Mg content or with the CL. With respect to the lack of correlation with Ti, they emphasize that:

> It does, however, point out one important limitation of the CL technique, the difficulty of identifying a specific activator. Although CL color readily distinguishes between minerals and between different textures of the same phase, the sensitivity of CL to poorly known factors, such as crystal lattice orientation and trace element concentration, often obscures the actual production process. . . .

5.5 Summary

The CL of feldspars has proved to be a very valuable petrographic tool. Sometimes part of the trace element composition of the feldspar can be identified from the CL spectra. Also the visual contrast between different feldspars, which is commonly observed, is useful in quickly distinguishing K-spar from plagioclase. However, there is certainly room for further research to make it even more useful. Many of the areas worth investigating were mentioned in the discussion but they are summarized here also.

(a) What is the dependence of CL intensity on concentration for Ti, Mn, and Fe, both when present singly and in combination?
(b) What are the quenching elements (if any) and what is their concentration dependence? Are the quenchers the same for the different activators?
(c) For a given activator and concentration, what is the variation in CL intensity for different possible feldspar hosts?
(d) Is there an intrinsic CL or CL due to lattice defects in feldspars? Why does it not appear in authigenic feldspars?
(e) How does the magnitude of the polarization effect vary for the different activators and hosts? Does the direction of the incident beam have an effect?
(f) Do authigenic feldspars show CL at higher excitation voltages, analogous to quartz?
(g) What levels of Mn, Ti, and Fe exist in authigenic feldspars?
(h) What is the mechanism by which Ti content influences CL in feldspars?

Answers to some of these questions will greatly enhance the use of CL in the study of the feldspars.

Chapter 6

The cathodoluminescence of quartz and some geological applications

The intensity of quartz CL is weak compared to the CL of feldspars, carbonates, and many other minerals. Nevertheless, it exhibits interesting and useful cathodoluminescence properties. Most examples of quartz from igneous and metamorphic rocks show CL. Detrital quartz also often luminesce but the CL of authigenic quartz is usually weak compared to that of the detrital grains. The colors of quartz CL that have been reported in the literature include gray, blue, brown, purple, rust, dull red, and pink.

Crookes looked for CL in quartz as early as 1879 but with indefinite results. Smith and Stenstrom (1965) included two examples of quartz CL in their paper and they commented on the observed differences between the luminescence of detrital quartz (blue) and secondary quartz (dull red). Long and Agrell (1965) observed dull blue luminescence of quartz in some of their samples.

Surprisingly, perhaps, the mechanisms responsible for the CL of quartz and other SiO_2 minerals are among the least well understood of those of the common minerals. The activators and quenchers, if any, have not been identified with certainty. Also effects such as irreversible fading of CL under prolonged bombardment, and CL changes apparently due to structural damage, are not well understood. The situation is similar to that for quartz ThL. Bernhardt (1982), updating an earlier statement by Wintle, states: 'Although "quartz is one of the most widely studied naturally occurring thermoluminescent materials" (Wintle 1975) the physical nature and structure of thermoluminescent centres in quartz is not clear. . . .'

6.1 Observations on the physical bases for CL in SiO_2

Peters (1930) saw red CL in glass and quartz and he references much earlier (1907) similar observations. Saksena and Pant (1951) used a gas discharge to excite samples of colorless quartz and amethyst and observed luminescence between 5,650 and 6,550 Å. The luminescence was weak.

Claffy and Ginther (1959) reported on the preparation of a red-luminescing (emission maximum at 6,500 Å) synthetic quartz which was made by the addition of 1.5 percent MnO and 0.1 percent Al_2O_3. This amount of MnO is much larger than that encountered in natural samples so this observation is of limited geological interest.

Medlin (1963b) carried out an extensive study of thermoluminescence in quartz and included some studies of X-ray and UV excited luminescence as part of this work. His synthetic samples incorporated any one of many (24) different possible activators, including Mn^{4+} but not Mn^{2+}. He also analyzed (by spectrographic analysis) some natural quartz samples and found that the intensity of the 165° K ThL glow peak correlated poorly with the Ti concentration. He found that UV excitation produced no photoluminescence but all specimens exhibited XL. The X-ray excited luminescence was about the same for all samples, independent of the Ti concentration, and he concluded that it was probably 'due to some sort of defect center rather than to the Ti ion.' He noted several glow peaks (ThL) including one at 350° K (77° C). The spectrum of this ThL emission and the spectrum of X-ray excited luminescence for some of his samples is shown in Figure 6.1.

6.2 Observations in geological samples

In addition to the two geological observations already mentioned, there are many more examples. Smith and Stenstrom (1965) observed both red- and blue-luminescing quartz. They noted that the CL of authigenic quartz is much duller than that of blue-luminescing detrital grains and thought the latter might have originated in igneous rocks formed at elevated temperatures. The CL from authigenic quartz changed from essentially nonluminescent, at low beam loading, to a dull red at high beam loading. They reported a shift from blue to red CL after prolonged bombardment of the detrital grains in some cases. These same authors also noted an example of a quartz grain which

Figure 6.1 Emission spectra for quartz. (a) The spectrum of a ThL glow peak at 330°K for an annealed single crystal specimen. (b) The spectrum of the 165°K ThL glow peak of a synthetic, Ti-containing sample. It is identical to the spectrum at 77°K of the X-ray excited luminescence. (Modified from Medlin 1963b.)

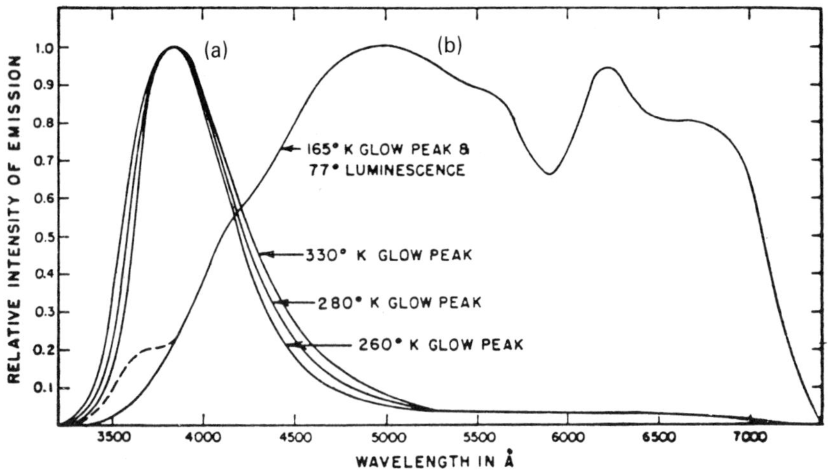

luminesced blue except for a central region with red CL. The red-luminescing region contained zircon and they inferred that the CL change was due to radiation damage from the thorium associated with the zircon.

Sippel pioneered the application of the CL observations to sandstone petrology and in 1968 he published a comprehensive article on the subject. He reported that red and dull blue colors were most common in quartz in sandstones. He pointed out the usefulness of CL to help locate and identify, in sandstones, minerals such as feldspars, zircon, apatite, and the carbonates. He reported that the CL of quartz ranged from one to three orders of magnitude lower in intensity than that of the potash feldspars. He observed the dull or nonluminescent properties of authigenic quartz and reported that blue-luminescing quartz was never seen as overgrowths. He also reported the presence of transient phenomena, e.g., a shift in color from blue to red after electron beam bombardment. Sippel made abundant use of the nonluminescence of authigenic quartz to distinguish cementation, pressure solution phenomena, reworked grains, etc.

The secondary (authigenic) quartz which grows in optical continuity with detrital grains has been identified historically by the presence of such features as bubbles or dust, collected along the rims of the detrital cores. These features are sometimes difficult to see or equivocal in interpretation. However, the CL contrast, when present, is much clearer and a routine CL examination to distinguish the two types of quartz is valuable. The study of fabric (the size, shape, and orientation of the detrital grains) is facilitated with CL. Sippel cautioned that the absence of luminescence contrast does not prove the absence of secondary quartz – sometimes both the overgrowth and the detrital core have the same dull CL color. He reported that in immature sandstones one frequently sees phenomena related to fracturing and healing by secondary quartz. Also he stated that 'most of the polycrystalline quartz grains found in sediments originate by brittle fracture with subsequent healing and various degrees of annealing and recrystallization.'

Sippel (1971) reported that quartz from lunar samples showed pink and bluish CL. The CL polarization effects were stronger than those in terrestrial quartz samples and rotation of a polarizer between the sample and the microscope produced a color change of CL from red to blue. He and Spencer (1970b) also found that some lunar quartz had an anomalously bright blue CL and there is evidence this is caused somehow by shock damage.

Sibley and Blatt (1976) utilized CL in their studies of the Tuscarora orthoquartzite. The recognition of intergranular pressure solution was facilitated because of the clear differentiation of authigenic from detrital quartz by CL. Some contacts which appeared to be pressure solution with polarized light were found not to be so with CL. In this formation they concluded that: 'Pressure solution can account for only 30–35% of the pore-filling quartz cement.' Plates 6a and b are examples from their work. They also found that extension fractures were often well delineated under CL because the secondary quartz which fills the fracture has a much duller luminescence and stands out clearly in contrast with the original grains (Plates 6a and b). Another interesting example of fracturing, also taken from the work of Sibley and Blatt, is shown in Plates 6c and d. This sample appears with polarized light to have three separate grains. However, with CL it is revealed to be a single grain which has been rotated and fractured and the wedge-shaped fractures are filled with nonluminescing quartz. Sibley and Blatt also noted the transient phenomena and Sibley (1975) states that, in some cases, if the electron beam is on the same spot for periods of 20 minutes or more, the red component of the luminescence increases. This variability of CL with bombardment history for some quartz is a complication because prolonged bombardment is often necessary for the photographic recording of quartz CL.

Sprunt *et al.* (1978) have studied the CL of two sequences of naturally deformed quartzites. They found that metamorphism appears to homogenize the luminescence and the 'color' is related to the metamorphic grade – low temperature causing red luminescence and high temperature causing blue CL. They also stated that strain is a factor in the CL of quartz.

Sprunt and Nur (1977) have used cathodoluminescence as a tool in their study of the 'Destruction of porosity through pressure solution.' They looked at thin sections of deformed and undeformed St Peter sandstone with CL. Grain boundaries and pore spaces, as well as overgrowths, are clearly revealed in many cases, but not in every case. They concluded: 'whereas cathodoluminescence is a very useful tool in studying solution transfer and possibly a useful method of detecting very low grade metamorphism by the absence of blue quartz luminescence, cathodoluminescence studies cannot be used to determine the amount of pressure solution.' These same authors (1979) also reported observations of CL in quartz from granites and pegmatites. They found that in 84 percent of the granite samples, most of the quartz exhibited blue CL. In the remainder the 'dominant quartz CL color is red.' The red and blue are believed by Sprunt and Nur to represent quartz introduced at low and high temperatures respectively. Within the blue-luminescing quartz, many samples have structures which show up as red with CL and these same structures are not usually visible with PoL or TrL. This red quartz seems to be deposited in pre-existing grains, i.e., it is healing fractures in grains formed at higher temperatures. They suggest that there may be a relationship between cracking and healing phenomena in quartz and fluid transfer in crustal processes.

Morton (1978) has provided an example of radiation damage of quartz in a uranium-containing sandstone. Distinct haloes, invisible in transmitted light or polarized light, show up as red-luminescing rinds on the quartz grains under CL. The haloes surround blue-luminescing cores (Plates 5a–d). The halo widths are of the order of 20 μm and the interpretation is that they are due to damage by alpha particles, emitted from the uranium minerals in the pore space. The range of alpha particles from uranium in a solid is of the order of 20 μm (Gentry 1973) and they apparently produce a radiation-damaged region with changed CL properties. The haloes are stable under heating to temperatures up to 300 °C. As Morton has suggested, these damage haloes may be useful indicators of previous uranium content in leached sandstone. This damage information would be retained after the leaching process, provided the rock had not been subjected to temperatures high enough to anneal it.

Zinkernagel (1978) has presented an extensive discussion of quartz CL, with well reproduced color plates. He defines three types of quartz, based on CL. They are:

(a) *violet-luminescing quartz* from plutonic, volcanic, and contact metamorphic rocks, rocks which have undergone fast cooling;
(b) *brown-luminescing quartz* from low-grade metamorphic rocks, and from high-grade metamorphic rocks which have been slowly cooled;
(c) *nonluminescing authigenic quartz.*

He did not observe zoning in quartz with brown CL and very rarely observed zoning in violet-luminescing quartz.

Zinkernagel (1978) recorded CL spectra from some of his samples and he also studied the effects of thermal treatments on the spectra. An example of his results is shown in Figure 6.2. Zinkernagel's comments on quartz colors and spectra may be of interest. They help to emphasize the contrast in preciseness of description of CL between a monochromator scan and a visual estimate of color. He states (1978, p. 18):

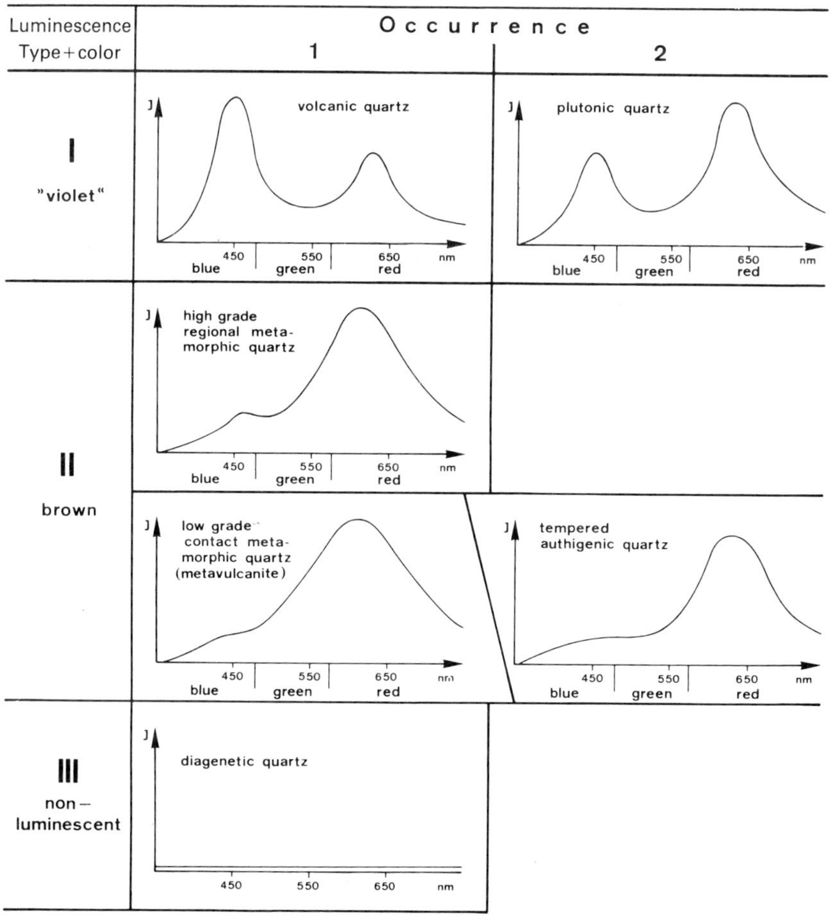

Figure 6.2 CL spectra of various types of quartz according to the classification proposed by Zinkernagel (1978).

> All luminescence spectra of 'violet' luminescing quartz described as blue, blue-violet, violet, or red-violet in igneous and some contact metamorphic rocks show narrow emission 'bands' in the blue and red range. Even blue luminescing quartz has always an additional emission maximum in the red range. For blue and blue-violet luminescing quartz the intensity in the blue range is higher than in the red range. For violet luminescing quartz the emission intensity in the blue and red range is almost identical. For red-violet luminescing quartz the emission in the red range is more intensive than in the blue range.
>
> Brown luminescing quartz has a broad emission range in the red. But most strikingly, a minor emission maximum in the blue range, too, is developed in all samples, which cannot be recognized by the eye.
>
> Thus, the superposition of red and blue light explains the 'violet' and brown luminescence of quartz, which is significant for the temperature history and is related to geological processes. Likewise, it becomes clear that a simple distinction between 'violet' and brown luminescence is only useful for rough examinations of sand and sandstones.

One of the conclusions he reached, in agreement with Sibley and Blatt (1976), is that the degree of pressure solution is overestimated if CL is not used. He also concluded that the CL of quartz is more likely due to the degree of ordering of the lattice than to trace amounts of activators.

Grant and White (1978), using an SEM operated at 30 kV, reported the same two major peaks that Zinkernagel reported, one between 400 and 500 nm and one at about 620 nm. The relative heights of the red and blue differed somewhat from those found in Zinkernagel's samples. These same authors also reported 'blocky textures' in some quartz overgrowths.

Dietrich and Grant (1985) presented an extensive discussion of the CL of syntectonic quartz fibres. Their observations were made with an SEM, operated at electron energies up to 30 keV, and equipped with a semi-ellipsoidal mirror for collecting CL. With this arrangement, they were able to study the CL emission of small areas of weakly luminescing material surrounded by material with high CL emission. They pointed out that prior studies had found that low-grade metamorphic quartz CL was uniform in color. However, at the higher voltages, up to 30 kV in the SEM that they employed, there were distinctive variations in the CL spectra of different specimens. These variations could be related to the different temperatures of formation and to deformation effects due to strain. They found two bands in the CL emission spectra of the low-grade quartz fibres they studied – a red and a blue. A relatively high intensity of the red band was related to 'a higher temperature during fibre growth, that is during the deformation of the rock unit containing the fibres.' A relatively higher blue band seemed to be 'related to lattice modifications induced by continuing deformation of the earlier formed fibres.'

6.3 The question of activators in quartz

Speculation on the possible rôle of activators in quartz CL has taken place for many years. There is still little firm information. In this respect the status is similar to the explanation for the color of smoky quartz. As Nassau (1978) points out in his discussion of the origins of color in minerals: 'reinterpretations continue to appear even in materials, such as smoky quartz, which have been extensively investigated over many years.'

Titanium is frequently mentioned as a possible activator. As noted previously, Medlin (1963b) did find a weak correlation between ThL intensity and Ti concentrations (in the range of 900–9,000 ppm) in his synthetic samples. However, the titanium content of natural quartz may be much lower than this. Dennen (1967), for example, reported Ti contents of only 44 ppma or less in a group of samples of quartz from the Perry Formation. He found the Fe contents were also in this range.

Sprunt (1981) looked at many different samples of quartz with differing water content and differing amounts of various trace elements. She concluded, among other things, that the cause of the CL color of these samples is not necessarily related to the cause of the color observed in natural lighting. She found that the amounts of Cu, Al, and water have no simple relationship to the CL color. She did establish a correlation between the CL color and the Ti : Fe ratio. She found that this ratio is high for blue-luminescing quartz and low for red-luminescing quartz. The absolute amounts of Ti and Fe were not given.

Mn may also be postulated as an activator, since it plays this rôle in so many other minerals. As noted earlier, Medlin did not include Mn^{2+} in the list of impurities he used to dope the synthetic quartz samples, so his results do not rule it out. Long and Agrell (1965) observed some pinkish-luminescing quartz and probed these samples for Mn. The Mn content was very low, less than 0.01 percent.

There is discussion about the possibility of aluminum, a common trace element in quartz, playing a rôle in the CL of quartz. Dennen *et al.* (1970) found between 10 and 260 ppm of Al in natural and synthetic quartz samples, formed over a temperature range of 90–1,000 °C. Scotford (1975) reported Al contents of up to 549 ppm for 18 natural quartz samples. Grant and White (1978) have commented that a 400–500 nm region peak is common in quartz CL emission spectra and may be associated with Al replacing Si. Sprunt (1981) comments on the possible rôle of Al impurities in smoky quartz and rose quartz but concluded that it is not important in these particular samples. Ramseyer and Mullis (quoted in Matter & Ramseyer 1985) reported an inverse correlation between CL intensity and the Al content of some natural quartz samples.

Owen (1984b) has made some interesting observations on the possible rôle of water in determining the CL of quartz samples. He found that if he heated quartz samples in the CMA by turning the beam up to high power, high enough to cause incandescence, the quartz grains soften and bubble, and after 10–20 seconds there is another color change. When they are allowed to cool to room temperature and are once again observed under more normal CL excitation conditions, they have acquired an apparently permanent, medium intensity, blue CL. He presumes that the gas lost during the bubbling is water vapor. As partial confirmation of this, he has found that quartz from extrusive, shallow intrusive, and high-grade metamorphic rocks all show CL which is purple or blue (Owen 1984a, c). These rocks are usually depleted in water. Low-grade metasedimentary rocks, which normally have abundant structural water, always luminesce red-brown.

One of the arguments which has been advanced against the presence of activators in quartz is the general absence of zonation of CL like that seen in carbonates, feldspars, apatites, garnet, and other minerals. If activator impurities were present, one would not expect that they would always be distributed uniformly, i.e., zonation should be seen. Actually there have been a few reported examples of zonation in quartz but this zonation is not nearly so dramatic in appearance as the zonation seen in carbonates (Ch. 7) and in several other minerals (Ch. 8). Sippel (1968) did report an example of zoning growth patterns in the red luminescence but it was not clear if this zonation took place during the formation of authigenic quartz or whether the zoned quartz was a detrital fragment of vein quartz. He suggested that the zoning might be due to either foreign ion activators or to impurity-induced defects. Zinkernagel (1978) saw slight zonation in occasional samples of 'violet'-luminescing quartz. He also included a color plate of a 'zoned quartz cement of greenish, brown, and non-luminescent zones indicating changing conditions during cementation . . .' (p. 41). He concluded, however, that 'the ordering degree of the quartz lattice' was the cause of CL and variations therein, not trace elements. Ramseyer (1982a) observed zoned CL in quartz from volcanic rocks. The overgrowths either exhibited green CL or were black (no CL). Other investigators (Tovey & Krinsley 1980, Grant & White 1978) have also noted some evidence of zonation or similar textures at higher excitation energies.

Another example of zonation in quartz CL is shown in a photograph by Zinkernagel in Stel's article (Stel 1981). The article is concerned with crystal growth in micro-cracks. These cracks may contain crushed fragments of various minerals and also may contain euhedrally terminated quartz crystals. There are growth bands visible in the quartz, even with polarized light, and they are believed to be duc to zones of small crystallographic misorientation. The same zones are revealed even more clearly with CL. This is another case of CL zonation which is not apparently associated with activators.

Hanusiak (1975) advanced another argument which tends to rule against the rôle of activators. He measured the CL intensity as a function of excitation current over a range of three orders of magnitude and found the relationship was still linear to within 5 percent at 10^{-7} amperes beam current (Fig. 6.3). He calculated the density of activators required to produce a linear relationship and arrived at a number of 0.3 mole percent or greater. This was much greater than the known abundance of any impurity in the quartz and hence tended to rule impurities out. He also noted, as have other authors, that the position of the emission maximum of the CL spectrum in the blue region is similar to that of many other silicates and this would be unlikely if an activator were involved.

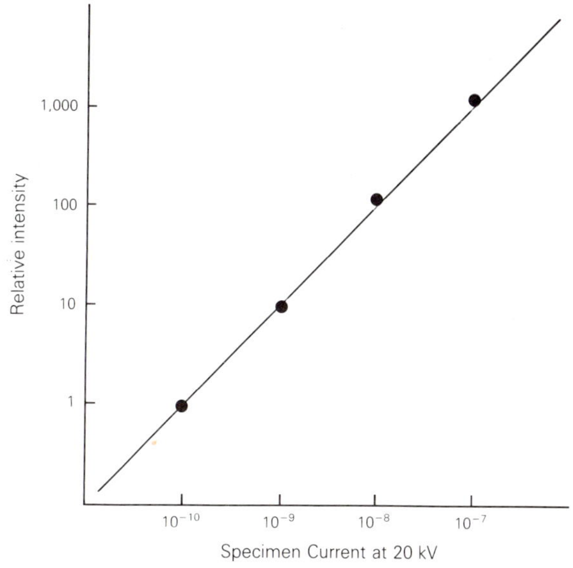

Figure 6.3 The variation of the intensity of CL of α-quartz with bombarding beam current. The data is fitted by a linear relationship to within ± 5% over this range (Hanusiak 1975.)

An example of a zoned quartz cement is shown in Plate 4b. The zonation in this sample shows up at electron energies of only 15 keV.

6.4 Damage and quartz luminescence

The concept of 'damage' and its effect on the CL of quartz seems to merit further study. Evidence from several directions suggests that crystal structure damage plays a more significant rôle in quartz CL than for most other minerals. The CL changes noted by Morton (1978), where alpha particle bombardment changes quartz CL from blue to red, is an example. Smith and Stenstrom (1965) observed a radiation damage halo around an included zircon and an example of this is also shown in Plates 4c and d. Sippel and Spencer (1970b) noted that shock damage affects the CL of both plagioclase and quartz. Krinsley and Hyde (1971) and Sprunt *et al.* (1978) have also commented on the apparent effects of strain on CL of quartz.

Hanusiak (1975) and Hanusiak and White (1975) have reported the results of certain 'damage' experiments on samples of natural and synthetic quartz. Damage mechanisms included sand-blasting and grinding. They found a difference in the CL of damaged and undamaged material, an increase in intensity in some cases and a decrease in others. At low specimen temperatures (−100 °C), they found the damaged quartz to be significantly more luminescent.

If damage is indeed important, then one must raise the question of what damage happens to a sample during the preparation procedures of lapping and polishing. Does the preparation procedure itself affect the CL and, if so, can the effect be reproduced?

The influence of surface damage effects may also explain the higher electron energies required to observe CL in quartz. Unlike many carbonates, feldspars, and other common minerals, quartz samples often do not develop appreciable CL unless the electron energy is 12–14 keV or higher. As noted in Section 3.5, electron penetration depths for 10–30 keV electrons are estimated to be in the few microns range, i.e., the primary CL excitation occurs relatively near to the surface. Perhaps the electrons must penetrate through the first few microns of 'damaged' quartz before they cause CL representative of the bulk of the thin section. If surface damage plays a rôle, then one might suspect that there would be a difference between CL observed on a normally prepared thin-section surface and one which has been etched with HF after preparation to remove the outer layer. This type of experiment has been done by Tovey and Krinsley (1980) who carried out a number of different studies on individual quartz grains, including studies of the effects of grinding, polishing, and etching. They observed the CL on an SEM and they noted changes in several cases. They also etched grains and produced CL changes in some cases but not all. The etching apparently removed a layer of about 10 μm.

It is interesting to note, also, that Charlet (1974) has done similar experiments, etching quartz grains for thermoluminescence studies. He found differences between the ThL of the surface and the interior of the quartz grains.

Further work in this area, e.g., looking at etched surfaces of polished sandstone slabs before and after etching 20–30 μm, looking at CL on cleaved, unpolished surfaces, or preparing surfaces by alternative procedures such as ion milling, would be worthwhile.

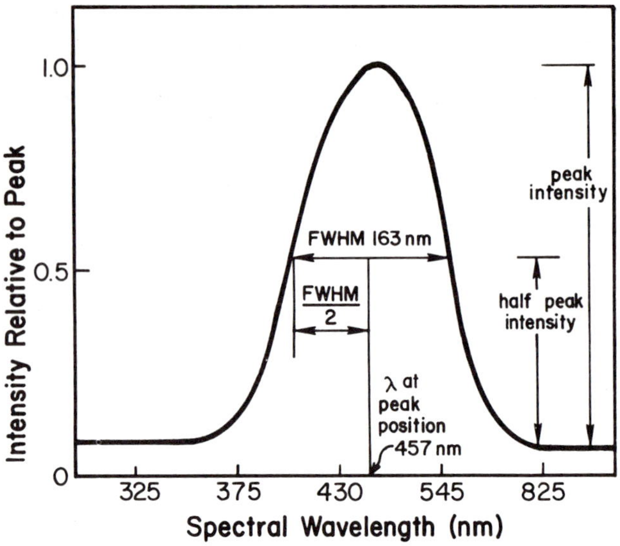

Figure 6.4 An example of the spectrum of CL of α-quartz at a sample temperature of −100 °C. Note that the wavelength scale is not linear. (Hanusiak 1975.)

6.5 Influence of temperature on quartz luminescence

The effect of the sample temperature on the CL of quartz is another area worthy of additional investigation. It is known (Leverenz 1968) that the luminescence of many materials (but not all) increases at lower temperatures and such an enhancement would be appealing for the case of a weakly luminescing material such as quartz.

Medlin (1963b), as part of the work mentioned earlier, found that the X-ray excited luminescence of quartz at 77° K is almost an order of magnitude more intense than the XL at room temperature.

Blair and Edgington (1968, 1970) noted that the spectrum of proton excited luminescence (10^8 protons/cm^2sec, 146 meV energy) of quartz from a granite showed blue and red components at 20 °C. They found that the intensity change at −80 °C was only slight but a dramatic increase occurred between −80 and −180 °C. At −196 °C the blue component intensity increased dramatically, 15× or more. Their low-power density in these experiments (2.4×10^{-5} W/cm^2) minimized damage.

Hanusiak (1975) and Hanusiak and White (1975) built a cooled stage for an SEM and observed the CL spectra and the variation of CL with temperature for several samples of natural and synthetic quartz, tridymite, cristobalite, and orthoclase. With the exception of the fused quartz and the orthoclase samples, all showed strong increases in intensity at −100 °C, by factors of 100 to 1,000. The CL spectrum of one of the quartz samples is shown in Figure 6.4. The rise and decay times were lengthened by 1.0–1.8 ms at lower temperatures. Peak maxima in the emission spectra remained at about the same location but the half-width decreased slightly for most samples. An example of the variation of the CL intensity with temperature for one of the samples is shown in Figure 6.5. Clearly there is a large increase in CL intensity at lower temperatures for this particular sample. Note that the onset of the intensity increase is at about −70 °C and occurs rather abruptly.

For a synthetic quartz sample, Hanusiak and White observed an increase in the polarization effect at low temperatures and postulated that this might occur if more than

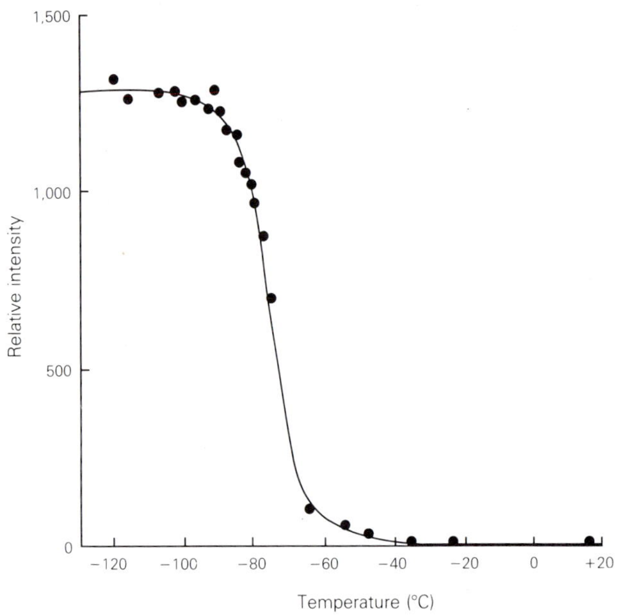

Figure 6.5 An example of the variation of CL intensity with sample temperature for α-quartz. Note the sharp onset of the intensity increase at about −70 °C and the magnitude of the overall increase which Hanusiak quotes as 1,300× between room temperature and −80 °C. (From Hanusiak 1975 and Hanusiak & White 1975.)

one mechanism is responsible for the CL, the different mechanisms having different temperature dependences.

Grant and White (1978), using an SEM, also reported an increase in CL intensity of quartz at liquid nitrogen temperature, the increase being a factor of 50× over the level of CL at room temperature.

The author (Marshall 1980a) has looked at a number of thin sections with a liquid nitrogen cooled stage mounted on a CMA. The dramatic increase in intensity was noted on some thin sections of natural, predominantly quartz, samples, but not for all (Plates 5c and d). It appears that the onset of the increased intensity is sharply dependent on temperature, as would be expected from Figure 6.5, and only slight differences in cooling capability – the thermal contact between the thin section and the cooling plate – may be critical in determining whether or not the sample reaches a sufficiently low temperature to develop the increased intensity.

The intensity increase at low temperatures is important, but the most important consideration is the contrast between authigenic and detrital quartz and between damaged and undamaged quartz. It has still not been clearly shown that these contrasts are retained at lower temperatures.

6.6 Time-dependent cathodoluminescence effects in quartz

Many authors have commented on the fact that some quartz samples show an apparent cathodoluminescence which changes with time. Characteristically, the CL is a light blue initially but fades in intensity and changes color to red later. The time period over which the change takes place varies from a few seconds to a few minutes. Sometimes part of the original blue luminescence is recoverable if the sample is set aside for a period.

To the author's knowledge, no one has as yet demonstrated the reason for this alteration of CL in quartz. It may represent the emission of thermoluminescence from a glow curve peak whose temperature can be reached with the power from the electron beam. Rémond (1977) has pointed out that natural quartz samples which showed strong ThL emission exhibited CL which varies with time. One of the characteristic ThL glow peaks in quartz is at 77 °C, a temperature which may be reached locally in some samples. The temperature reached on the portion of the sample in the center of the electron beam is a highly variable quantity, because of variations in the excitation available (depending upon the system design and the mode of operation – current density – etc; see Secs. 3.5 & 3.7), the length of duration of the observation, and the thermal conductivity of the sample. As already noted, weak quartz CL can sometimes be enhanced by the use of high excitation voltages and high current densities. The thermal conductivity of a sandstone is highly variable. If grains are well connected, thermal conductivity is relatively high and local heating is minimized. In other cases, if a single grain absorbs a high proportion of the incident beam energy, it may become incandescent, even at low beam powers. Because of the wide variability in the performance of observation systems, some authors may use photographic exposure times of several minutes and the time-dependent effects will be more important. Others may use exposure times of 30 seconds or less (Plates 4a–f) for quartz and the effects will be less important. Until a better understanding of the reasons for the time-dependent effects is achieved, a thorough reporting of experimental procedures will be helpful to other investigators.

6.7 Techniques for observing quartz CL

The comments in Chapter 3 on the benefits of using a microscope with good transmission are particularly apropos when studying quartz. All authors agree that the CL of quartz is relatively weak, compared to that of carbonates, feldspars, and many other common minerals. A darkened room, or preferably an actual dark room, is recommended for observations.

Another important observation is that the intensity of quartz CL is more strongly dependent on the initial electron energy than the CL of any of the other common minerals. Whereas some carbonates and feldspars can be studied with electron energies of only 5–7 keV, one usually needs 12 keV or higher for quartz. And there are occasional samples with features which are only revealed if the electron energy exceeds 20 keV. Smith and Stenstrom (1965) noted the effect of higher beam loading to develop the CL of authigenic quartz. And Zinkernagel (1978) has described his design of a CL apparatus, specially suited to quartz studies, with electron energies up to 30 keV. The reasons for the increased energy requirements for the study of quartz CL are not completely known although one possibility, the influence of surface damage, was discussed in Section 6.4. It may also be that an important part of the benefit from increased energy is simply increased intensity and to some extent this can also be achieved with better microscope transmission. The example already cited in Plate 4b was photographed on a microscope system with good transmission and zonation at 15 keV electron energy needed only a 30 second exposure.

Some investigators (Sippel 1965) have favored the use of He gas to support the cold cathode discharge when quartz is studied. The beam conditions are easier to adjust at higher beam voltage when helium is used; it is also possible that the sample remains cooler because helium molecules transport heat away more effectively than air molecules. Sippel also used hydrogen gas, but this is rarely, if ever, used now because of the additional safety considerations.

The photographic conditions for quartz CL that have been reported in the literature vary widely. Some investigators now use as short a time as 30 seconds. Others have reported using exposure times of 20 minutes or more. The differences can usually be traced to the several factors already mentioned in Sections 3.9 and 3.13. It is very important, for a relatively weakly luminescing mineral such as quartz to have a simple, high-transmission microscope, high-excitation power density, and fast film.

6.8 Geological applications

There are several areas in which information derived from quartz CL can contribute to the solution of geological problems, especially in the study of sandstones. These include mineral identification, provenance studies, and diagenesis studies.

6.8.1 Mineral identification

Grains of different mineral species in a sedimentary sample will often show diagnostically different CL. A typical assemblage might contain brightly luminescing feldspars and carbonates, weakly luminescing or nonluminescent quartz, and a number of less common minerals, e.g., zircon and apatite, with varying CL properties. With a little experience, the user of CL will be able to quickly distinguish between feldspars and quartz, even when they both show blue CL, because of the wide difference in intensity. The feldspars are nearly always much brighter, and sometimes there is difficulty in photography because they will be badly overexposed when quartz is correctly exposed. Similarly the red CL of some quartz will not be confused with the orange-red CL of the carbonates because of both the differences in the color and also the much greater intensity of the carbonate CL. Modal analyses for quartz, feldspar, and other minerals is greatly facilitated with CL (Thomas 1974). Clearly this type of mineralogical analysis can also be done by more conventional light optical methods but the differences in the CL of the various minerals are sometimes so great that the analysis can be done faster with CL (see Plates 3c, d, f, 4e, f, 6g, and h, for examples). Often the preliminary CL examination can be done on loose grains or on a slab, without the need for preparation of a thin section. Furthermore there are often cases where trace minerals may be present in very small amounts which are very difficult to locate and identify without time-consuming operations. When they show bright CL, such as zircon, apatite, and some others do, they can be immediately spotted, even if only one grain were present in an entire thin section. The identification may not always be possible from the CL observation and then the other techniques would be used, once the grains have been located.

6.8.2 Provenance studies

The relative amounts of different minerals in a sedimentary sample bear a relationship to the source area rock types. By comparing the results of mineralogical analysis of a suite of samples collected over an area, one can often observe trends which furnish evidence on the source rocks. This mineralogical analysis is facilitated by CL. Thomas (1974), for example, has applied these techniques in his study of the Vanoss Formation in south-central Oklahoma.

Total mineralogical analysis is one approach used. Another approach is to look at the trace element content of a particular mineral, such as Dennen (1967) has successfully done with quartz derived from granites. Another approach is to concentrate on the relative amounts of quartz grains of differing CL characteristics.

There is at least partial agreement that the color of luminescing quartz grains in a sedimentary sample can be related to the type of source rock. There is agreement that blue-luminescing quartz is usually of a high-temperature origin and red-luminescing quartz is of a low-temperature origin. Authigenic quartz is nonluminescent or only weakly luminescent. Quartz which has been intimately associated with radiation-producing elements will show a crust with changed CL properties from the interior of the grain. Recycled grains can sometimes be distinguished by their compound CL characteristic. Recycled authigenic quartz grains, for instance, might show dull red CL in the core and dull red CL in the overgrowth but the two dull red CL colors would not appear identical. These rules are not absolute but, when correlated with other evidence, they can be useful guides to the identification of source rock types.

Owen (1984a) and Owen and Carozzi (1983, 1986) used the CL colors of quartz, specifically the ratio of brown- to blue-luminescing grains, to determine the source of the Jackfork Sandstone in Arkansas. They made the assumption that the CL color of the quartz grains they observed was unchanged from its original source rock CL color, even after transportation, deposition, and diagenesis. Statistically significant differences in the ratio of 'brown to blue' were observed in various samples, furnishing an important guide to the source area. Photomicrographs of samples containing both brown- and blue-luminescing quartz grains (from a different sandstone) are shown in Plates 4c and d.

This study is illustrative of the problems associated with the use of quartz CL at this time. One might argue that all the reasons for CL in quartz need to be known before the colors can be used for this type of study. Unfortunately, they are not known, and there is apparently little, if any, work going on now to provide the answers. So can one be sure that the CL colors are a usable characteristic – one which has meaning and does not change during the processes of transportation, deposition, and diagenesis? There is some uncertainty. On the other hand one can argue that one of the ways to obtain the supporting information is by more studies of this type. If their conclusions stand the test of time, i.e., stand up in comparison to results from other methods of study of the same areas, then this is at least indirect support for the hypothesis of constant detrital quartz CL characteristics.

Matter and Ramseyer (1985) have presented an expanded discussion of the uses of CL in provenance studies, and their publication includes color plates of quartz CL. They use a much more detailed subdivision of high-temperature quartz into several origins, based on the CL color. Their categories are:

(1–3) Light blue to blue to blue violet (mauve): plutonic quartz as well as quartz phenocrysts in volcanic rocks.
(4) Bluish-black: plutonic quartz crystals which have been plastically deformed.
(5) Red: matrix quartz in volcanic rocks.
(6) Brown: quartz from regional metamorphic rocks.

This detailed breakdown illustrates the advances being made in the sophistication of the application of CL information to provenance studies. It should be emphasized, however, that while a more elaborate scheme such as this may be quite useful to a particular investigator(s), there is still a difficulty in transference to other investigators because of the problems of subjective color description. Also it should be emphasized that this breakdown associates the quartz grains of different CL color with their possible geological origin but does not provide the underlying solid-state physics principles that are controlling the CL properties.

6.8.3 *Diagenesis*

Diagenesis refers to the processes which take place in a sediment after deposition. These processes may include cementation, dissolution, pressure solution, and fracturing and healing of grains. CL can contribute valuable insights to the presence or absence of each of these. Sippel (1968) was the first to make extensive use of CL in studying sandstone diagenesis and his paper is still the most valuable discussion of the technique. He demonstrated with many examples that overgrowths, original fabric, evidence of pressure solution, and fracturing and healing phenomena are all more clearly revealed with CL than with PoL or TrL.

The important observation on which this is based is the fact that quartz cement and quartz overgrowths will have either no CL or a noticeably weaker CL than that of the detrital grains. In either case there is usually a marked contrast. Even though there may be overgrowths, optically continuous with the detrital core, the two regions are frequently easily distinguished with CL. The contrast is greatest when the core is a blue-luminescing quartz grain.

Plates 4a and b show one such example. In this particular instance, the original shapes of some grains are also revealed with TrL, but this is not always the case. The zonation of the quartz cement in the pore spaces is also revealed only with CL in this example. Plates 6g and h have many examples where the overgrowth is more clearly revealed with CL.

Unlike the case for carbonates (Ch. 7), in which a detailed record of changing pore water chemistry during diagenesis may exist in the CL characteristics of the cement, the quartz CL record is more likely associated with the lower temperatures of formation of the quartz overgrowths and cement than with the chemistry of the pore solutions.

Olaussen *et al.* (1984) used CL as one of their tools to study depositional environments and diagenesis in a Jurassic sandstone. Following the scheme of Zinkernagel (1978), the CL colors of the quartz grains implied two major and one minor sources, these being regional metamorphic rocks, low-grade or unmetamorphosed sedimentary rocks, and plutonic rocks. In their diagenesis study, they found that two generations of authigenic quartz were revealed by CL in some cases. Their publication also contains a color plate.

Another discussion of pressure solution and cementation in quartz sandstones is the article by Houseknecht (1984) which includes information on the utilization of both CL and SEM techniques. He used combined CL and point counting 'to estimate volumes of detrital quartz, overlap (dissolved) quartz, quartz cement, brittle nonquartz grains, ductile nonquartz grains, nonquartz cement, and porosity.'

Ruppert *et al.* (1985) examined the quartz content in facies of the Upper Freeport coal bed. The grain size was small ($<15\,\mu m$) and they elected to use an SEM and EMP, rather than a CMA for their CL studies. The EMP was fitted with a monochromator and photomultiplier to permit measurement of the CL emission spectrum. Detrital grains luminesced 'in a range from extremely faint orange to bright blue . . .' (p. 335). Authigenic quartz did not luminesce. The majority of the quartz that they looked at in separates from the coal did not luminesce and is interpreted to be authigenic in origin.

6.9 Suggestions for additional research

Further investigation of quartz CL with synthetic samples would be useful. Both impurity-doped and undoped samples, as well as samples synthesized at high and low temperatures, might help to unravel some of the questions that remain.

Further evaluation of the rôle of various types of structural damage is needed, both in synthetic and natural samples.

The studies of quartz CL at very low temperatures should be expanded.

Important questions which are still unresolved include:

(a) Is there more than one mechanism for producing blue CL in quartz?
(b) Is one of the causes of blue CL actually thermoluminescence, thus accounting for its transient nature in some samples?
(c) Is the change in the CL color from blue to red under prolonged electron bombardment associated with only one of the causes of blue CL?

6.10 Summary

Quartz displays CL which is low in intensity among the common minerals but the observations still provide very useful information. Many different colors, ranging from blue, violet, and gray to red and brown, are commonly seen. It is believed that quartz with blue CL is formed at high temperatures whereas red-luminescing quartz is of low-temperature origin. Authigenic quartz is nonluminescent or dully luminescent. The CL of quartz is believed to be associated primarily with structural defects rather than activators, but questions remain. Instrumental techniques and conditions are especially important in observations of quartz because of the low CL intensity. Both high-transmission optical microscopy and high-excitation power density are useful. The excitation voltage is especially important because the threshold voltage for quartz is high. Additional features may be revealed in some samples as the voltage is increased from 15 kV to 20 or even 30 kV.

Transient effects occur in the CL of some quartz and result in a change of CL emission from blue to red during bombardment.

Chapter 7

Cathodoluminescence properties of the carbonate minerals and some geological applications

The luminescence properties of feldspars and quartz are interesting and proving to be of valuable assistance in petrological examination of samples. But when one turns to the use of CL to study carbonates, and especially the sedimentary carbonates, the appearance of the sample with CL, as well as the information which is added, is often nothing short of spectacular. The revelation of complex zonation of cements, clarification of the growth history of crystals, and beautiful resolution of details in cements and in fossils are some of the surprises awaiting someone who has not had the opportunity to study carbonate samples with CL. Of course, this is not true for every such sample but it does apply to a surprisingly large number, and CL has become firmly established as a tool in this field of study.

Most of the carbonate minerals, when activated, can exhibit intense cathodoluminescence. In most instances, the CL is attributed to the presence of Mn^{2+} as an activator ion. Fe^{2+} is believed to be the most important quencher ion. The CL of the various carbonate minerals ranges in color from green to red.

7.1 Early observations of the physical–chemical bases for the CL of the carbonate minerals

In this section we will discuss the observations that took place in the period roughly from the late 1940s to the early 1970s. There was renewed interest and a rapid development in the subject after the publications by Meyers and coworkers and others in the mid-1970s. Both the usefulness of the applications which they described, and the more general availability of CL apparatus, helped to foster the renewed and continuing interest in carbonate CL. The data which have resulted from this spurt in interest will be discussed in the later sections of the chapter.

7.1.1 Calcite

The typical cathodoluminescence of activated calcite has been variously described as orange-red (Smith & Stenstrom 1965), yellow-orange (Martin & Zeegers 1969), and orange (Sippel & Glover 1965). These observations are of activated calcites. It is generally accepted that the luminescence of calcite in natural samples is due to the presence of Mn^{2+} in the crystal and that the ferrous ion acts as a quenching agent. In both cases, the impurity is structurally incorporated.

Schulman *et al.* (1947) studied the UV and electron-excited luminescence of calcite. They prepared synthetic calcites containing various amounts of manganese, lead, cerium, or thallium. They found that the presence of only Mn was sufficient to produce CL. The incorporation of Mn with any one of the other three elements was necessary for the appearance of photoluminescence (UV excitation) in calcite. The second impurity (Pb or Tl or Ce) apparently acts as a sensitizer. They observed the same behavior in some natural calcites. One such sample from Guanajuato, Mexico, showed, under UV excitation, a brightly luminescing outer rim and a nonluminescing center. The Mn content was uniform throughout the crystal at 0.1 percent, but Pb was present (several tenths of a percent) only in the rim. They did not report on the CL from this specimen.

One of the more frequently quoted references to calcite activation is a paper by Medlin (1959) which is actually entitled 'Thermoluminescent properties of calcite.' The ThL process is related to, but not identical to, CL. Medlin prepared a number of samples of synthetic calcite with different impurities. He found that the most important activator impurities are Mn^{2+} and Pb^{2+} and he presented data on the variation of the intensity of ThL with the concentration of these activators. The maximum ThL intensity occurred at an Mn concentration of 0.003 mole fraction. For Pb, the maximum occurred at 0.0004 mole fraction. At concentrations above these values, concentration quenching set in and the intensity of ThL decreased.

Occasionally one finds Iceland spar crystals with a dull blue luminescence. Sippel and Glover (1965) reported that several Iceland spar samples showed a uniform blue luminescence. Because of the uniformity, they suggested, tentatively, that it was not due to an activator ion but rather to a defect center. They also noted an irregular green luminescence in occasional samples which they suggested might be due to an as yet unidentified activator. Barsanov and Sarsembaeva (1962) stated that the green ThL in Iceland spar is due to interstitial Mn.

When we turn our attention to the information on quenching elements in calcite, the available data are even less complete. Most authors attribute the quenching in natural samples to the presence of ferrous iron and they usually cite Medlin (1959). Medlin in fact reported in this particular paper that the effective quencher in calcite is ferric iron (and possibly also Co^{2+} or Ni^{2+}). In a later paper (Medlin 1961b), he reported that ferrous iron is the quencher in dolomite. Medlin (1975, pers. comm.) later pointed out that the crystal sizes he used were small in comparison to those available with the crystal-growing techniques of today. As a result, the area-to-volume ratio of his samples was relatively large. This may have resulted in appreciable amounts of Fe^{3+} being present as adsorbed ions rather than being structurally incorporated. The presence of an appreciable amount of ferric iron in this form might have obscured a small amount of ferrous iron incorporated substitutionally, when the samples were analyzed.

Further, Medlin pointed out that the quantities accessible to measurement were indirect – the concentration of ferric or ferrous iron in solution before crystal formation and the same quantities after the crystal was later dissolved for analysis. This introduced additional uncertainty. It would be better to carry out direct (in-crystal) measurements of the ferrous and ferric iron concentration in large homogeneous crystals and relate these results to the CL intensities. This type of measurement may be possible with the techniques of electron paramagnetic resonance (Wildeman 1969/70, Schindler & Ghose 1970, Marfunin 1979).

In 1968, Medlin presented a summary of the results of his study of the physics of the ThL of calcite and other carbonates. He presented an analysis of the energy levels involved, and identified the transition responsible for the characteristic yellow-orange emission from calcite. This work has been outlined in part in Chapter 1.

Long and Agrell (1965), using an electron probe, studied a zoned calcite crystal and found that brightly luminescing regions contained 1 percent Mn and 0.1 percent Fe; nonluminescent regions contained 1 percent Fe and 0.1 percent Mn; regions with intermediate luminescence intensity contained intermediate amounts of Fe and Mn. They noted another calcite example, with 'blue-black' CL, which contained less than 0.03 percent of both Mn and Fe.

Martin and Zeegers (1969) reported electron probe results for Mn in a natural calcite and noted values of Mn of 0.03 ±0.02, 0.14 ±0.03, and 0.35 ±0.07 (all weight percent) for regions that showed low, medium, and high intensities of CL, respectively. The Fe content was not reported.

Conventional staining techniques will reveal the presence of iron, at least at levels greater than 0.5–1.0 percent (5,000–10,000 ppm) (Dickson 1966, Evamy 1969, Lindholm & Finkelman 1972). A combination of staining and CL might help to establish whether the lack of luminescence in a particular sample is due to the presence of iron or to the lack of manganese.

It is worthwhile emphasizing that Mn and Fe, to be effective activators and quenchers, must be substitutionally incorporated (or perhaps interstitial in some instances). The mere presence of iron staining on the surface of a sample, for example, does not necessarily mean that there is Fe^{2+} incorporated in the structure, quenching luminescence. It could be present only as a separate iron oxide phase.

In the practical applications of CL to carbonate petrography, there is great interest in calcites containing varying amounts of magnesium. However, there is little data on the manner in which the intensity and spectra of calcite CL vary with the magnesium content. The thesis by Sommer (1969) contains some data on this subject which will be presented later.

Although most investigators ascribe the quenching action in carbonates to ferrous iron, we should note that Glover (1977) has considered this problem, using EPR studies of natural and synthetic samples, and concluded that ferric iron is responsible for quenching in calcite. This question needs further investigation.

Table 7.1 Preliminary values for the intensity of luminescence for calcites of varying manganese content. All samples are believed to have low iron contents.

Lum. intensity: Bright	Dull	None or low	Mn content	Lum. technique	Reference
×			0.35 ±0.07 wt.%	el. probe	Martin & Zeegers (1969)
	×		0.14 ±0.03 wt.%	el. probe	Martin & Zeegers (1969)
		×	0.03 ±0.02 wt.%	el. probe	Martin & Zeegers (1969)
×			0.2% wt.%	CMA	Freeman (1971)
	×		0.1% wt.%	CMA	Freeman (1971)
		×	<0.1% wt.%	CMA	Freeman (1971)
×			0.55 mole %	ThL	Medlin (1959)
	×*		0.03 mole %	ThL	Medlin (1959)

* Chosen by this author as the point where intensity is 10% of the maximum.

Table 7.1 is a compilation of the few bits of data that existed in the early 1970s on the manner in which the CL intensity varies with the Mn concentration in calcite. For comparison the table also contains some ThL data from Medlin. The conclusion reached from this very limited amount of data was that the effective manganese concentration for causing luminescence in calcite is between a few hundred and a few thousand parts per million. This represented the state of knowledge in the mid-1970s. Later we will look at more recent data on this subject.

CALCITE CL SPECTRA

The reported colors of calcite CL, based on visual examination, are variable. Examination with a calibrated monochromator is a reliable measure of the emission spectrum,

Figure 7.1 Examples of the emission spectra of synthetic calcite with added Mn^{2+}. The spectra shown are for ThL glow peaks and X-ray excited luminescence, at the different temperatures indicated. The particular sample contained 2,000 ppm Mn^{2+} impurity but similar spectra were seen in other samples containing widely differing amounts of activator or quencher (see text). (From Medlin 1963a.)

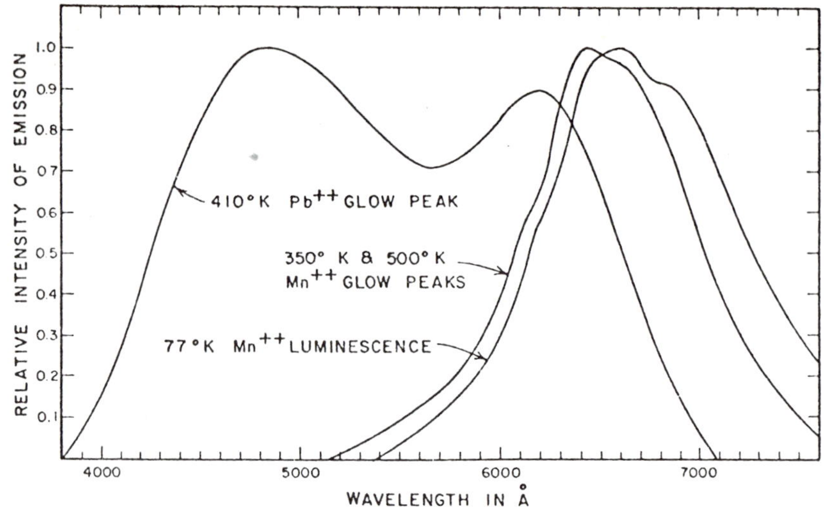

whereas to the eye, and even to some color film, there may be apparent color shifts, especially if there are large variations in the intensity of the emission.

Schulman *et al.* (1947) reported that the calcite spectrum, for UV excitation, extends from 5,780 Å to higher values. The results were similar for CL. The UV-excited emission, in this instance, occurs when Pb is present as a sensitizer with the Mn but it was reported that the spectrum was not affected by the amount of Pb present.

Medlin (1963a) recorded spectra of the glow peaks (ThL) of synthetic calcite containing 2,000 ppm Mn^{2+} and observed a broad maxima between 6,400 and 6,700 Å (with a measurement precision of ±400 Å). Spectra of ThL for other concentrations of Mn, in the range between 50 and 30,000 ppm were the same. He also saw no change in the spectra with quencher contents up to 500 ppm. XL spectra were also recorded on some samples and they were closely similar to the spectra of the ThL peaks.

Coy-Yll (1969/70) used an interference filter to measure spectra and reported a maximum for the calcite CL spectrum at 5,800 Å.

Sommer (1972a) used both an interference filter system and a grating monochromator system and reported a CL emission spectra maximum at 5,900 Å for calcite.

In all of the above, the presumption is that the activator is Mn^{2+} and that the spectrum is the typical broad band (Fig. 7.1). However, we re-emphasize that there are examples of REE-activated calcite, and the CL of some REE-activated calcite, to the unaided eye, appears identical to that produced by Mn^{2+}. Careful monochromator examination may be necessary to distinguish between the activators. See Chapter 8, Section 8.2.4, for further discussion of this subject.

7.1.2 *Dolomite* $CaMg(CO_3)_2$ *Trigonal*

Cathodoluminescence in dolomite is also usually considered to be controlled by the concentrations of Mn^{2+} as an activator and Fe^{2+} as a quencher (Medlin 1961b, 1968).

Medlin (1961b) prepared synthetic dolomites, doped with impurity elements, and studied their thermoluminescence. Fifteen different elements which occur as impurities in natural samples were tried but only Mn^{2+} was found to be effective as an activator. Fe^{2+}, Co^{2+}, and Ni^{2+} all acted as quenchers. For the various ThL peaks, the behavior of relative efficiency with Mn concentration followed similar curves, showing a peak at approximately 0.3 mole percent (3,000 ppm) and a self-quenching or concentration-quenching effect at higher concentrations. These results are at least an indication of the expected results for CL.

In a later paper (Medlin 1963a), he published the emission spectra of the glow peaks from five synthetic samples, with Mn content ranging from 160 to 15,000 ppm. All had the same spectra, with a maxima at 6,700 Å. One sample, containing 500 ppm Mn and 500 ppm Fe, had the same spectrum as the others with the addition of a small peak at 3,900 Å. The X-ray excited luminescence of these samples had the same emission spectra as the ThL glow peaks. An example of a dolomite glow peak spectrum is shown in Figure 7.2.

Long and Agrell (1965) noted weak blue CL in an exsolved dolomite in a carbonatite. This dolomite contained 0.07 ±0.02 percent Mn and 0.8 ±0.1 percent Fe. The associated calcite luminesced orange-red, especially in the neighborhood of the dolomite where Fe was depleted in the calcite and enriched in the dolomite. Figure 7.3 is a profile across this zone showing the strong fine-scale variations in the amount of the quencher, variations which are mirrored in the CL intensities.

Smith and Stenstrom (1965) described a zoned dolomite rhomb in which the center luminesced but the iron-rich (quenched) border was dark (nonluminescent).

According to Sippel and Glover (1965), the extremes of dolomite CL are from brilliant yellow to deep red. They noted a total range of 1,000 Å in the dolomite spectra of their samples.

Sommer (1972a) described dolomite CL as showing two peaks – one behaving like calcite at 597 nm and one behaving like magnesite with a peak at 675 nm.

Amieux (1982) has noted some examples of pale blue CL in dolomite and has suggested that the cause may be Zn activation of an otherwise relatively pure dolomite. He has also seen some examples of yellow-green CL in dolomite which,

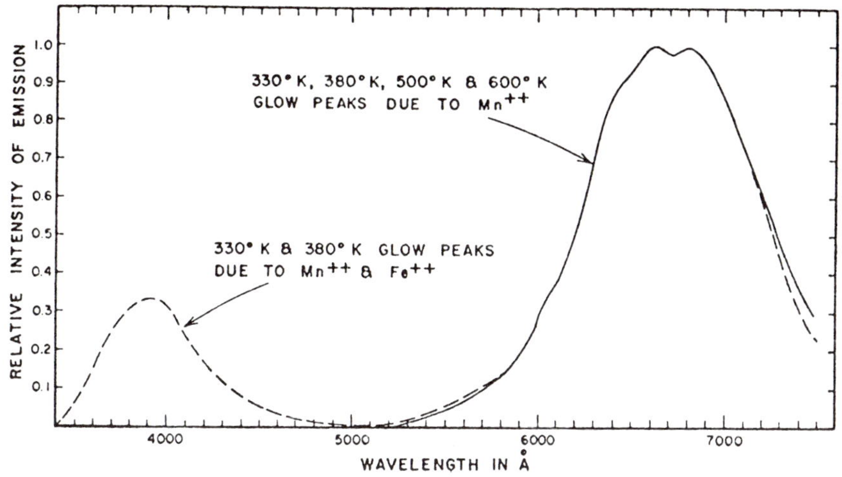

Figure 7.2 Examples of the emission spectra of synthetic dolomite with Mn^{2+} impurities. The spectra shown are for ThL glow peaks but the spectra of X-ray excited luminescence were found to be identical. The solid-line spectrum is the same as that observed for five different synthetic samples containing Mn^{2+} in the range of 160–15,000 ppm and one natural dolomite with 200 ppm of Mn. The dotted line spectrum is for a sample containing 500 ppm Mn^{2+} and 500 ppm Fe^{2+}. (From Medlin 1963a.)

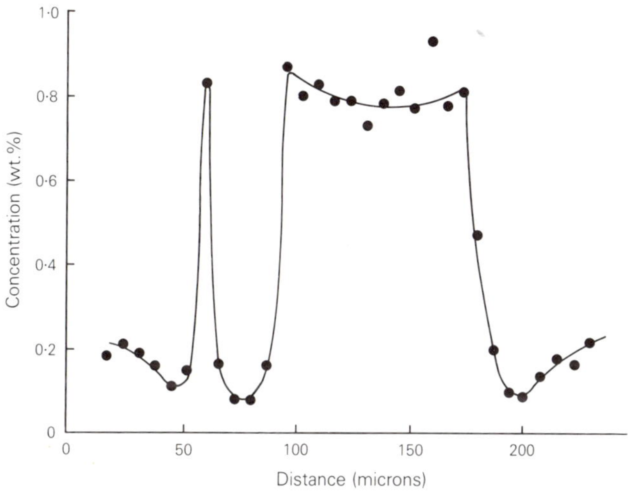

Figure 7.3 The variation in the iron concentration in a calcite, near boundaries with exsolved dolomite. The sample is from the Palabora carbonatite. The calcite CL intensity is greater near these boundaries because the iron is preferentially taken up by the dolomite. (From Long & Agrell 1965.)

he suggests, may be due to interstitial Mn or to rare earth activation. These suggestions merit further examination.

7.1.3 Aragonite $CaCO_3$ Orthorhombic

Aragonite is another important carbonate mineral. It is thought to be the original precipitate in some carbonate rocks and is also an important constituent of the shells of many types of organisms.

Medlin (1961a, 1963a) studied the ThL properties of synthetic aragonite and found that Mn^{2+} (50–480 ppm Mn^{2+}) is an activator, while Fe^{2+}, Co^{2+}, and Ni^{2+} are quenchers, i.e., the same situation as for calcite. He noted that Mn does not substitute easily for Ca in aragonite because of differences in ionic radii and coordination number, and its concentration in natural aragonite is low. The intensity of aragonite ThL was reported to be weak compared to that of calcite. The ThL glow was described as yellow and the spectrum was very broad, with a peak at about 6,300 Å.

Sommer (1972a) looked at the CL and XL of both synthetic and natural aragonites. In the Mn-doped synthetic samples, he found the emission maximum at 540 nm and described aragonite CL as 'yellow-green.' A number of synthetic samples were prepared and from measurements on these he prepared the curve relating intensity of X-ray excited luminescence to Mn concentration which is shown in Figure 7.4. He found that aragonites which contained Sr had a lower CL intensity than Sr-free aragonites, but even 17 mole percent of $SrCO_3$ reduced the intensity by only 50 percent. The peak maxima were shifted to longer wavelengths as the Sr content increased, reaching a value of 590–600 nm for the case of 100 percent $SrCO_3$ (Fig. 7.5).

Richter and Zinkernagel (1981) have reported green-luminescing aragonite in natural samples. Amieux (1982) has also reported a green CL in natural aragonite which he says is due to Mn^{2+} in an interstitial position.

7.1.4 Magnesite

Medlin (1961a, 1963a) found that Mn^{2+} (50–2,000 ppm) is an activator and Co^{2+} and Ni^{2+} are quenchers for ThL in

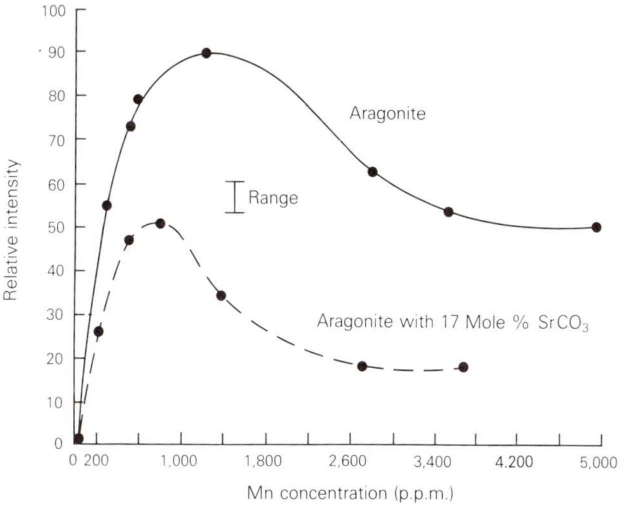

Figure 7.4 The intensity of X-ray excited luminescence from synthetic aragonites, doped with varying amounts of Mn. The solid line is for Sr-free aragonite and the dotted line is for an aragonite which contains 17 mole percent $SrCO_3$. There is a reduction in intensity and a shift of the most effective concentration to a lower value in the latter case. (From Sommer 1972a.)

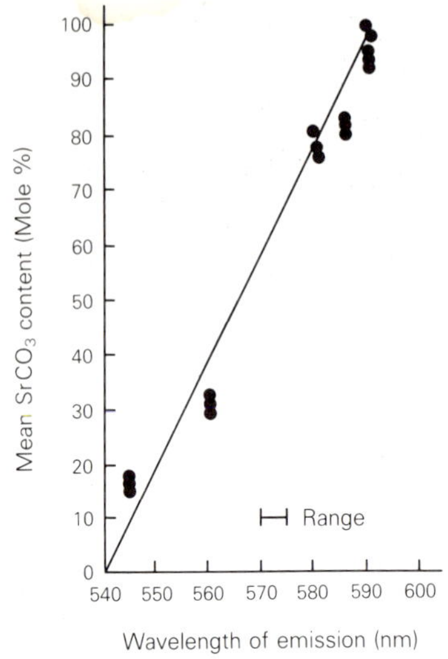

Figure 7.5 The relationship between the CL emission spectrum peak maximum and composition for the aragonite–strontianite solid solution series based on CL observations on synthetic samples (means of 131 replicates). The Mn contents are 400–500 ppm; the Fe contents are 100–200 ppm. (From Sommer 1969, 1972a.)

synthetic magnesites. The ThL emission spectra showed a peak between 6,600 and 6,800 Å.

Sommer (1972a) reported that the CL spectrum of Mn-activated magnesite peaked at 675 nm, the same value he found for one of the two peaks in dolomite. He studied the manner in which the peak emission wavelength varies in the calcite–magnesite solid solution series and found the relationship shown in Figure 7.6. He emphasized that there are deviations from a simple linear relationship in the data.

Interestingly, Sommer also found that magnesite CL was easily changed by the electron beam and, unless much reduced current density was used, the luminescence emission could be lost in less than a second.

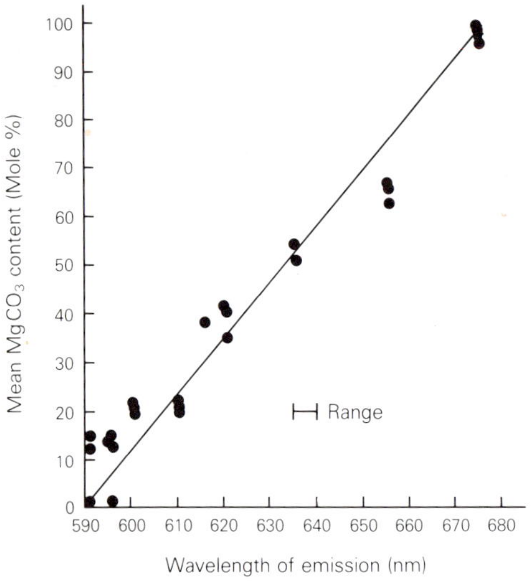

Figure 7.6 The relationship between the CL emission spectrum peak maximum and composition for the calcite–magnesite solid solution series based on CL observations on synthetic samples (means of 173 replicates). The Mn content of these samples ranges from 800 to 2,800 ppm. The iron content is between 200 and 430 ppm. (From Sommer 1969, 1972a.)

Table 7.2 The position of the peak maximum in the emission spectrum (CL, XL, ThL, or PhL as noted) as reported by several authors for several of the carbonate minerals. All values are in nanometers.

Author	Calcite	Dolomite	Aragonite	Magnesite
Schulman *et al.* (1947)	578 & up (PhL) 578 & up (CL)	—	—	—
Medlin (1963a)	640 (ThL) 670 (XL)	670 (ThL & XL)	630 (ThL)	660–680 (ThL)
Medlin (1968)	650 (ThL) (broad)	660–680 (ThL)	560–620 (ThL)	—
Coy-Yll (1969/70)	580 (CL)	—	—	—
Sommer (1972a)	590 (CL)	597 & 675* (CL)	540 (CL)	677 (CL)
Pierson (1981)	—	620–690 (CL); max. about 650	—	—
Amieux *et al.* (1983)	578 (CL)	—	—	—
Mariano (1983b)	605 (CL)	640 (CL)	—	640 (CL)
Robbins (1983)	640 (PhL)	—	—	—

* He found the spectrum to be the combination of the spectrum of Mn in a calcite host and Mn in a magnesite host. If the Mn is primarily on the Ca site, the color tends towards the yellow. If the Mn is on the Mg site, the color tends towards the red.

7.1.5 CL spectra of some carbonate minerals

For easy reference and comparison, the various values for the CL emission spectra peak maxima of the four carbonate minerals discussed are collected in Table 7.2. The agreement among these values is not good and a comprehensive study of this problem would be a worthwhile contribution. One possible reason for the variation in results is a true variation among calcite samples, perhaps due to variations in the magnesium content or other causes. One should not overlook the possible existence of calibration differences among the various instruments used. In many published articles calibration curves for the standards used are not furnished. There is also a possibility that some intensely bombarded samples show a combination of incandescence and CL spectra, biasing the spectra towards the red.

Figure 7.7 is a presentation of the spectra of four carbonate minerals, taken under identical conditions on the same instrument. The fact that these several minerals are all activated principally by Mn^{2+} but have different spectra is another illustration of the fact that luminescence spectra are controlled not only by the activator ion but also by the host mineral.

7.2 Some early observations of the CL of carbonate minerals in geological samples

Despite the fact that only a limited amount of information was available on the relationship between CL intensity and the concentrations of manganese and iron, interest in the

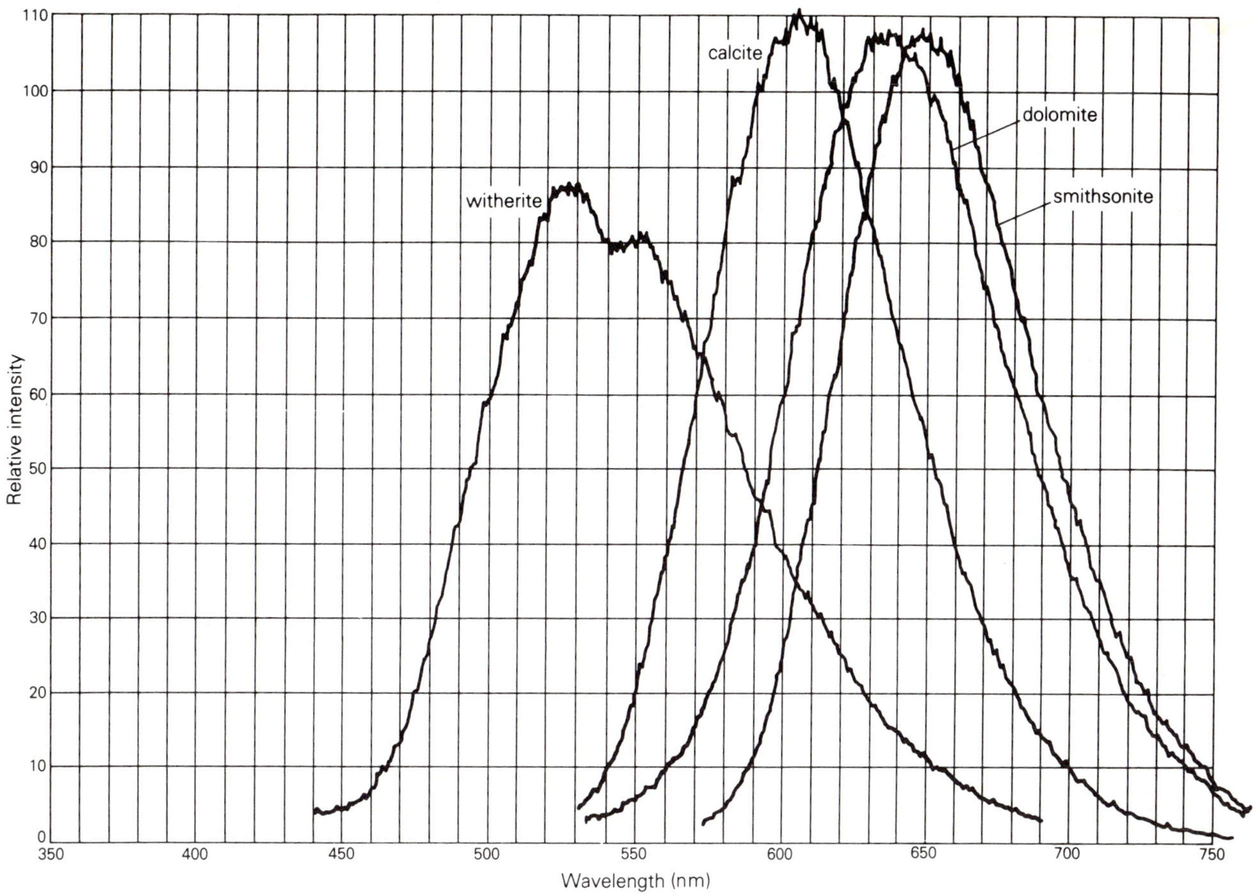

Figure 7.7 The spectra from four natural carbonate mineral samples, recorded on the same spectrophotometer under nearly identical conditions. (Spectra recorded by A. N. Mariano.)

use of CL expanded in the early 1970s and luminescence petrography became well established as a technique for the study of carbonates, particularly in sedimentary rocks. Its use frequently helped to reveal information which was not obtainable by any other readily available technique.

7.2.1 *Calcite*

One of the applications mentioned in the oft-quoted paper by Smith and Stenstrom (1965) is the ready detectability of calcite cement in sandstones and shales and the ease with which its distribution can be evaluated.

Sippel and Glover (1965) demonstrated the use of CL to reveal the detailed growth history of crystals. One of their examples was of a calcite in-filling in dolomite in which 50 growth bands were distinguishable by their differing CL intensities. They commented that the 'Variations in the growth rate along the leading edges of the growing crystal have been recorded in minutest detail' (p. 3).

Freeman, in 1971, reported the results of a study of the Fernvale Limestone (Ordovician) from northern Arkansas. The calcite cement showed no evidence of discontinuities with plain polarized light or with crossed polarizing prisms. Nor were differences revealed with potassium ferricyanide staining. With CL, however, there is a clear contrast between two generations of cement (Plates 7a & b). The earlier nonluminescing cement formed as microstalactites on echinoderm fragments, indicating cementation under vadose conditions. The second cement is brightly luminescent which indicates that there was a change in the chemistry of the pore fluids. Freeman carried out an electron probe analysis for iron and manganese both in cement zones and in the echinoderm fragments. He found that the iron content was below the detection limit (less than 0.1 percent) in all three regions. Manganese concentrations were below the detection limit in the earlier, nonluminescent cement, at the detection limit (0.1 percent) in the echinoderm fragments, and readily detectable (0.2 percent) in the later, brightly luminescing, cement.

Region	Fe	Mn	
echinoderm fragments	$< 0.1\%$	$< 0.1\%$	dull luminescence
early cement	$< 0.1\%$	$\sim 0.1\%$	nonluminescent
late cement	$< 0.1\%$	0.2%	brightly luminescent

The variation in the CL intensity reflects the variation in manganese content.

Meyers, in 1974, published the results of a comprehensive study of the Lake Valley Formation (Mississippian) in New Mexico. Using CL, he identified five different cement zones in samples from this formation. Three of these zones correlated within 200 foot (60 m) stratigraphic sections and correlated laterally for at least 10 miles (16 km). Thus the luminescence characteristics could be used as a stratigraphic indicator and he suggested the term 'luminescence stratigraphy.' Photomicrographs of two of his samples are shown in Plates 8a–d. The width of some of the very brightly luminescing bands within certain zones is only of the order of 10 μm – nevertheless they are clearly revealed by CL. All but zone 5 of these samples are apparently free of iron so the variations in luminescence are due to variations in the manganese content, not to Fe quenching. Based on the data available at that time, he concluded that about 1,000 ppm of Mn was necessary for bright CL in calcite. This conclusion has been changed with newly acquired data (Section 7.3.3).

Possible activation of sedimentary carbonates by REE has been discussed (Amieux 1982, Machel 1983) but, so far as the author knows, this has not yet been supported by actual analyses of sedimentary carbonates exhibiting CL. Gies (1975) has reported on the photoluminescence of REE-activated carbonates. Mariano (Section 8.2.4) presents two relatively rare examples of REE activation in a hydrothermal carbonate and in calcite associated with a laterite.

7.2.2 Dolomite

CL in dolomite was reported by Smith and Stenstrom (1965). They presented a photograph of a zoned dolomite with iron-rich, nonluminescing borders. Long and Agrell (1965) also reported an example of calcite and exsolved dolomite in a carbonatite. In this example, the dolomite CL was blue and very weak. They attributed this lack of CL to Fe quenching and provided a profile of Fe content in the calcite and dolomite showing a strong concentration of Fe in the dolomite near the calcite–dolomite boundaries (Fig. 7.3), in agreement with the CL observations.

Some of Meyers' samples contain small redder-luminescing grains which may be very small dolomite crystals. These may be difficult to see under TrL but those which luminesce are easily located with CL (Plates 8c and d). The subject of microdolomite-rich cements is treated in subsequent papers (Lohmann & Meyers 1977, Meyers & Lohmann 1978). Further comments on the CL and origin of microdolomites are contained in a discussion between Oglesby and the authors (Meyers & Lohmann 1979, Oglesby 1979). It is worth pointing out that this is another instance where CL is useful as a locating technique for small grains. Even though it may not be able to provide an unequivocal identification, the use of CL facilitates the location of the small grains which can then be identified by other techniques.

7.2.3 Aragonite

In addition to his work with synthetic aragonites discussed in Section 7.1.3, Sommer (1972b) also observed CL in aragonite in a variety of organisms. Some of the aragonite contained no detectable Mn and was nonluminescent. In other cases, the Mn content was found to be as high as 660 ppm and the characteristic green CL was exhibited. CL was found to be a useful tool to locate regions of calcite in aragonite shells because of the distinct CL contrast. Sommer (1972b) also reported on a pelecypod which contained small areas of light orange luminescence. These areas were rich in $SrCO_3$ and were probably strontianite.

Richter and Zinkernagel (1981) published a comprehensive article on the applications of CL in carbonate petrography. They reported yellowish-green CL in aragonite, some of which had Sr contents as high as 2.5 percent. Their publication contained several color plates including two plates with the yellowish-green-luminescing aragonite. They stated that additional investigation of the reasons for CL in aragonite is necessary.

7.2.4 Nonsedimentary samples

Many granites, basalts, and other nonsedimentary rocks contain carbonate minerals which show the characteristic carbonate luminescence; however, very little has been published on this subject (Plates 2d, f & 10a). Presumably the carbonate mineral can either be original, or an alteration product, or can be introduced later as a void filling of some sort. The general impression that the author has gained is that the carbonate (usually calcite) in these types of rocks almost always shows luminescence. The luminescence is usually uniform without zonation, although there are a few notable exceptions (Plate 12c).

7.3 Present-day uses of CL in the study of geological carbonate samples

The previous two sections have introduced the basic information on CL in carbonates. In the remainder of this chapter the geological applications will be expanded upon. The background of the author is not that of a carbonate or sedimentary petrologist, and the treatment of the subject which is adopted here is therefore more from the viewpoint of the physicist or physical chemist. Many interesting and important questions can be profitably discussed from this vantage point and their eventual answering will further the application of CL in carbonate petrology.

To supplement the material in this section there are many extensive articles on CL in carbonates, written by experts in the general field of sedimentary petrology and the specific field of carbonate petrology. Among these comprehensive articles are the following: Amieux (1982)*, Dickson (1985), Fairchild (1980, 1983), Frank (1981), Frank *et al.* (1982), Grover and Read (1983), James and Klappa (1983)*, Lohmann and Meyers (1977), Meyers (1974, 1978), Meyers and Lohmann (1985), Moldovanyi and Lohmann (1984), Richter and Zinkernagel (1975, 1981)*, and Wilkinson *et al.* (1985). These articles, read hand-in-hand with the chapter material, will provide a balanced, comprehensive view of the present status of the applications of CL to sedimentary carbonates and will suggest some possible directions for future investigations. The articles marked with an asterisk include color plates of luminescing carbonates and therefore are especially useful supplements to the plates in this text. However, it should be noted that many important examples of sedimentary carbonate luminescence can be illustrated nicely in black and white reproductions and most of the other articles in the list include such photographs.

7.3.1 *Mineralogical analysis*

One of the very useful applications of CL is to facilitate mineralogical analysis. Calcite has a distinct color and intensity of CL in many samples and the amount and distribution are quickly revealed. And this is true of calcite in both sedimentary and nonsedimentary rocks.

Smith and Stenstrom (1965) mentioned the use of CL, to make an instantaneous evaluation of the distribution of carbonate cement in sedimentary samples. The brightly luminescent calcite (or dolomite) usually stands out in sharp contrast to the other components such as feldspar and quartz which either luminesce weakly or luminesce blue. Plates 3c, d, and 6g and h are examples of sediments containing quartz, feldspar, and carbonates, and they illustrate this use.

Similarly in nonsedimentary samples, the carbonate content can be evaluated quickly with the aid of CL. This is illustrated in Plates 2c–f.

We should caution that there are many cases of nonluminescing carbonates, so, more properly, when we speak of a carbonate we should speak of two (or more) phases; it is only the luminescent phase that is revealed.

7.3.2 *Distinction between calcite and dolomite*

Mineralogical analysis is further facilitated in some samples because of the CL contrast between calcite and dolomite. As noted earlier, dolomite CL is nearly always of longer wavelength than that of calcite. The relationship is not absolute, and some dolomites have been reported to luminesce in the yellow region of the spectrum (Sippel & Glover 1965, Carrasco 1973). Nevertheless, even though the CL color of both minerals can vary, there is apparently always a contrast present in natural samples.

Plate 10a shows one example of a sample which contains dolomite exsolution laths within the regions of calcite. These laths are very difficult to isolate with PoL or TrL but they stand out very well with CL. Plates 7c and d show another example in a sedimentary sample.

Another important example of this use is shown in Plate 8c. This particular limestone sample contains microdolomites. These microdolomites are usually not seen at first with TrL or PoL. Even with CL they may be overlooked initially because the calcite CL is so striking. But with more careful observation they are seen and then, because of the contrast in color, their distribution can be quickly evaluated. Some of the microdolomites show CL zonation which is revealed in the higher magnification photograph shown in Plate 8d. After these microdolomites have been found with CL, one can usually go back to TrL or PoL and find their outlines, although they are not nearly so sharply defined.

The microdolomites are not always luminescent: Meyers and Lohmann (1978) reported both luminescent and nonluminescent examples, and in the latter case other means must be resorted to for their identification. The significance of the microdolomites has been discussed in the publications by Meyers and Lohmann (1978, 1979) and by Oglesby (1979).

7.3.3 *Estimation of Mn and Fe contents of calcites and dolomites*

One of the possible applications of CL in carbonates is as a qualitative analytical method to set some limits on the Fe and Mn content, based on the observed CL intensity. If this can be done with CL, then one has a relatively simple, nondestructive, analytical method, requiring minimum sample preparation. One can then go on to make some judgments about possible geological situations which existed during the formation of the carbonates. This application underlines the need for better knowledge about the effective concentration of Mn and Fe (and possibly other trace elements) required to produce a given intensity of CL in carbonates. This need was responded to by Oglesby (1976), Richter and Zinkernagel (1975, 1981), Frank *et al.* (1982), Pierson (1981), Fairchild (1980, 1983), and ten Have and Heijnen (1985), among others. All of these authors have made significant contributions to the database on CL intensity versus Mn and Fe concentration in calcite and dolomite.

Oglesby's interest was in determining the distribution of Ca, Mg, Mn, and Fe in calcite and dolomite cements in the Upper Smackover Formation in Eastern Mississippi (Oglesby 1976). He tested a hypothesis that the amounts of these elements could 'be used as a means of recognizing various types of diagenetic environments.' The electron probe measurements had a detection limit for Mn at the 0.05 mole percent (500 ppm) level. Analyses were made with a spot diameter of 15 μm or smaller, so finely zoned cements could be studied in detail.

Oglesby found many examples of multiple stages of cementation. One of his more striking examples was a single calcite crystal which appeared to be uniform with TrL; inspection with CL revealed nine separate identifiable zones (Plates 10e & f). Each of these zones was analyzed for Fe and Mn with the results shown in the plate caption. Some of his data is included in that summarized in Figure 7.8.

He called the pattern of alternating zones in a cement or crystal the 'luminescent signature,' and showed examples where calcite and dolomite adjacent to one another had the same luminescent signature, implying that they had formed contemporaneously, under nearly identical conditions.

In a subsequent paper (Frank *et al.* 1982), this work and much additional work done at the University of Missouri–Columbia was reported. The analytical results are summarized in Figure 7.8. The conclusions stated in this latter paper include:

> Cathodoluminescence petrography coupled with microprobe analysis shows that luminescent zones in the calcite cement correspond to distinct compositional zones.
>
> [some data] does suggest that calcites can luminesce with concentrations of manganese less than 0.05 mole percent $MnCO_3$. . . .
>
> Luminescent intensity displayed by a particular calcite zone appears to be controlled by the Fe/Mn ratio in the zone, not by the absolute concentrations of either cation.
>
> Trends . . . suggest that there may be no critical amount of Fe^{2+} in calcite above which there is no luminescence. Data suggest that regardless of the amount of Fe^{2+}

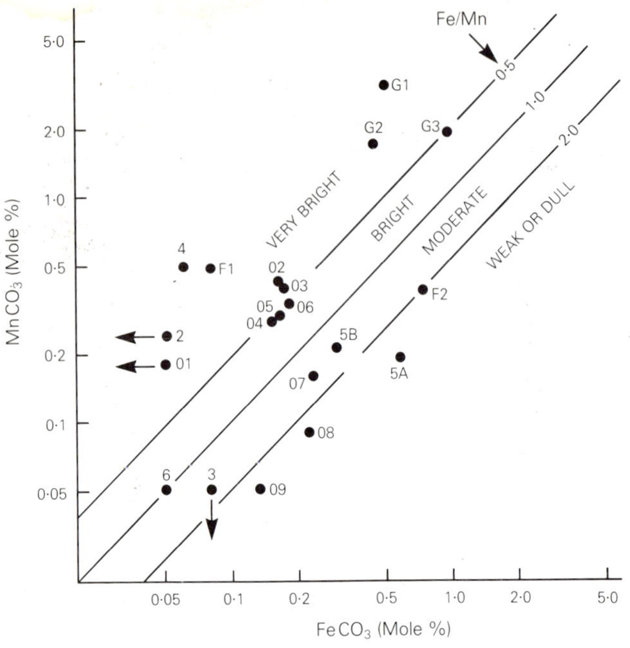

Figure 7.8 Relationship between concentration of $FeCO_3$ and $MnCO_3$ in calcite and the CL intensity (Frank *et al.* 1982). The authors stated that these data suggest that 'the ratio of Fe/Mn, not the absolute amounts of either cation, controls the luminescent intensity displayed by calcite.'

present, a calcite zone may luminesce if sufficient Mn^{2+} is present.

Pierson (1981) made a detailed examination of 86 samples of dolomite from North America and Europe. He analyzed them for Fe and Mn and observed the CL intensity and color for each. Intensities were judged by eye from the microscope observations and rated relative to the brightest sample. The color of the CL was described by comparison to a set of color charts in Pauling's text (Pauling 1970) on general chemistry. Dolomite CL fell in the range of 620–690 nm (orange to deep red). Occasional samples showed pink or red-purple CL.

Pierson's chemical analyses were based on the use of arc spectroscopy, atomic absorption, and the electron probe. Fe contents varied between 500 ppm (0.05 percent) and 10 weight percent. Mn concentration ranged from less than 100 ppm to more than 2,500 ppm. Traces of Pb, Ti, Cr, Ni, Co, and Sr were found. Pierson observed that the variations of the concentration of Mn and Fe were correlated, although the total range of variation was greater for Fe. Samples with very low iron content sometimes luminesced brightly with only 100 ppm or less of Mn. Samples with more than 15,000 ppm of iron showed no CL, regardless of the amount of Mn present. His results are summarized in Figure 7.9.

In a practical sense, the CL of Pierson's dolomite samples was controlled by the Fe content, since all of the samples contained enough manganese (more than 100 ppm).

Fairchild (1980) investigated the rôles of cementing versus replacement textures formed during the dolomitization process, using a combination of CL and electron probe analysis. His probe detection limits were quite low (20 ppm for Fe and Mn in dolomite). Three types (1, 2, 3) of dolomite were recognized petrographically. Type 1 had uniform, dull red CL. Type 3 had either dull red or no CL. Cathodoluminescence proved to be especially useful in recognizing the growth zones in Type 2 dolomite, zones which were revealed only by CL (or by laborious point-by-point electron probe analyses for Fe and Mn).. In a subsequent paper, Fairchild (1983) reviewed the work which had been done to date on Mn and Fe contents and CL intensities of calcite and dolomite and presented additional data based on his own CL observations and his analytical work in dolomites and calcites from the Bonhaven Formation. These results are presented in Figures 7.10 (calcite) and 7.11 (dolomite).

Richter and Zinkernagel (1981) also present an extensive discussion of CL in calcite and dolomite. They give a lower limit of 20–40 ppm of Mn as an effective activator concentration in both calcite and dolomite, with quenching effects by iron setting in, even at low Fe concentrations. However, they continued to observe a dull CL in some dolomites containing as much as 10–15 weight percent of $FeCO_3$. The Fe-rich dolomites apparently contained appreciable Mn. One of the interesting figures in their paper relates the CL color of dolomites to the $FeCO_3$ concentration, based on visual descriptions of the color. The trend is from orange-red, at low Fe content, to dull red brown to very dark brown at high Fe content.

ten Have and Heijnen (1985) prepared synthetic calcites with Mn^{2+} incorporated and observed CL if the Mn^{2+} content exceeded 15–30 ppm. For Fe contents up to about 150–200 ppm, the Fe had no quenching effect. Analyses by atomic absorption spectrometry (AAS) on some natural calcites with low Fe contents showed CL when Mn is greater than 20–25 ppm. For Fe contents less than about 200 ppm,

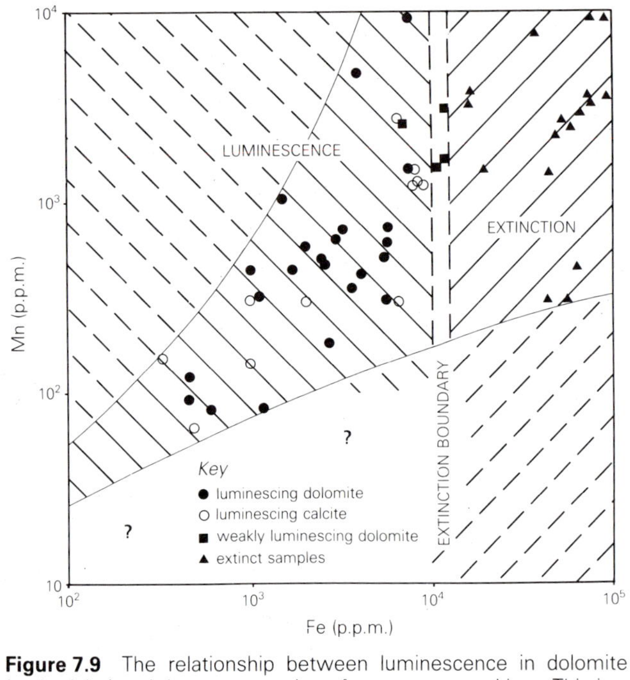

Figure 7.9 The relationship between luminescence in dolomite (and calcite) and the concentration of manganese and iron. This is a summary of results obtained with the electron probe and atomic absorption analysis. The intensities were judged by eye with reference to a standard brightly luminescing sample. (From Pierson 1981.)

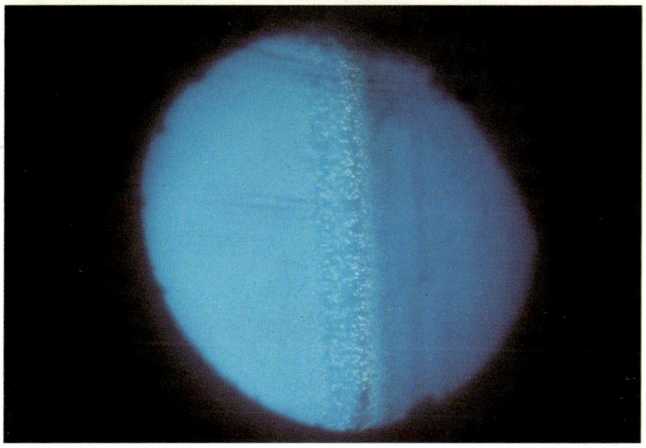
(a)

(b)

(c)

(d)

(e)

(f)

Where a matched pair of transmitted light (TrL) and cathodoluminescence (CL) photos are provided, the photographic conditions given are for the CL photograph. Unless otherwise noted, the sample used for each plate was collected by the same person who did the photography. The comment 'App' after the photo credit means that additional CL or photographic conditions are summarized in the Appendix. (Note: text has cross-references to plate numbers but not vice versa.)

Plate 1a Yellow-luminescing giant platelets' in a growth band in a natural Type 1a diamond. CL. See Collins and Woods (1982) for a detailed discussion. *CL conditions:* JEOL electron probe, approximately 35 kV, specimen current density of 1–10 μA/mm^2. *Scale:* the field is 300–400 μm in diameter. (Photo by G. S. Woods.)

Plate 1b Synthetic diamond exhibiting growth sectors. CL. The overall size of the diamond is approximately 5 × 6 mm. The blue-luminescing linear defects, 0.3 mm from the top edge, are believed to be dislocations. Their CL is strongly polarized. The phosphorescence decay time of the CL from this sample is less than 0.1 seconds. For a complete discussion see Woods and Lang (1975). This was published (in black & white) as Figure 4a of their paper. (The orientation in this plate is 90° CCW, relative to Figure 4a of their paper.) *CL conditions:* Cambridge Inst. Ltd. SEM, 20 kV, beam current density approximately 1 μA/mm^2. (Photo by G. S. Woods.)

Plate 1c Cassiterite. CL. This is Type II cassiterite (Hall & Ribbe 1971) with yellow CL. These samples contain appreciable amounts of Fe and Ti and usually do not contain significant amounts of W, Ta, or Nb. The zoning revealed by CL is not visible under TrL or PoL conditions. *CL conditions:* electron probe, beam current approximately 0.5 μA. *Photographic conditions:* exposures on the electron probe were quite long, 15–30 minutes, on Anscochrome 500 film. *Scale:* the luminescent spot is 300 μm in diameter. (Photo by M. R. Hall & P. H. Ribbe.)

Plate 1d Cassiterite. CL. This is Type III cassiterite (Hall & Ribbe 1971) with blue CL. The blue CL is associated with moderate amounts of W, Fe, and Ti. Ta and Nb are not present in significant amounts. *CL conditions:* as for Plate 1c. *Photographic conditions:* as for Plate 1c. *Scale:* the luminescent spot is 300 μm in diameter. (Photo by M. R. Hall & P. H. Ribbe.)

Plate 1e Emerald from Carnaiba area, Bahia, Brasil. CL. The zonation which is revealed with CL is not visible with either PoL or TrL. *CL conditions:* not available. *Photographic conditions:* 1 s. *Scale:* 35 mm on transparency ≘ 3.8 mm on sample. (Sample, A. N. Mariano 1974. Photo by A. N. Mariano, App. (I).)

Plate 1f A macro-photograph of an 11.7 metric carat synthetic corundum with CL. This sample appears brown in natural light. *CL conditions:* 5 kV, 1 mA. *Photographic conditions:* macro-photograph of sample in Luminoscope Bulk Specimen Chamber. (Photo by J. Ponahlo.)

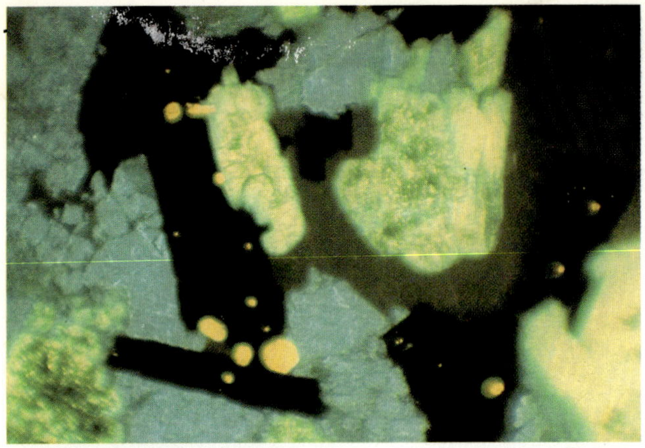

(a)

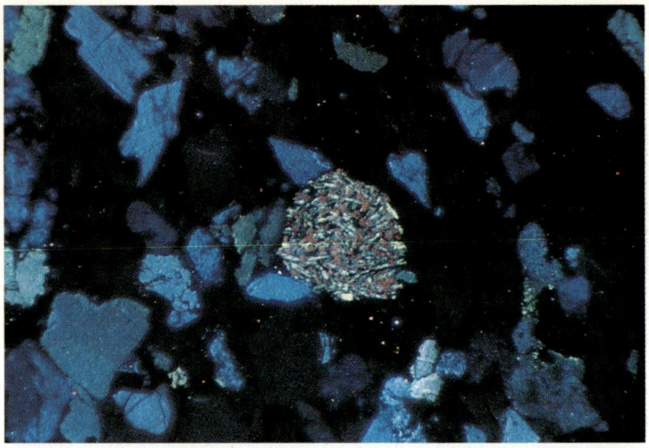

(b)

(c)

(d)

(e)

(f)

Plate 2a Granite, Peggy's Cove, Nova Scotia. CL. The two feldspar phases, a blue-luminescing K-spar and a yellow-green luminescing plagioclase, stand out clearly with CL. The bright yellow grains are apatite. The dull brown-luminescing areas are quartz. *CL conditions:* not available but typically 12 kV, 0.5 mA beam current, beam focused to approximately 4 mm diameter. *Photographic conditions:* 15 s, ASA 160. *Scale:* 35 mm on transparency $\equiv$ 3.8 mm on thin section. (Photo by A. N. Mariano, App. (I).)

Plate 2b Tuffaceous Miocene sandstone from the Back-arc, Japan. CL. This sandstone contains quartz, plagioclase, alkaline feldspars, and rock fragments. The complex grain in the center of the photo is approximately 0.7 mm in diameter. Even though it is the only such fragment in the entire thin section, it stands out clearly from the other components and is easily located with CL. Identification of the components within the fragment would require use of the SEM or other more sophisticated instrumentation. *CL conditions:* 15 kV, 0.5 mA electron beam current, electron beam focused to approximately 5 mm diameter spot. *Photographic conditions:* 45 s, ASA 400. *Scale:* 35 mm on transparency $\equiv$ 3.5 mm on thin section. (Sample courtesy of T. Uchida, Japan Petroleum Exploration Co. Ltd.) (Photo by D. J. Marshall.)

Plates 2c and d Westerly, Rhode Island granite. 2c: TrL. 2d: CL. The bright blue-luminescing mineral in the upper left-hand corner is K-feldspar. The plagioclase feldspar in the lower right-hand corner luminesces a grayish-green which is easily distinguished from the K-feldspar by eye although the colors appear similar in this print. The deep blue-luminescing mineral in the center of the photo, running diagonally from the lower left to the upper right, is quartz, which appears more purplish in color to the eye than in the print. A considerable amount of calcite (orange CL) and apatite (yellow CL) shows up well also. There are some nonluminescing (black) opaque minerals. *CL conditions:* 16 kV, 0.7 mA beam current, 5 mm diam. spot. *Photographic conditions:* 25 s exposure on ASA 400 Ektachrome. (Photo by D. J. Marshall.)
Scale: 35 mm on transparency $\equiv$ 5 mm on thin section.

Plates 2e and f A porphyritic basalt from Sam Goosly Lake, British Colombia. 2e: PoL. 2f: CL. CL clearly reveals that there are multiple generations of feldspars. The brightest phenocrysts are slightly overexposed (white) to bring out the feldspars in the groundmass. To the eye, these plagioclase phenocrysts appear pale yellow-green. The blue CL in the groundmass is K-feldspar and, based on color differences, at least two other feldspars can be seen in the groundmass. The four nearly circular grains to the SW of the large phenocryst are anhydrite which is lanthanide-activated and appears pale blue. A thin vein of dolomite appears as a vertical line on the extreme right of the photograph. The CL of the dolomite is redder than that of the calcite. There are many very small grains of yellow-luminescing apatite and orange-luminescing calcite. The distribution of both of these is nicely revealed with CL. *CL conditions (typical for this kind of sample):* 10 kV, 0.5 mA, unfocused (10 mm diameter) beam. *Photographic conditions:* 20–40 s, ASA 400 film. *Scale:* 35 mm on transparency $\equiv$ 3.8 mm on thin section. (Photo by A. N. Mariano, App. (I).)

(a)

(b)

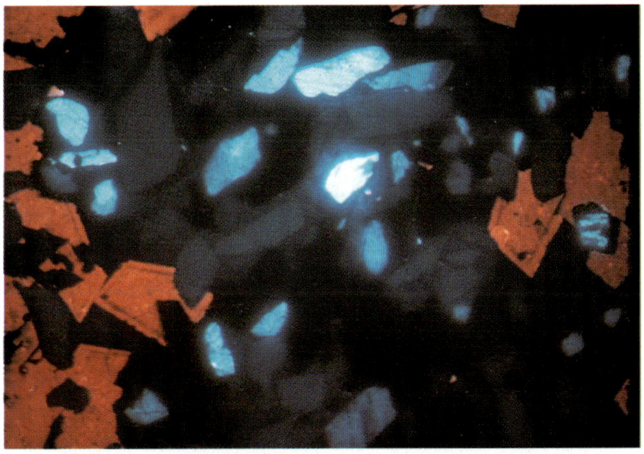

(c)

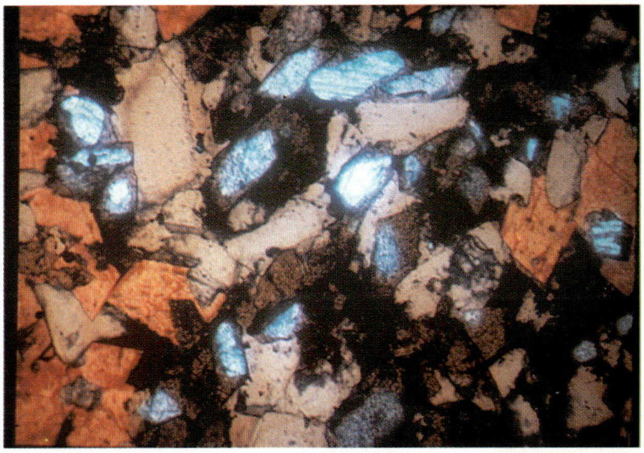

(d)

(e)

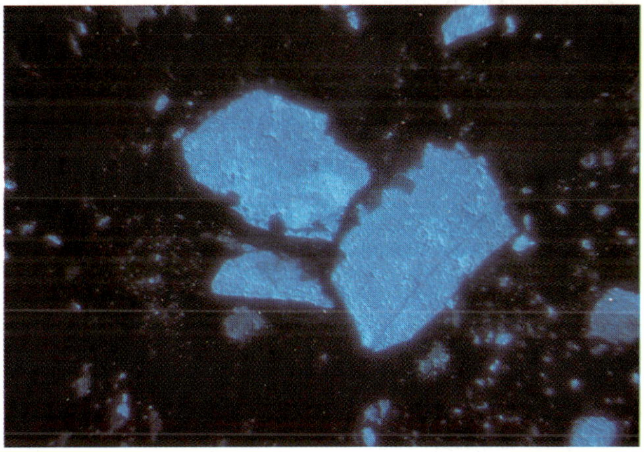

(f)

Plates 3a and b An example of authigenic feldspars in a calcrete from the Triassic Sherwood Sandstone, East Budleigh, Devon, UK. 3a: TrL. 3b: CL. The orange luminescence is the calcite of the calcrete nodule. The two bright blue grains are detrital potassium feldspar cores which are surrounded by a narrow region of nonluminescing (black) authigenic feldspar. The boundaries between the detrital and authigenic feldspar are visible on the TrL photo but they are located much more quickly and unequivocally with CL. CL and photographic conditions not available. (Photos by A. Leonard.)

Plates 3c and d Tribes Hill Dolomite from Mohawk Valley, northern New York State. This particular thin section contains detrital quartz with overgrowths, dolomite, and detrital feldspars with overgrowths. 3c is a conventional CL exposure. The quartz and the quartz overgrowths luminesce slightly different shades of purple. The dolomite shows characteristic red CL with some dark, Fe-rich, quenched zones. The detrital K-feldspars luminesce blue. 3d is a conventional CL exposure with a very short transmitted light exposure superimposed. The thin section has been stained for feldspars and the stained feldspars appear yellowish to the eye but as a mottled brown on the photos. Where there is a detrital grain with authigenic feldspar, the luminescing detrital core appears blue on both plates whereas the authigenic portion appears black (nonluminescent) on Plate 3c and a mottled brown on Plate 3d. (The violet halo which appears around the detrital feldspars on Plate 3d is apparently a photographic artifact.) Thus in a single transparency one has information immediately displayed on both the authigenic and the detrital feldspar composition of the sample. The areas which are black on both plates are probably clay or organic materials. This pair of photos appeared in black and white as Figure 4 and 5 of Buyce and Friedman (1975). *CL conditions:* not available. *Photographic conditions:* 5 minute CL exposure and less than 1 s TrL exposure. *Scale:* 35 mm on transparency ≡ 0.6 mm on thin section. (Photos by M. R. Buyce, App.)

Plates 3e and f A weathered granite from the Boscobel Quarry, Richmond, Virginia. 3e: TrL. 3f: CL. The bright blue-luminescing mineral is K-feldspar which is altering to clay (kaolinite and illite). The feldspar grains show up much more clearly with CL than with TrL, especially some of the very small residual fragments from almost completely weathered grains. Other nonluminescing minerals which are present include iron-rich biotite and quartz. The sample is a lapped thin section. *CL conditions:* 15 kV, 0.18 mA, 8 mm diameter beam. *Photographic conditions:* 15 s, ASA 400. *Scale:* 35 mm on transparency ≡ 0.5 mm on thin section. (Photos by T. G. Powell, W. T. Thompson, and J. B. Thomas, App.)

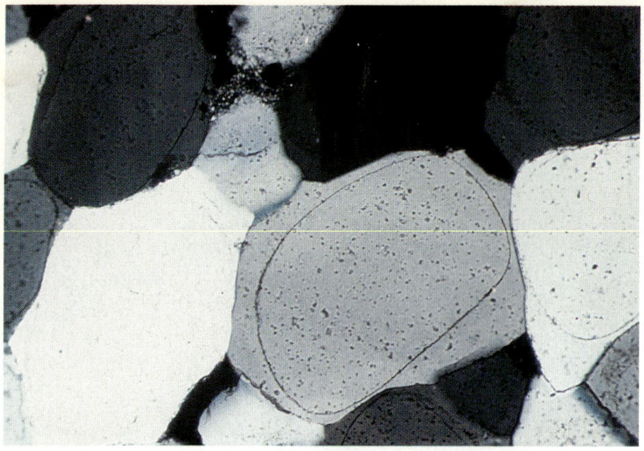
(a)

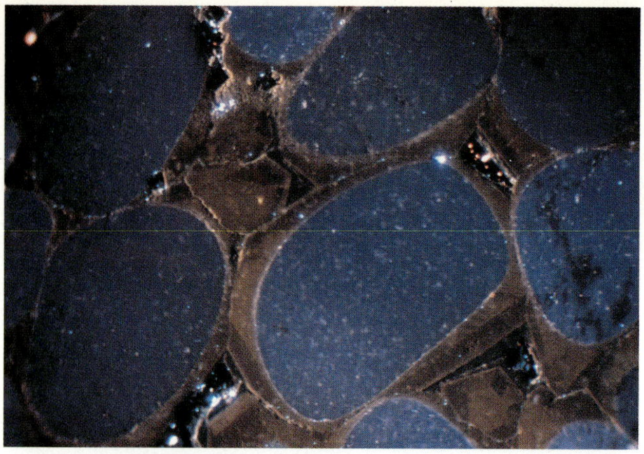
(b)

(c)

(d)

(e)

(f)

Key

1 K-feldspar
2 quartz grain with (**3**) zircon inclusions
3 zircon inclusions
4 plagioclase
5 rutilated quartz
6 calcite
7 K-feldspar (microcline)
8 quartz

Plates 4a and b St. Peter Sandstone (Ordovician?), Strawberry, Arkansas. 4a: PoL. 4b: CL. Overgrowths on many grains are revealed by dust rims in PoL, and the CL boundary matches these exactly. The authigenic quartz cement, however, shows zonation under CL (16.5 kV) which is not revealed under PoL (Owen 1984a). *CL conditions:* see App. *Photographic conditions:* 30 s exposure on ASA 400. *Scale:* 35 mm on transparency ≡ 1 mm on thin section. (Photos by M. R. Owen, App.)

Plates 4c and d Caseyville Sandstone (Pennsylvanian), southern Illinois. 4c: PoL. 4d: CL. This sandstone contains both brown- and blue-luminescing quartz. The very bright circular region in the center is a luminescing zircon grain and surrounding this is a radiation damage halo in the quartz grain which is revealed only with CL. Just above the zircon is a small region of overgrowth which is not revealed in PoL. On the left-center is a grain with extensive secondary quartz, again not revealed with PoL. The bright blue region in the lower right-hand corner of 4d is kaolinite. *CL conditions:* see App. *Photographic conditions:* 30 s exposure on ASA 400. *Scale:* 35 mm on transparency ≡ 1 mm on thin section. (Photos by M. R. Owen, App.)

Plates 4e and f Fountain Formation (Pennsylvanian), Climax, Colorado. 4e: PoL. 4f: CL. The minerals are identified in the accompanying sketch. The bright yellow-luminescing inclusions in the quartz in the lower left-hand corner may be rutile. The quartz in this sample shows a purple CL. The quartz grain at top-center contains two circular regions which show up dimly with CL. These are radiation damage haloes around zircon. *CL conditions:* see App. *Photographic conditions:* 30 s exposure on ASA 400. *Scale:* 35 mm on transparency ≡ 1 mm on thin section. (Photos by M. R. Owen, App.)

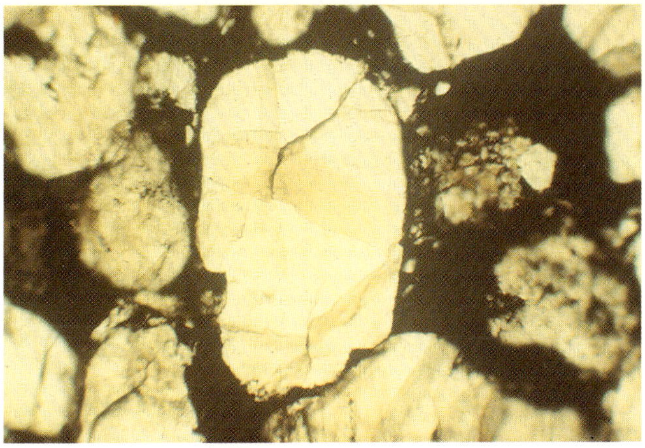
(a)

(b)

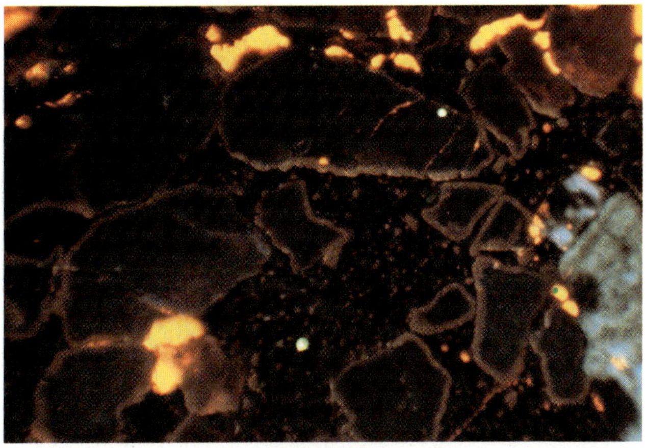
(c)

(d)

(e)

(f)

Plates 5a and b Sandstone from Great Slave Lake, northern Canada. 5a: TrL. 5b: CL. The pore spaces are occupied by uranium-containing minerals. Alpha particles from the uranium have a penetration depth in quartz of the order of 20 µm and they have 'damaged' the quartz grains, producing a damaged region with CL properties different from those of the bulk. Note several grains, including the one in the lower right-hand corner, where radon has penetrated the crack region and radiation damage extends from both sides of the crack. Neither TrL nor PoL reveals any indication of the rims (Morton 1978). CL and photographic conditions not available. *Scale:* 35 mm on transparency ≡ 2.2 mm on thin section. (Sample from R. D. Morton. Photo by A. N. Mariano, App. (II).)

Plates 5c and d Sandstone, from Moab, Utah, containing quartz pebbles, calcite, and uraninite. Both photomicrographs are with CL taken at (5c) room temperature and (5d) at low temperature with all CL conditions, photographic conditions, and exposure times identical. The low temperature was established with the sample in contact with a liquid nitrogen-cooled plate, but exact sample temperature is not known. The intensity of quartz CL increases at the lower temperature but there is little or no change in the CL intensity of the calcite or the feldspar grain in the lower right-hand corner. This sample also shows the radiation damage rinds due to alpha particles from the uraninite. *CL conditions:* 15 kV, 0.8 mA, unfocused (1 cm diameter) beam. *Photographic conditions:* 20 s, ASA 400 Ektachrome. *Scale:* 35 mm on transparency ≡ 2.2 mm on thin section. (Photos by D. J. Marshall and A. N. Mariano. Other conditions A. N. Mariano, App. (II).)

Plate 5e CL. Zircons separated from a sample of Precambrian paragneiss from the Bighorn Mountains, Wyoming (sample no. H67–66). After separation the grains were mounted with adhesive on a glass slide. *CL conditions:* 15.5 kV, 0.1 mA, 3 mm beam diameter. *Photographic conditions:* 35 s, ASA 160 daylight film. *Scale:* 35 mm on transparency ≡ 2.2 mm on thin section.

Plate 5f CL. Zircons separated from a sample of Pennsylvanian Sharon Conglomerate from northeastern Ohio (sample no. 735). After separation the grains were mounted with adhesive on a glass slide. *CL conditions:* 15.5 kV, 0.5 mA, 3 mm beam diameter. *Photographic conditions:* 15 s, ASA 160 daylight film. *Scale:* 35 mm on transparency ≡ 2.2 mm on thin section.

The grains in 5e are relatively uniform, although many exhibit dark cores with clear overgrowths. The grains in 5f are not uniform and they represent contributions from multiple source rocks. The rocks from which these samples were derived are discussed in papers by Malcuit & Heimlich (1972) and Heimlich *et al.* (1975), respectively. (Samples from R. A. Heimlich. Photos by A. N. Mariano, App. (II).)

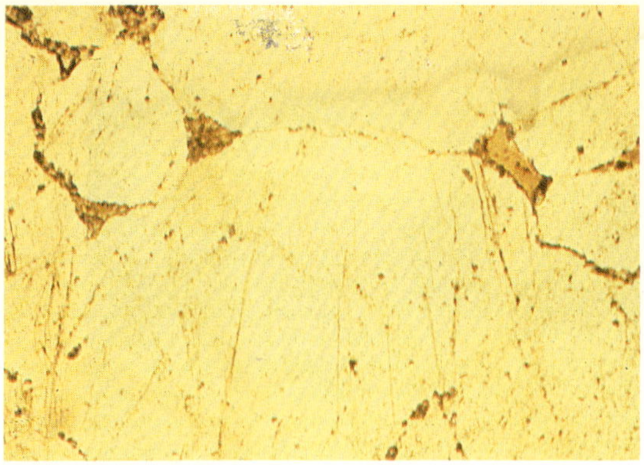

(a)

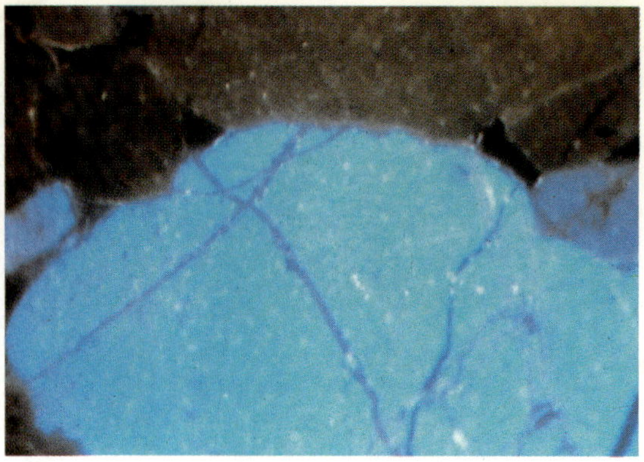

(b)

Plates 6a and b Tuscarora orthoquartzite. 6a: TrL. 6b: CL. Quartz grains exhibit two different CL colors. CL also reveals the presence of fractures which have been filled with nonluminescing (black), secondary quartz. The fractures are barely or not at all visible in the TrL photograph, and are not distinguishable from the sample preparation marks, whereas they are clearly revealed in the CL photograph (Sibley & Blatt 1976). *CL conditions:* Sippel type CMA (Sippel 1965). *Scale:* 35 mm on transparency ≡ 0.92 mm on sample. (Photos by D. F. Sibley.)

Plates 6c and d Tuscarora orthoquartzite. 6c: PoL. 6d: CL. PoL seems to show three grains but CL reveals this to be a single grain, rotated during fracturing, and the fractures are filled with secondary, nonluminescing quartz (Sibley & Blatt 1976). *CL conditions:* Sippel type CMA (Sippel 1965). *Scale:* 35 mm on transparency ≡ 0.74 mm on sample. (Photos by D. F. Sibley.)

(c)

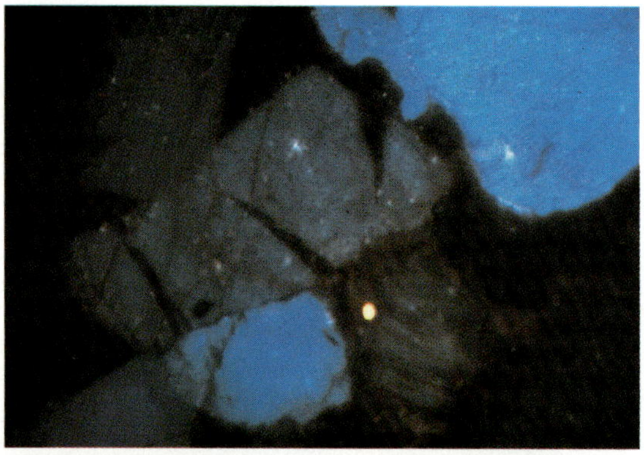

(d)

(e)

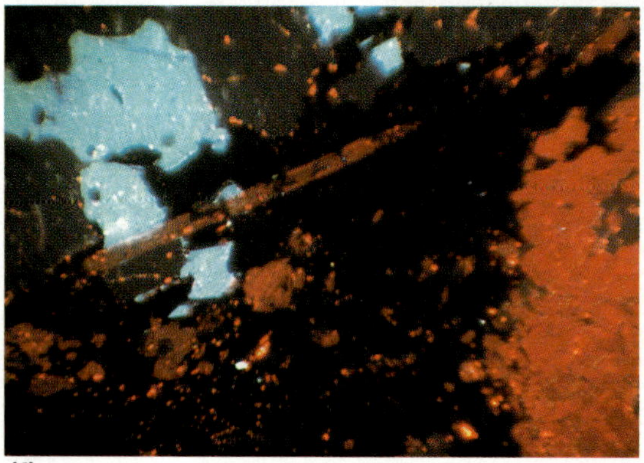

(f)

Plates 6e and f Sillimanite gneiss, Carlisle, Massachusetts. 6e: TrL. 6f: CL. The red-luminescing grains are sillimanite. This particular section does not contain any carbonates but it should be stressed that the color of the sillimanite CL is distinctly different from that of the common carbonates and would be readily distinguishable. The homogeneous blue-luminescing grain is a feldspar. The dull-luminescing grains on either side of the larger feldspar grain are quartz. *CL conditions:* 15.5 kV, 0.9 mA, 4 mm beam diameter. *Photographic conditions:* 7 s, ASA 400. *Scale:* 35 mm on transparency ≡ 3.8 mm on thin section. (Photo by A. N. Mariano, App. (I).)

Plates 6g and h A sandstone from the Sussex Formation, Upper Cretaceous, Wyoming. 6g: TrL. 6h: CL. The orange-luminescing mineral is carbonate cement. The dull purple and dull brown grains are quartz, and some of these, especially near the center of the field, have a dark blue-luminescing outer region of authigenic quartz. The bright blue (almost white)-luminescing grains are potassium feldspar, some of which are surrounded by dark authigenic material. The occasional greenish-luminescing grain is plagioclase. In this type of specimen, the determination of the relative amounts of the various minerals by point counting is easily done with the aid of CL. *CL conditions:* 16 kV. Beam current and beam diameter not available. *Photographic conditions:* 480 s, ASA not available. *Scale:* 35 mm on transparency ≡ 5.9 mm on sample. (Photo by R. L. Brenner.)

(g)

(h)

(a)

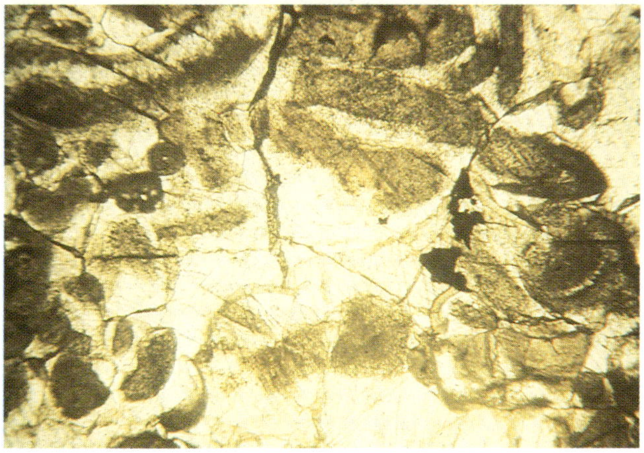

(b)

(c)

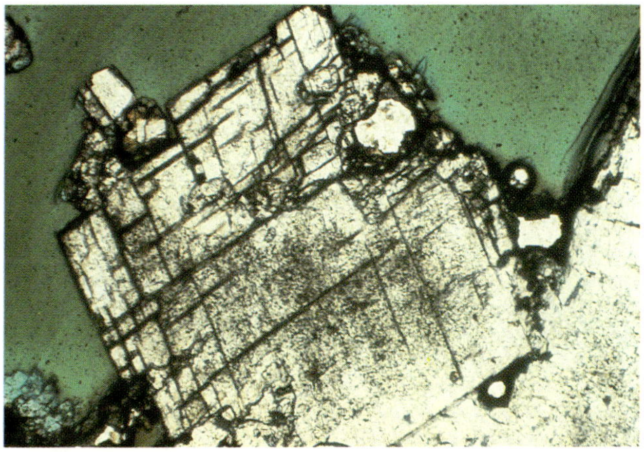

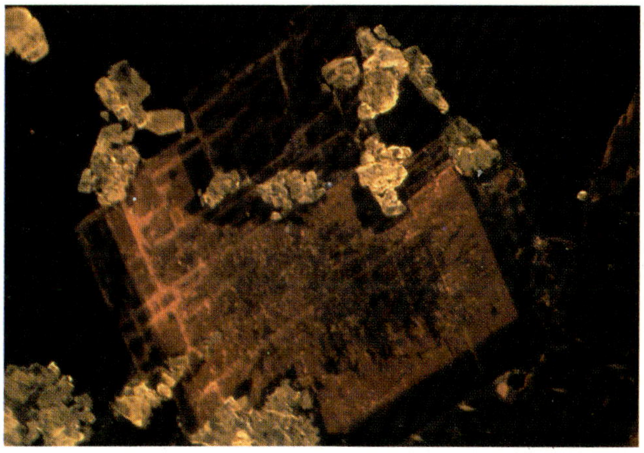

(d)

(e)

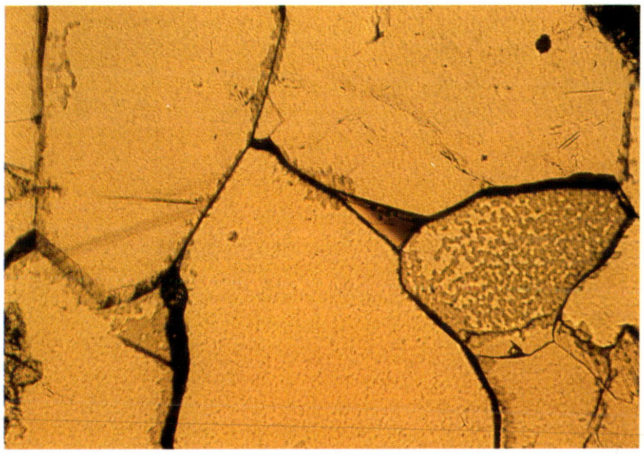

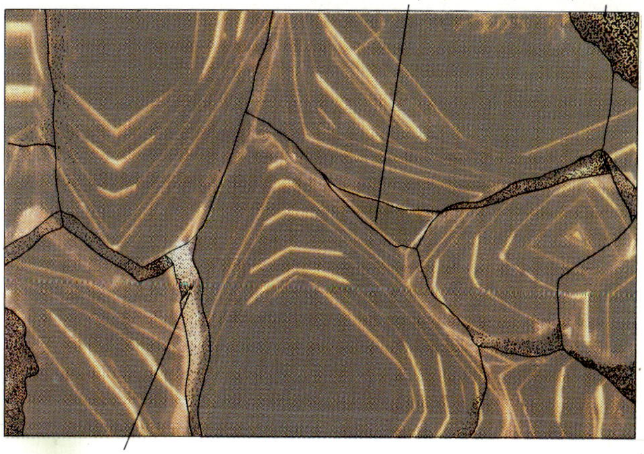

(f)

Plates 7a and b Fernvale Limestone (Ordovician) from northern Arkansas. 7a: TrL. 7b: CL. The TrL photomicrograph shows calcite cement between the crinoid fragments which looks more or less uniform. CL reveals two cement generations. The first, nonluminescing, cement formed as microstalactites on the crinoid fragments, indicating cementation under vadose conditions. The second cement is brightly luminescent, reflecting a change in the chemistry of the precipitating fluids (Freeman 1971). *CL and photographic conditions* are not available. *Scale:* 35 mm on transparency ≡ 6.8 mm on thin section. (Photos by T. Freeman.)

Plates 7c and d A dolomitized algal mound from Paradox Stage oil reservoir, Paradox Basin, southeastern Utah. 7c: TrL. 7d: CL. The orange-luminescing calcite is original. The dolomite is a late stage fill in shelter porosity. Leaching of the original aragonite (and calcite?) generated the pore space in which the dolomite developed. The zoning in the red-luminescing dolomite is not revealed with TrL. *CL conditions:* 12 kV, 0.14 mA, 5 mm beam diameter. *Photographic conditions:* 1 minute exposure, ASA 400. *Scale:* 35 mm on transparency ≡ 0.5 mm on thin section. (Photos by T. G. Powell, W. T. Thompson, and J. B. Thomas, App.)

Plates 7e and f Lower Magnesian Limestone, Permian, Yorkshire, UK. 7e: TrL. 7f: CL. This sample exhibits prominent intercrystalline boundaries. The area on the slide contains orange-luminescing calcite, and dark, nonluminescing calcite. The centers of the crystals have a slight blue cast which is visible to the eye but not well reproduced photographically. The diffuse orange areas are thought to be the result of intense CL from adjacent dolomite, reflected from the intercrystalline boundaries (see sketch). The growth outlines of the several crystals and the chronology of the growth are nicely revealed with CL. The use of CL to assist in studying crystallization phenomena has been presented in a paper by Dickson (1983). *CL conditions:* 20–25 kV, 0.65 mA, approx. 2 mm beam diameter. *Photographic conditions:* 38 s, ASA 400 Ektachrome daylight film. *Scale:* 35 mm on transparency ≡ 1.25 mm on original thin section. (Photos by J. A. D. Dickson, App.)

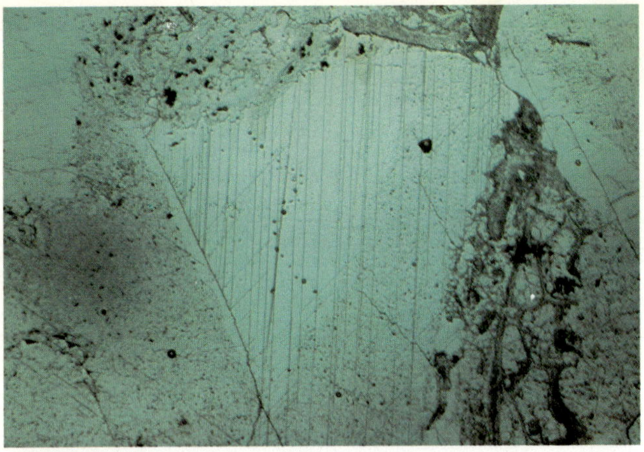

(a)

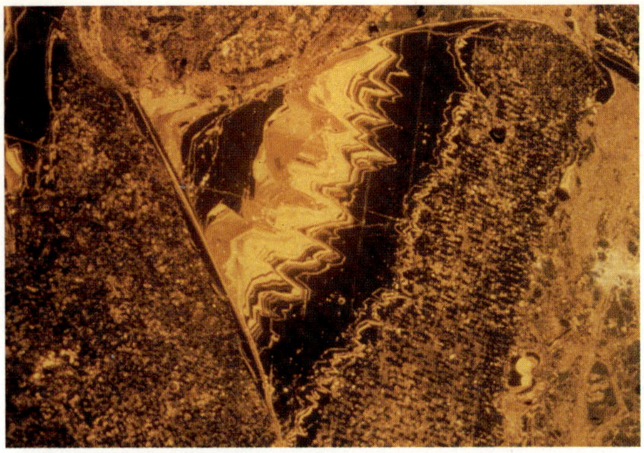

(b)

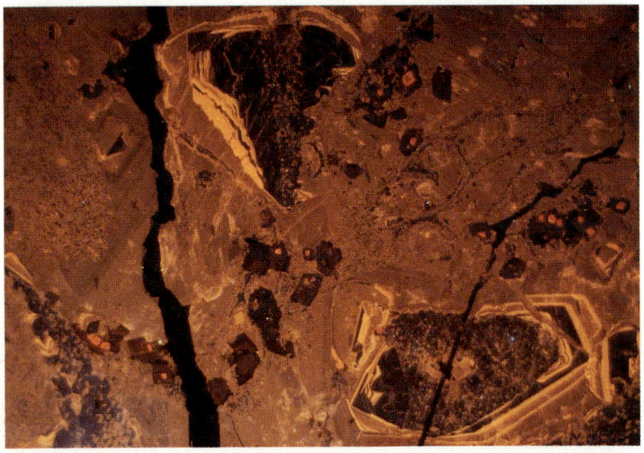

(c)

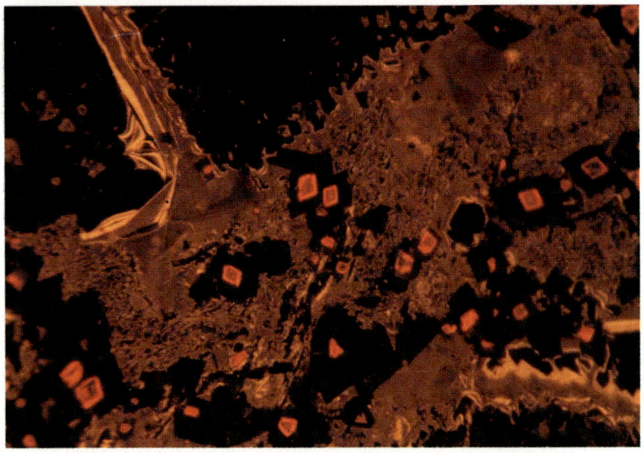

(d)

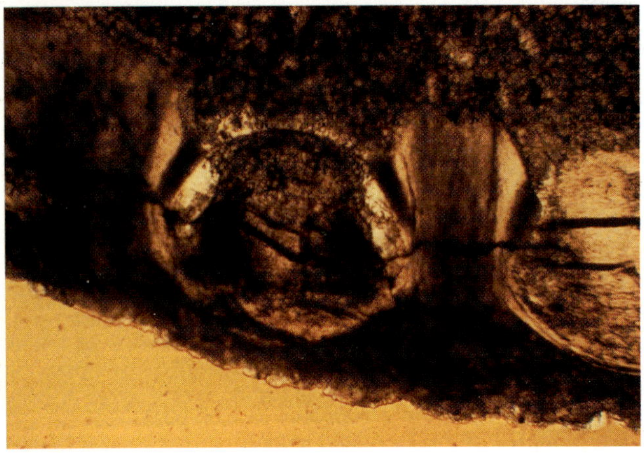

(e)

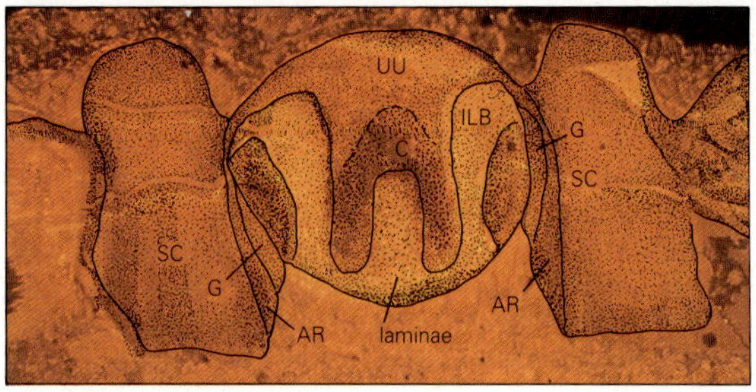

(f)

Key

UU upper unit of Huygensian lens doublet
ILB intralensar bowl (lower unit of lens doublet)
G girdle
C pyriform core (Fe-rich after Mg calcite)
SC scleral pillar
AR alveolar ring

Note the fine laminae seen in the lower unit of the lens and also indications of radially arranged fiber crystallites of calcite seen with SEM

Plates 8a and b Crinoidal grainstone from the Lake Valley Formation (Mississippian), San Andres Mountains, New Mexico. 8a: TrL. 8b: CL. The cement region between the three crinoidal fragments exhibits the complex zonation in cement which is sometimes revealed so clearly with CL. The region on the right is a bryozoan. The series of small black dots seen clearly in the TrL photo are from an electron probe traverse made across the cement region to measure Mg, Fe, and Mn concentrations. W. Meyers work in this geographical area is discussed in several papers, e.g. Meyers (1974, 1978) and Meyers and Lohmann (1978, 1985). Polished unstained thin section. *CL conditions (typical):* Luminoscope; 12 kV, 0.8 mA beam current. *Scale:* 35 mm on transparency ≡ 2.5 mm on thin section. (Sample and photographs by W. J. Meyers.)

Plates 8c and d Crinoidal limestone from the Lake Valley Formation (Mississippian), Sacramento Mountains, New Mexico. Both photos are with CL. 8c exhibits another example of complex cement zonation which is not seen with either TrL or PoL. Throughout the section there are areas of the order of 100–200 μm in longest dimension which are darker than the other regions. Within these dark areas, there are many small red-luminescing grains. These are luminescing dolomites and the dark region surrounding them is non- or weakly-luminescing dolomite. 8d is a magnified view of another region of the same sample. At this higher magnification, the dolomites are seen to have a zoned luminescence. *CL conditions (typical):* Luminoscope; 12 kV, 0.5 mA beam current, 30 s exposure, ASA 400 film. *Scale 8c:* 35 mm on transparency ≡ 1.75 mm on original thin section. *Scale 8d:* larger dolomite rhombs are approximately 50 μm in longest dimension. (Samples from W. J. Meyers. Photos by D. J. Marshall (8c) and Institute of Petroleum, Beijing (8d).)

Plates 8e and f Trilobite eye structures, *Phacops rana milleri*, showing the additional preservation of detail that is revealed when viewed with CL. 8e: TrL. 8f: CL. The section is somewhat oblique and off lens axis. UU = upper unit of Huygensian lens doublet; ILB = intralensar bowl (lower unit of lens doublet); G = girdle; C = pyriform core (Fe-rich after Mg calcite); SC = scleral pillar; AR = alveolar ring. Note the fine laminae seen in the lower unit of the lens and also indications of radially arranged fiber crystallites of calcite seen with SEM. Discussion of this work appears in the article by Miller and Clarkson (1980). *CL conditions:* 18 kV, 0.35 mA, 4 mm beam diameter. *Photographic conditions:* 10 s (or less). ASA 200 film. *Scale:* diameter of eye is approximately 0.6 mm. (Photos by J. Miller, App.)

Plates 9a and b Unaltered bryozoans, genus *Monotrypa*. 9a: TrL. 9b: CL. The wall structure shows up well with either TrL or CL but some additional details are revealed with the latter (Ferrigno & Walker 1973). CL and photographic conditions not available. (Photos by K. F. Ferrigno.)

(a)

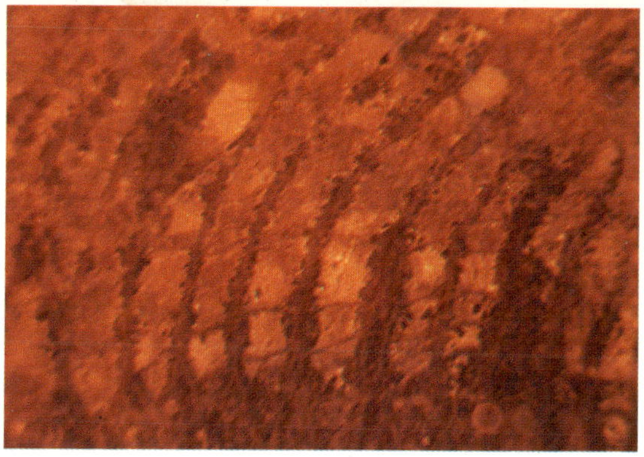

(b)

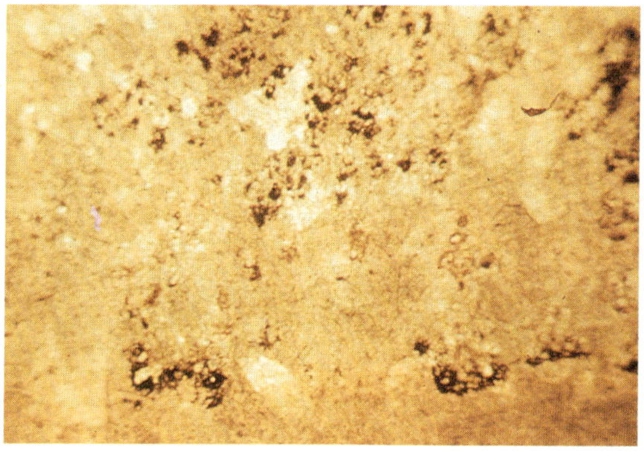

(c)

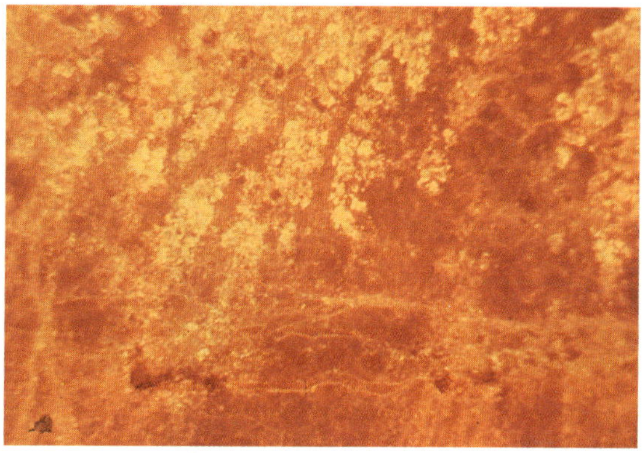

(d)

Plates 9c and d Very badly altered bryozoans, possibly genus *Monotrypa*. 9c: TrL. 9d: CL. The wall structure is barely detectable with TrL but is still readily visible with CL. (Ferrigno & Walker 1973.) CL and photographic conditions not available. (Photos by K. F. Ferrigno.)

Plates 9e and f Hidden breccia in Knox Dolostone, Right Fork Area, Central Tennessee zinc district. 9e: PoL. 9f: CL. 'Using crossed polarizers, the material appears to be a fairly uniform, crystalline dolostone. Using cathodoluminescence, the breccia fragments are outlined because of differences in the amounts of activators and/or quenchers' (Kopp 1981, p. 111). *CL conditions:* 10.5 kV, 0.5 mA, 4–5 mm beam diameter. *Photographic conditions:* 4 minutes, ASA 400 Ektachrome daylight. *Scale:* 35 mm on transparency ≡ 2.8 mm on thin section. (Photos by O. C. Kopp, App.)

(e)

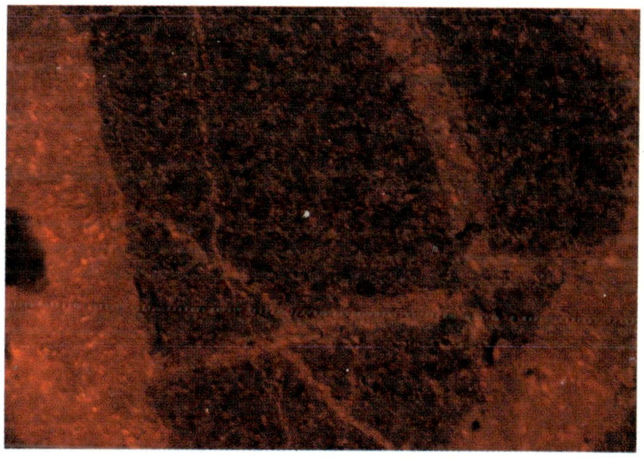

(f)

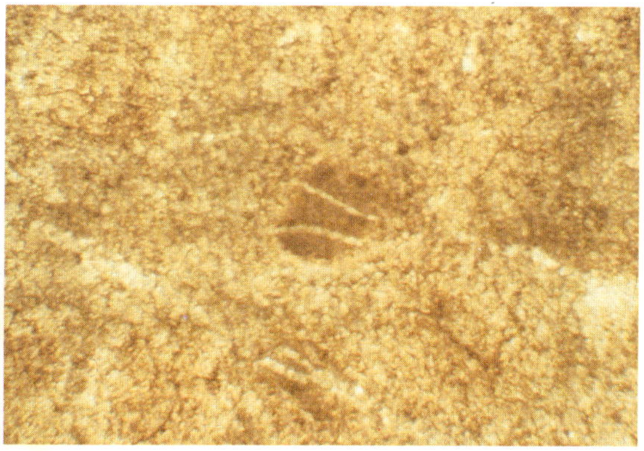

(g)

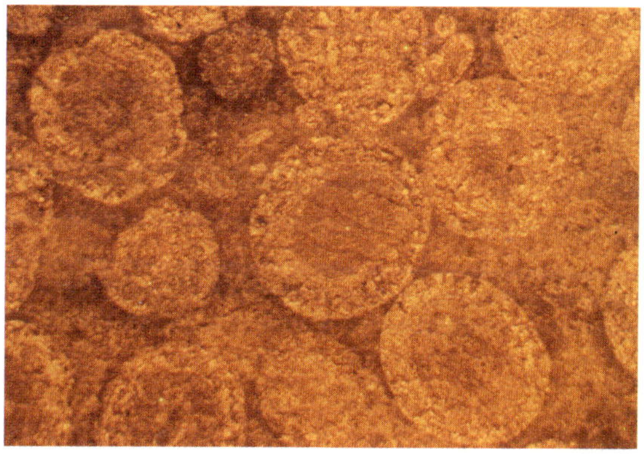

(h)

Plates 9g and h Conasauga Limestone, east Tennessee. 9g: PoL. 9h: CL. The presence of oolites is barely discernible with ordinary light or with polarized light whereas the oolites are obvious with CL. *CL conditions:* 10.5 kV, 0.5 mA, 4–5 mm beam diameter. *Photographic conditions:* 2 minutes, ASA 400 Ektachrome daylight film. *Scale:* 35 mm on transparency ≡ 2.8 mm on thin section. (Photos by O. C. Kopp, App.)

(a)

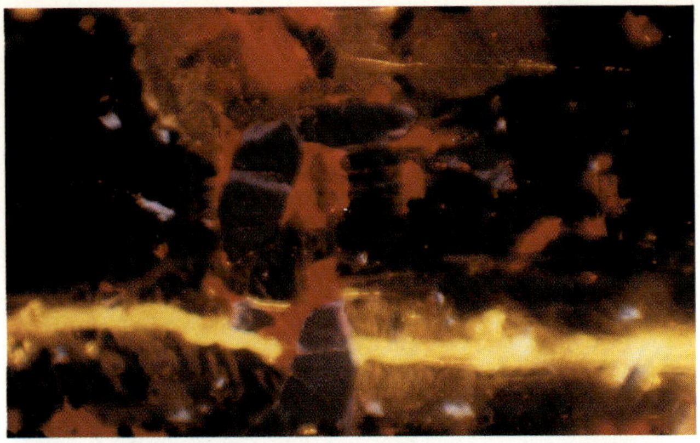

(b)

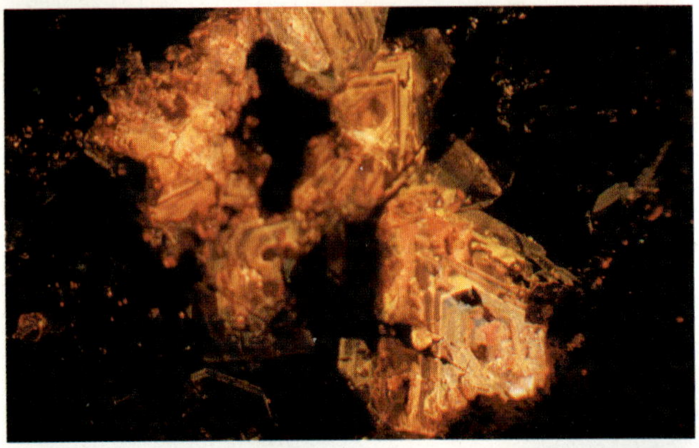

(c)

(d)

(e)

(f)

Plate 10a Carbonatite from Oka, Quebec. CL. The yellow orange is calcite and the dark red regions within the calcite are dolomite exsolution laths which stand out quite clearly with CL. With TrL or PoL, the dolomite is difficult to resolve from the calcite. *CL conditions:* Luminoscope. Other conditions not available. *Photographic conditions:* not available. *Scale:* 35 mm on transparency ≡ 3.8 mm on thin section. (Photo by A. N. Mariano, App. (I).)

Plate 10b Okaite from Husereau Hill, Oka, Quebec. CL. The deep blue-luminescing prisms are apatite with strong Eu^{2+} and Sm^{3+} activation. The reddish-orange grains are calcite which is Mn^{2+}-activated, confirmed by a spectral scan which showed a peak at 605 nm. The late yellow vein that transects nonluminescing melilite is wollastonite which is a product of late calcium metasomatism that sometimes is observed in carbonatite environments. In some cases at Oka, wollastonite metasomatism is accompanied by vesuvianite which displays bright red CL. This type of metasomatism should not be confused with fenitization. *CL conditions:* 10 kV, 0.8 mA, 4 mm beam spot diameter. *Photographic conditions:* 45 s, ASA 160. *Scale:* 35 mm on transparency ≡ 3.8 mm on thin section. (Photo and caption by A. N. Mariano, App. (I).)

Plate 10c Garnets in a skarn from Ok Tedi, New Guinea. CL. An extremely complex growth pattern is revealed. *CL conditions:* 12 kV, 0.8 mA, 4 mm beam diameter. *Photographic conditions:* 45 s, ASA 160. *Scale:* 35 mm on transparency ≡ 3.8 mm on thin section. (Photo by A. N. Mariano, App. (I).)

Plate 10d Birmingham, Alabama Iron Ore. CL. The iron ore has a high phosphorous content (about 0.3 percent). The phosphorous is present as a carbonate-fluorapatite (CF) and these grains cannot be seen in a TrL photo which is predominantly opaque (hematite) with occasional light areas of calcite. However, most of the CF grains are luminescent and quickly revealed with CL. CL also reveals calcitic fossil fragments that have been partially replaced by hematite and are effectively concealed to TrL observations. Knowledge of the modes of occurrence of the various apatites is important in the development of ore beneficiation methods. (Hagni 1984.) *CL conditions (typical):* 12–14 kV, 0.7–0.9 mA, 2–4 mm beam diameter. *Photographic conditions:* 14 minutes, ASA 400, pushed to 1,600. *Scale:* The elliptical fossil fragment in the center of the Plate is approximately 125 μm in length. (Photo by R. D. Hagni, App.)

Plates 10e and f A zoned calcite crystal which appears to be a single uniform crystal under TrL. CL reveals nine identifiable zones. The Fe content of all zones is less than 0.05 mole percent. The CL intensities of the various zones correlate with the Mn content. (Oglesby 1976. See also Frank *et al.* 1982.) *CL conditions:* Luminoscope. Other CL and photographic conditions not available. *Scale:* Brightest zone, upper left-hand corner, about 50 μm wide. (Photo by T. W. Oglesby.)

Zone	Mn (mole percent)	CL intensity
1	<0.05	dead
2	<0.05	dull
3	<0.05	dull
4	<0.05	dull
5	0.06	bright
6	0.18	very bright
7	0.12	bright
8	0.18	very bright
9	<0.05	dull

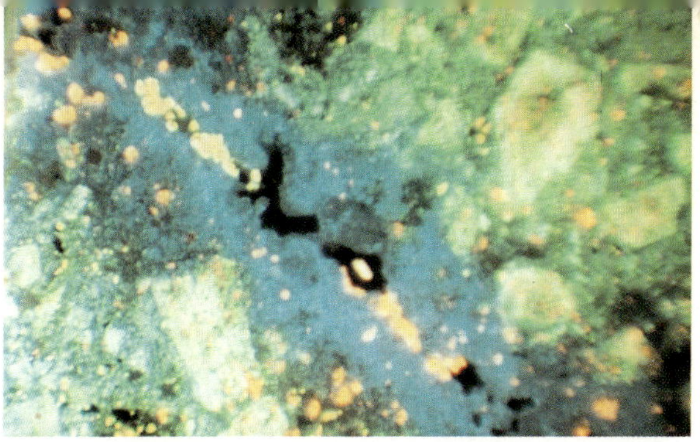
(a)

(b)

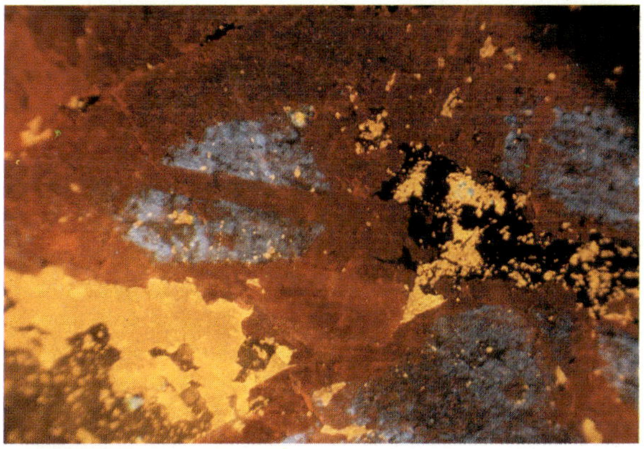
(c)

(d)

(e)

(f)

Plate 11a Blue-luminescing K-feldspar veining in quartz andesite from Safford, Arizona. CL. K-feldspathization is one of the hydrothermal alteration mechanisms which frequently accompanies base metal mineralization in porphyry copper and stockwork molybdenite deposits. Unlike the K-feldspathization which occurs during fenitization, and which exhibits red Fe^{3+}-activated CL, Safford hydrothermal alteration K-feldspar veins show only blue, Ti^{4+}-activated CL. Within the blue veins are calcite, apatite, and chalcopyrite. The quartz andesite groundmass shows bleached feldspars with Mn^{2+} and/or Fe^{2+} activation, both of which produce yellow-green luminescence. *CL conditions:* not available. *Photographic conditions:* 30 s, ASA 160. *Scale:* 35 mm on transparency ≡ 3.8 mm on thin section. (Caption and photo by A. N. Mariano, App. (I).)

Plate 11b Quartz diorite from Ste. Honoré Carbonatite, Chicoutimi, Quebec, showing incipient fenitization of country rock. CL. Fenite K-feldspar displays red CL from Fe^{3+} activation. Within the K-feldspar is light blue-luminescing apatite, cogenetic with the fenitization, and LREE-activated. In contrast, the country rock feldspar shows blue, Ti^{4+}-activated feldspars and yellow, Mn^{2+}-activated apatite. Orange-luminescing calcite has accompanied the introduction of fenite solutions. *CL conditions:* 18 kV, 0.8 mA, 4 mm beam diameter. *Photographic conditions:* 15 s, ASA 400. *Scale:* 35 mm on transparency ≡ 3.8 mm on thin section. (Caption and photo by A. N. Mariano, App. (I).)

Plate 11c Advanced incipient fenitization of country rock from the Fen, Norway. CL. The photomicrograph shows fenite feldspar with red CL from Fe^{3+} activation which is engulfing primary feldspar with blue CL. Orange-luminescing calcite accompanies the fenitization. *CL conditions:* 17.5 kV, 0.8 mA, 3 mm beam diameter. *Photographic conditions: 7 s, ASA 400. Scale:* 35 mm on transparency ≡ 2.2 mm on thin section. (Caption and photo by A. N. Mariano, App. (II).)

Plate 11d Brecciated rock from the Fen Carbonatite, Norway. CL. The red fragments are Fe^{3+}-activated orthoclase. The blue fragments are Ti^{4+}-activated country rock feldspar. The brilliant white-luminescing grain is zircon. Orange calcite specks and dull greenish-brown apatite occur as interstitial disseminations. The apatite luminescence is due to complex LREE activation. *CL conditions:* 15 kV, 0.8 mA, 3 mm beam diameter. *Photographic conditions:* 7 s, ASA 400. *Scale:* 35 mm on transparency ≡ 2.2 mm on thin section. (Caption and photo by A. N. Mariano, App. (II).)

Plate 11e Internal fenitization of hornblende syenite, McClure Mountain, Colorado. CL. The McClure Mountain syenite crystallized under conditions of low alkalinity as indicated by the presence of hornblende and magnetite. In the late stages of crystallization, residual fluids became enriched in alkalis resulting in the stabilization of Fe^{3+}. At this time the primary feldspars with blue CL were subjected to peripheral replacement by fenite feldspar with Fe^{3+} activation. This is the cause of the red-luminescing rims around the feldspar grains. Calcite grains with orange CL are obvious and at least two blue-luminescing apatite grains also appear. They are LREE-activated. *CL conditions:* Luminoscope. Other conditions not available. *Photographic conditions:* 10 s, ASA 400. *Scale:* 35 mm on transparency ≡ 2.2 mm on thin section. (Caption and photo by A. N. Mariano, App. (II).)

Plate 11f Secondary green-luminescing plagioclase in mineralized quartz andesite, Safford, Arizona. CL. The large rectangular prisms with light blue CL are primary plagioclase crystals. The grains with moderate blue CL are K-feldspar. The quartz andesite has been subjected to intense hydrothermal alteration which includes green CL dendritic oligoclase that is predominantly Fe^{2+}-activated. The euhedral grains showing tan CL with some green CL zones are anhydrite. At least two yellow grains of Mn^{2+}-activated apatite also appear in this micrograph. *CL conditions:* Luminoscope. Other conditions not available. *Photographic conditions:* 60 s, ASA 160. *Scale:* 35 mm on transparency ≡ 2.2 mm on thin section. (Caption and photo by A. N. Mariano, App. (II).)

(a)

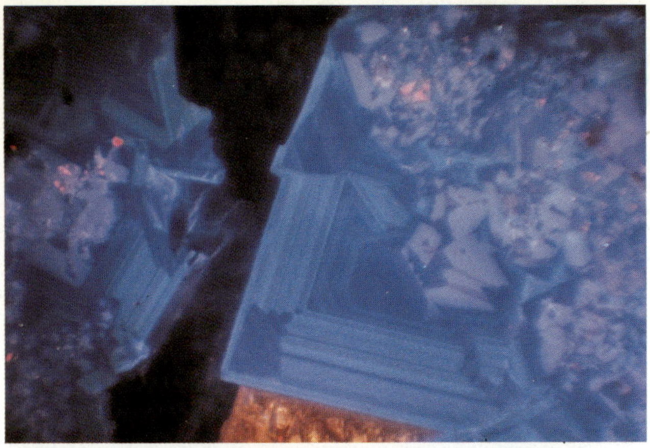
(b)

(c)
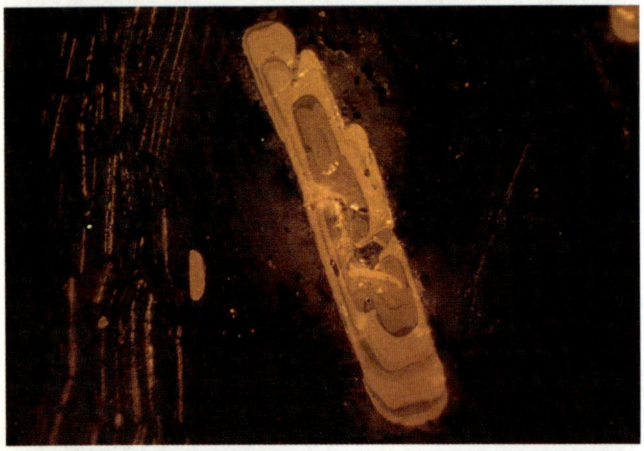

(d)

(e)

(f)

Plate 12a Cyclic zoning of Mn^{2+}-activated apatite from diabase, Ray, Arizona. CL. No probe analyses have been done on this sample but variations in CL are assumed to reflect variations in impurity composition and possible variations in Mn^{2+} activator content. *CL conditions:* Luminoscope. Other CL conditions not available. *Photographic conditions:* 75 s, ASA 160. *Scale:* 35 mm on transparency ≡ 3.8 mm on thin section. (Photo by A. N. Mariano, App. (I).)

Plate 12b Hydrothermal fluorite from the Okorusu carbonatite, Namibia. CL. Cyclic zoning is often observed in hydrothermal fluorites and may be mostly due to intrinsic effects. The violet CL specks in the fluorite are apatite with dominant Sm^{3+} activation. *CL conditions:* 9 kV, 0.8 mA, 4 mm beam diameter. *Photographic conditions:* 60 s, ASA 160. *Scale:* 35 mm on transparency ≡ 3.8 mm on thin section. (Caption and photo by A. N. Mariano, App. (I).)

Plate 12c Porphyritic calcite lava – Kerimasi Volcano, Tanzania. CL. The cyclic zoning exhibited in the calcite phenocrysts is only visible under CL. The aphanitic groundmass material is quenched calcite. The calcite zoning shows continuity in phenocrysts that have been fractured and faulted during flow, a strong indication that the zoning is a primary feature. In addition, Sr isotopic ratios for the phenocrysts and matrix calcite agree with those measured in mantle-derived crystallization. *CL conditions:* 17 kV, 0.8 mA, 3 mm beam diameter. *Photographic conditions:* 45 s, ASA 160. *Scale:* 35 mm on transparency ≡ 2.2 mm on thin section. (Caption and photo by A. N. Mariano, App. (II).)

Plate 12d Hydrothermal mineralization in quartz andesite, Safford, Arizona. CL. Cyclic zoning in anhydrite is revealed by CL. Green CL is from Mn^{2+} activation. Tan CL is from Sm^{3+} and Dy^{3+} activation. *CL conditions:* 14 kV. Other conditions not available. *Photographic conditions:* 60 s, ASA 160. *Scale:* 35 mm on transparency ≡ 2.2 mm on thin section. (Caption and photo by A. N. Mariano, App. (II).)

Plate 12e Hydrothermal mineralization in quartz andesite, Safford, Arizona. CL. Cyclic zoning of anhydrite. The light zones are mainly Sm^{3+} and Dy^{3+}-activated. The subtle variations in color are due to variations in Dy : Sm content. The dark zones are assumed to be areas of Mn^{2+} activation. The individual zones could not be isolated for spectral analysis of the CL emission spectrum. *CL conditions:* 14 kV. Other conditions not available. *Photographic conditions:* 60 s, ASA 160. *Scale:* 35 mm on transparency ≡ 2.2 mm on thin section. (Caption and photo by A. N. Mariano, App. (II).)

Plate 12f Beforsite from Fen, Norway. CL. This example shows the cyclic zoning in baddeleyite – brought out by CL. The central grain with dark blue and black zonation is baddeleyite (ZrO_2). The blue CL is intrinsic luminescence. The black zones are assumed to represent iron quenching. The red-luminescing groundmass is dolomite and the orange-luminescing material is calcite. *CL conditions:* 16 kV, 0.9 mA, 3 mm beam diameter. *Photographic conditions:* 15 s, ASA 400. *Scale:* 35 mm on transparency ≡ 2.2 mm on thin section. (Caption and photo by A. N. Mariano, App. (II).)

(**a**)

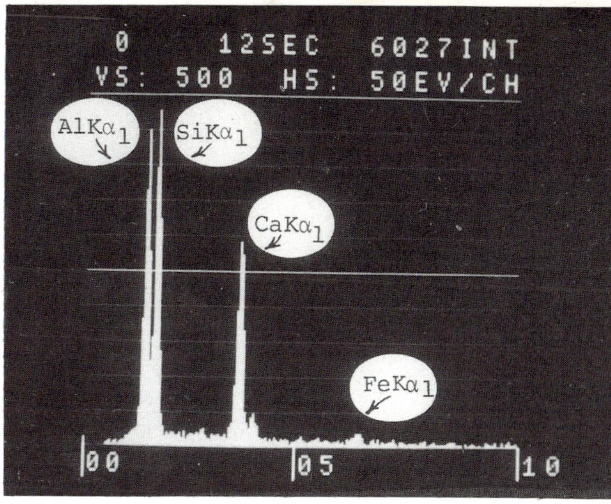

(**b**)

Plates 13a and b (a) Secondary electron image of synthetic anorthite. (b) Energy dispersive X-ray spectrum of synthetic anorthite.

The secondary electron image is of a crystalline aggregate of anorthite, synthesized by Jun Ito, with the composition $Fe_{0.1}Ca_{0.9}Al_2Si_2O_8$. The EDS spectrum was generated from the central flake with the rhombic cross section. All areas consistently give the same EDS results, indicating one phase with a homogeneous distribution of Fe. An X-ray diffraction scan of this material gave a pattern consistent with anorthite.

they concluded that the CL intensity is a function of the Mn content only, not the Fe:Mn ratio. Natural dolomites with Fe contents between 30 and 200 ppm exhibited CL if the Mn content exceeded 30–35 ppm.

The results of each of the authors which have been presented in Figures 7.8–11, standing alone, indicate a reasonably well defined field for the various levels of CL intensity. It was considered worthwhile, however, to combine all of these data into a single figure to evaluate the overall consistency of the data and to add the data from a number of other authors who have made a contribution to the subject. A similar summary has been prepared by Fairchild (1983).

To prepare these composite figures, the published data have been converted, if necessary, from the author's units to units of mole percent of the elemental carbonate. The conversion has been done using the exact chemical formula for either calcite or dolomite and assuming that only Ca, Mg, Fe, and Mn are present, all as divalent cations. It is assumed that the only anion present is carbonate. Some authors provide Ca and Mg contents while others give only the values of Fe and Mn. If the Mg content is not stated, it is assumed that the Mg content is zero for calcite and that Ca and Mg are equal, in molar amounts, in dolomite.

Fairchild (1983) has included analyses of some inhomogeneous samples and presented the results as a range of values. These data are important, but it has not been possible to incorporate them within the framework of the computer plotting routine used for the composite figures.

All of the samples represented in the figures are of sedimentary carbonates except those from Long and Agrell (1965).

All other trace elements are neglected in the calculation of the molar amounts of Mn and Fe. This assumption will have an insignificant effect on the calculated molar values of Fe and Mn in these carbonates. However, the presence of other trace elements could have an important bearing on the CL intensities observed. As noted earlier, there are situations where REE produce significant CL in calcite, Co and Ni may be quenchers in some carbonates, and Sr has been shown to depress the CL intensity in synthetic aragonite. As one example of the possible concentrations of other trace elements, Pingitore (1978) analyzed corals and found that Sr contents were comparable to the magnesium contents (500–3,000 ppm) and Zn contents were comparable to Mn (16–37 ppm).

Finally, there is the problem of relating the CL intensity values used by various authors, all of whom estimated intensity visually, sometimes with reference to a standard sample (Pierson 1981). Some use only the simple scale of luminescent or nonluminescent. Others use several categories to define the luminescence intensity. For the present purpose, the three categories chosen are nonluminescent, dully luminescent, and brightly luminescent. Any other categories have been placed into one of these three.

The results of plotting the data from the several authors are shown in Figures 7.12 (calcite) and 7.13 (dolomite).

The initial conclusion that can be drawn from the figures is that the boundary between the fields of luminescence intensity is a very broad one, when all results are considered, whereas the data from any single publication produce a much better defined field and narrower boundary. This may not be quite so surprising when one considers the variety of analytical techniques used, that the data have been reported over a period of 20 years or more, that different methods for observing the CL were used, etc. If CL intensities are to be used to estimate or to set limits on the Fe and Mn amounts then there is clearly more work to be done. The establishment of intra-laboratory standards for Fe and Mn analysis in carbonates, as well as standards for CL intensities, would seem to be important requirements to further the value of this particular application of CL.

It is tempting to try to draw in 'best fit lines' on Figures 7.12 and 7.13 to indicate at least the regions where there is agreement among the several authors. However, the author

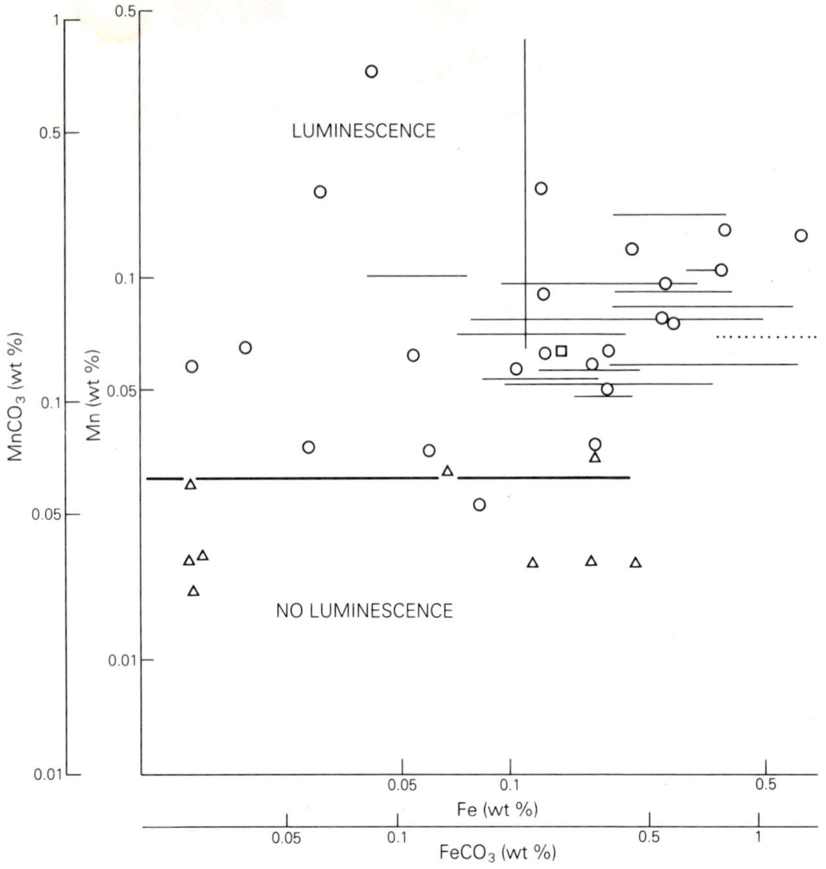

Figure 7.10 The relationship between the Mn and Fe contents and the luminescence intensity of calcites from the Bonahaven Formation of Islay, Scotland. Mn and Fe analyses were made by electron probe with detection limits for $FeCO_3$ and $MnCO_3$ of 0.004 weight percent. (From Fairchild 1983.)

Key

	Homogeneous chemistry	Inhomogeneous chemistry
approximate limiting line of luminescence		▬▬▬ (thick line)
bright luminescence	○	———
dull luminescence	□	– – – – – –
no luminescence	△	··············

Figure 7.11 The relationship between the Mn and Fe contents and the luminescence intensity of dolomites from the Bonahaven Formation of Islay, Scotland. Mn and Fe analyses were made by electron probe with detection limits for $FeCO_3$ and $MnCO_3$ of 0.004 weight percent. (From Fairchild 1983.)

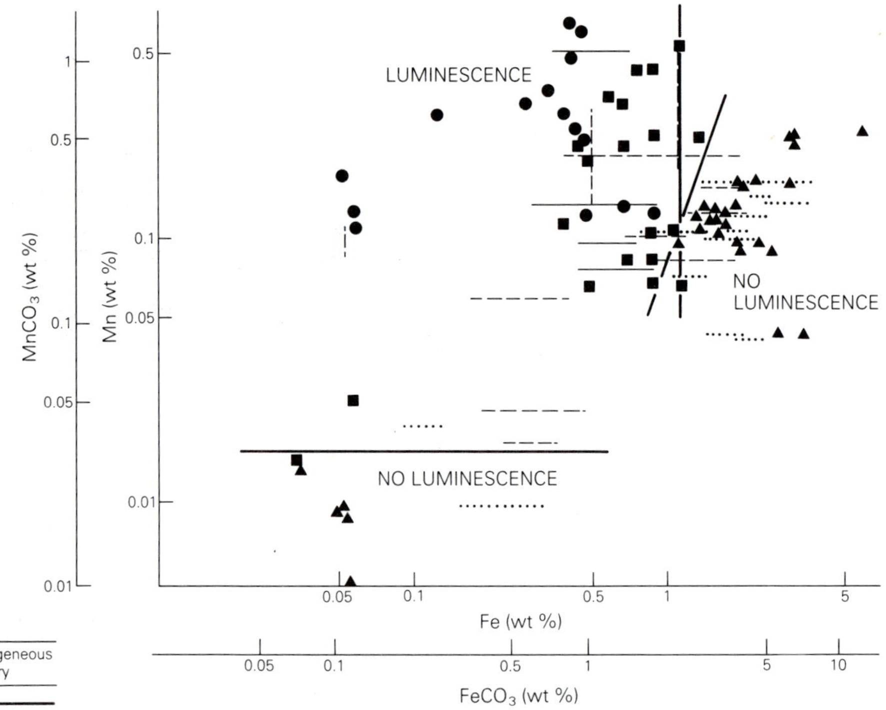

Key

	Homogeneous chemistry	Inhomogeneous chemistry
approximate limiting line of luminescence		▬▬▬ (thick line)
bright luminescence	●	———
dull luminescence	■	– – – – – –
no luminescence	▲	··············

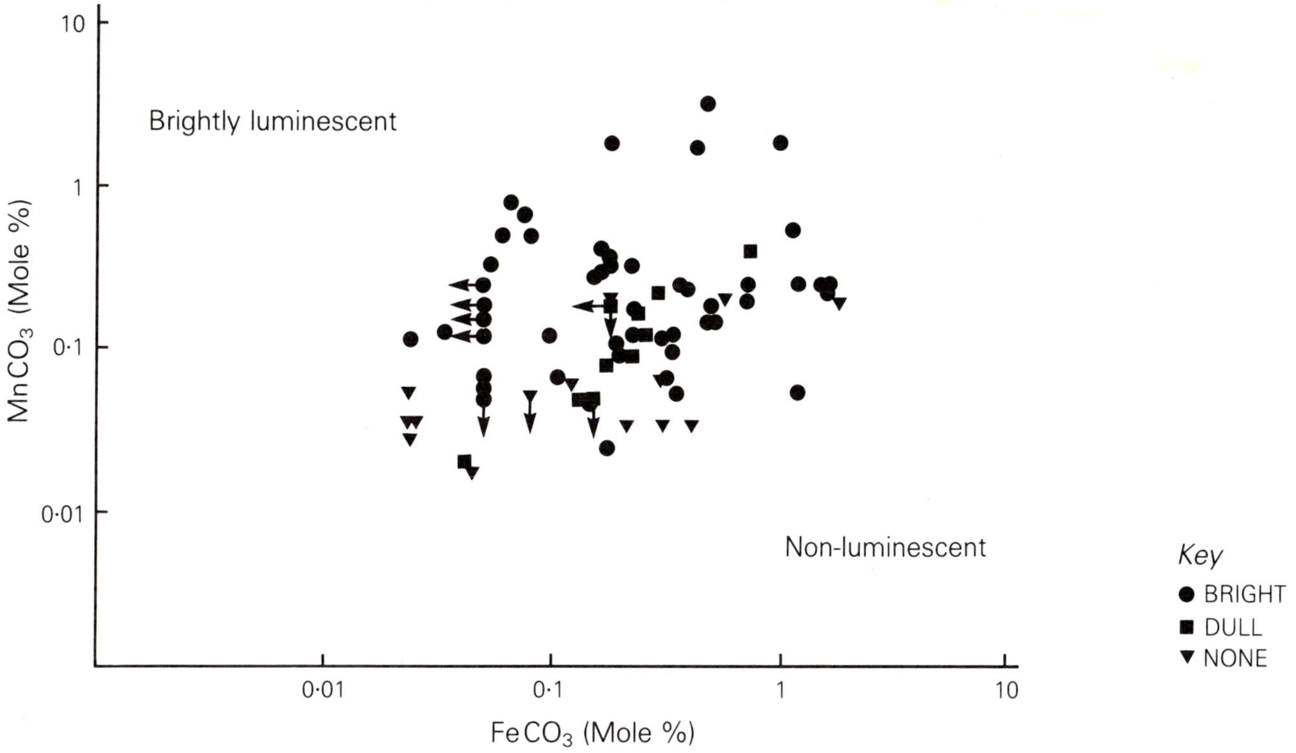

Figure 7.12 Summary of published data on calcite luminescence as a function of Mn and Fe contents. The data of Long and Agrell (1965), Martin and Zeegers (1969), Freeman (1971), Oglesby (1976), George (1977), Fox (1979), Pierson (1981), Frank *et al.* (1982), Grover and Read (1983), and Fairchild (1983) are included. Some authors have used several categories for intensity and we have reassigned these to one of the three shown – bright, dull, or none.

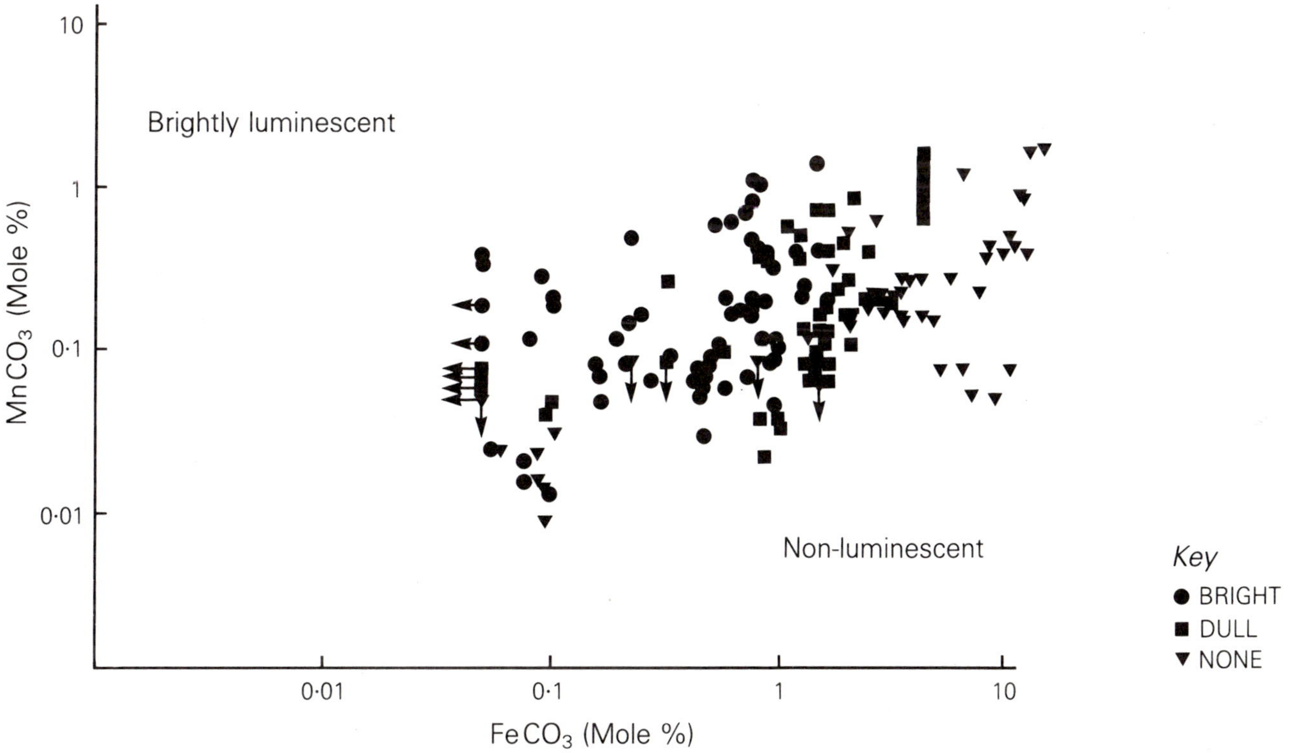

Figure 7.13 Summary of published data on dolomite luminescence as a function of Mn and Fe contents. The data of Long and Agrell (1965), Oglesby (1976), Ebers and Kopp (1979), Choquette and Steinen (1980), Pierson (1981), Churnet and Misra (1981), Fairchild (1983), and Niemann (1984) are included. Some authors have used several categories for intensity and we have reassigned these to one of the three shown – bright, dull, or none.

prefers to leave the interpretation of these diagrams open and to hope that additional work will better define the CL intensity fields.

Not only are the boundaries poorly defined in Figures 7.12 and 7.13 but it is also difficult even to estimate the shape. Is it a straight line or a curve on a log–log plot? Is it a straight line of constant slope, as proposed by Frank *et al.* (1982)? Is the boundary line horizontal at low Fe concentrations as proposed by ten Have and Heijnen (1985)? A theoretical model might help to answer this but the author is unaware of the existence, at this time, of a model or theory of the manner in which CL intensity should vary with the Fe and Mn concentrations. At least two effects need to be considered – the quenching effect of the Fe^{2+} and the self-quenching or concentration-quenching effect (Dexter & Schulman 1954) of increasing amounts of Mn^{2+}.

Some work has been done on the subject of concentration-quenching or self-quenching of a carbonate containing only Mn as an impurity activator. Medlin (1959, 1960) found a concentration-quenching effect in the calcite samples which he synthesized for ThL studies. The maximum intensity occurred in the range of 0.55 mole percent of Mn. Sommer's data (1972a) clearly show a self-quenching effect in aragonite, the maximum CL intensity occurring at about 1,000 ppm of Mn. Marfunin (1979, p. 174), on the other hand, states that many activator impurities show self-quenching but Mn does not. Each author's results and the composite presentations in Figures 7.12 and 7.13 do not appear to furnish evidence of concentration-quenching of CL up to $MnCO_3$ concentrations of a few percent molar.

Theoretical work in this area would be of value and might establish at least guidelines for the expected shapes of the boundaries between fields of varying luminescence intensity. If concentration-quenching takes place in the manner usually predicted, then it would be expected that the boundary lines would turn back toward the *y* axis at higher Mn concentrations.

All of the data in Figures 7.12 and 7.13 are for natural calcites and dolomites. Some work has also been done with synthetic crystals, an important example being that reported by ten Have and Heijnen (1985). They produced calcite crystals with Mn contents between 4 and 8,646 ppm using two different methods of crystal growth. All of these synthetic crystals had low (less than 157 ppm) Fe contents. The concentrations of all other trace elements which might be effective as activators, quenchers, or sensitizers were minimized. They concluded that 15–30 ppm of Mn is sufficient to activate CL in calcite, and that the CL intensity increased with increasing Mn concentration up to concentrations of about 200 ppm. Some of their synthetic calcite showed CL zonation and they concluded from their experiments and other information that the uptake could be controlled by both the Mn available and the growth rate. They cautioned that the use of CL zonation (cement stratigraphy) can be complicated because it might sometimes reflect growth rate variations, rather than trace element concentration variations.

7.3.4 *Factors which may influence the Mn and Fe contents of carbonates*

If we accept the conclusion that the concentrations of substitutional Mn^{2+} and Fe^{2+} are the most important variables controlling the intensity of CL in most sedimentary carbonates, then we can ask what conditions influence these concentrations. The list is long and includes both physical chemistry and geological factors. Biological selectivity may occur in some organisms also.

PHYSICAL–CHEMICAL FACTORS CONTROLLING THE Mn AND Fe CONTENT DURING THE FORMATION OF THE ORIGINAL SOLID PHASE

These physical–chemical factors are:

(a) the original mineralogy of the mineral;
(b) the concentration of Fe and Mn in the associated solutions from which the carbonate mineral formed;
(c) the partition coefficient for the particular set of conditions of temperature and other variables;
(d) the oxidation states of Mn and Fe in the solution;
(e) possible influence of other trace elements.

The simplest case is that of a major phase such as calcite, forming from a solution which contains both the constituents of the major phase and also a trace element ion such as Mn^{2+}. The extent to which the trace element ion is incorporated in the solid phase, relative to the concentration in the solution, is described by the partition coefficient (McIntire 1963). The partition coefficient is a thermodynamic quantity and its value will depend upon the specific nature of the major phase, the trace element, the temperature, and possibly other variables such as the presence of other trace elements.

The actual concentration of Mn^{2+} in the calcite will, therefore, depend on the value of the partition coefficient and on the concentration of Mn^{2+} in the solution. Depending upon the value of the partition coefficient, the trace element can be either enriched or depleted in the solid phase, relative to its concentration in the solution.

For each possible major phase (e.g., calcite, aragonite, dolomite), and for each possible trace element ion, there is a different value of the partition coefficient. If this value is known, and if the concentration of the trace element is known, the amount which is incorporated can be calculated.

If the system is open, i.e., if there is a continuous supply of solution of the same composition, then some uniform amount of the trace element will be incorporated throughout the major phase and the CL will be homogeneous.

If the system is closed, then the trace element concentration will vary with time as the formation of the major phase proceeds. Depending upon the value of the partition coefficient, the trace element concentration will either increase or decrease with time in some regular fashion, presumably producing a CL intensity which varies from the center to the outer boundary of the crystal (Churnet & Misra 1981).

It must be emphasized that when we speak of the concentrations of the trace element(s), we are really speaking of the concentrations of ions in a particular oxidation state. This is the divalent state in the case of Fe and Mn in carbonates, since these are the specific ions which substitute for Ca^{2+} and control CL. Therefore the quantity which is usually referred to as the redox potential of the solutions is important.

When all of the above quantities are specified, then, in a strict thermodynamic sense, the CL intensity of the resulting solid phase is specified. In the real world, of course, the solutions are more complex, for example, containing many trace element ions, containing different oxidation states of the same element, and possibly forming two or more solid phases simultaneously. Even here, in principle, one can predict the manner in which the trace elements will be distributed as the crystallization proceeds, if the appropriate partition coefficients and concentrations are known and the open or closed nature of the system is defined.

The references by Pingitore (1978), by Brand and Veizer (1980), and by Churnet and Misra (1981) will guide the interested reader into this subject. Much of the work of Medlin which has been referred to frequently in Section 7.1 falls into this category also.

The thermodynamic approach to the estimation of the Mn^{2+} and Fe^{2+} contents implicitly assumes that the process is at equilibrium. In a real situation, equilibrium in the thermodynamic sense may not be closely approached, and the conclusions can be modified. ten Have and Heijnen (1985) have observed that the manganese content of synthetic laboratory-grown calcites can be influenced by the crystal growth rate. They observed luminescence zonation in such crystals due, apparently, to variations in the growth rate.

Meyers (1974) was among the first to consider the use of CL to estimate Mn contents of the original pore solutions. His approach was based in part on the work of Michard (1968). Michard's experimental work on the coprecipitation of Mn^{2+} in carbonates indicated that the Mn content of calcite, precipitated from normal sea water, could not exceed 10–50 ppm. Meyers considered this, and the data on the Mn contents of natural nonmarine waters, and suggested that 'shallow groundwater and surface meteoric waters are more likely media for precipitating Mn^{2+} bearing calcites (i.e. luminescent calcites) than deep subsurface waters, and all three are clearly more likely than normal sea water.' At the time of this paper, it was still believed that an Mn concentration in the neighborhood of 1,000 ppm was required to cause bright CL in calcite. As noted in Section 7.3.3, this figure has now been reduced to less than 100 ppm for iron-free calcites and therefore his conclusions may need modification.

In connection with the precipitation of luminescent calcites from sea water, we note that Adams and Schofield (1983) present what they believe is the first published example of a luminescent marine cement. Their particular example was somewhat unusual, the cement being associated with the wreck of a ship containing metal machinery.

GEOLOGICAL FACTORS WHICH MAY INFLUENCE THE Mn AND Fe CONTENT OF CARBONATE MINERALS

It appears that the most important geological factor is the nature of the pore solutions; in particular, their Mn and Fe concentration and respective charge states. There is some agreement that the single most important factor is usually the oxidizing–reducing nature of the solutions, since apparently enough Fe and Mn are nearly always present in natural waters.

A model by Carpenter (Carpenter & Oglesby 1976, Frank *et al.* 1982) suggests that the CL intensity reflects the pH, redox potential, and total sulfur content of the pore fluid. This model assumes that there is always sufficient Mn and Fe available and emphasizes the importance of the oxidation states of Mn and Fe in the different regions of the Eh–pH field. At high oxidizing potentials, Mn is in the 3+ or 4+ state, and Fe is in the 3+ state. These ions are strongly excluded from the carbonate lattice and the characteristic CL is not produced. At somewhat lower oxidation potentials, the Mn is reduced to Mn^{2+} and can substitute for Ca^{2+}. However, the Fe remains in the 3+ state and is not incorporated as a quencher. The CL intensity is high. At even more reducing conditions, Mn and Fe are both in the 2+ state and both can substitute for Ca, leading to a dull, partially quenched CL. There are some further refinements of the model involving total dissolved sulfur and the precipitation of some of the Fe as sulfides.

The oxidizing nature of the solutions can be associated with the geological condition as illustrated nicely by Freeman's example (Section 7.2 & Plates 7a & b) where the first-stage nonluminescent cement formed under vadose conditions and the later-stage, brightly luminescent cement formed under other conditions. Meyers (1974) also pointed out the absence of CL in known vadose cements.

Grover and Read (1983) made a detailed study of regional cementation patterns in Middle Ordovician carbonates in Virginia. CL was one of their important tools. Their results supported Meyers' conclusion that CL zones can be used to define regional cementation patterns. Following Carpenter's hypothesis, they say 'The distribution of nonluminescent cement in Middle Ordovician subtidal facies defines the regional distribution of oxidizing portions of the paleoaquifer system' (p. 1,275). They found that carbonates undergoing progressive burial showed a sequence of nonluminescent, bright, then dully luminescing cement, as conditions changed progressively from oxidizing to reducing. Superimposed on the large-scale changes may be frequent small changes leading to more complex CL zoning. A number of black & white photomicrographs of their luminescence results are included.

Another important set of geological conditions might be lumped under the very general heading of hydrological controls, the details of fluid flow through the pores and the transport of the trace element ions in the fluid phase (Meyers 1974). The geometry of the pore spaces and their interconnection will control the movement of the pore fluids and define the open or closed nature of the system. The kinetics of the process of forming the major phase, diffusion in the liquid phase (Pingitore 1982), etc. may play a rôle. There may be situations where the dominant reactions take place in thin film reaction zones (Brand & Veizer 1980), rather than with the bulk fluids. And it seems certain that most of the above conditions would not remain constant, but would change as the cement-forming processes continue.

The original mineralogy, as compared with the present

mineralogy, of the carbonate must also be considered. As noted, for a given set of conditions (solution composition, temperature, etc.), the amount of Fe^{2+} and Mn^{2+} incorporated in the solid phase will depend upon the nature of the solid phase being formed. As an example, aragonite normally incorporates much less Mn than does calcite (Bencini & Turi 1974). If a particular present-day calcite formed from a precursor aragonite, then this calcite might have a relatively low Mn content and a correspondingly weak CL. The CL intensity of the calcite would not correctly reflect the composition of the original solutions. Pingitore (1982) has presented a detailed analysis of the manner in which a trace element such as Mn may be enriched or depleted in such a transformation. Meyers (1974) has interpreted some cements with blotchy CL as 'having resulted from Mn uptake during recrystallization of the original aragonite to calcite.'

GEOLOGICAL FACTORS INFLUENCING THE Mn AND Fe CONTENT SUBSEQUENT TO THE FORMATION OF THE ORIGINAL SOLID PHASE

The luminescence observations provide some evidence that the manganese and iron are introduced into the original cement and their concentrations are not significantly affected by diffusion after the solid phase is formed. This is suggested by the sharp boundaries in the photographs of cements published by Freeman (1971), Meyers (1974), and many others. It is difficult to envision a mechanism for their formation by a later diffusion of manganese in the solid state. The possible rôle of solid-state diffusion in influencing the Mn, Fe distribution has been mentioned but is not considered to be of primary importance (Meyers & Lohmann 1979, Oglesby 1979, Churnet & Misra 1981, Meyers 1978). The diffusion coefficient for Mn, Fe in carbonate minerals is believed to be too small to change the initial distribution, even over geological timescales. The very fine nature of many examples of zoned cements is, in itself, evidence of little solid-state diffusion (Marshall 1975).

Likewise it is difficult to conceive of a model which would lead to the formation of the zones by solid-state diffusion at some time after the original carbonate cement had been formed.

RELICT STRUCTURES

Cathodoluminescence provides further interesting evidence of the limited mobility of impurities. Smith and Stenstrom (1965) pointed out the use of CL to detect relict structures. Apparently in some recrystallization or other alteration processes the trace elements are not appreciably mobilized and the CL image approximates the original trace element distribution.

Another example of this is the altered and unaltered Bryozoans studied by Ferrigno and Walker (1973) and shown in Plates 9a–d. This is a clear case where information on original wall structure is preserved in the CL image, implying that the impurities have not been appreciably mobilized.

Kopp (1981) has presented an example of a hidden breccia (Plates 9e & f). This rock was apparently fractured at an early stage in its history and the fractures were filled with carbonates whose CL differs from that of the host. Subsequent recrystallization, without appreciable impurity redistribution, converted it to the present dolostone with uniform fabric. Without the use of CL, the history could not be unraveled so easily. Another of Kopp's examples (Plates 9g & h) is an oolite, barely discernible with TrL or PoL, but obvious with CL.

7.3.5 *The use of CL to identify temporal uniformity and to develop time relationships between cementation events*

Cathodoluminescence observations show that some cement zones and pore fillings are uniform in their CL characteristics, implying that they formed under conditions which remained uniform over a period of time. In many other cases, however, the CL observations provide immediate evidence of the presence of many zones of varying luminescence intensity (e.g., Plates 8a & b), indicating that the carbonate mineral has been formed under varying conditions. Some of these examples of zoned carbonates, with luminescence, are spectacular and for most there may be little or no hint given of the existence of the zonation by other commonly used techniques.

As an example of the possible complexity, in one of the very early papers Sippel and Glover (1965) published a photograph of a calcite infilling in dolomite which had 50 distinguishable growth bands.

The relevant conditions which may be changing are those discussed in earlier sections, i.e., the amounts of Fe and Mn available, the oxidizing–reducing nature (Eh) of the fluids, the temperature, and possibly the original mineralogy. It is generally agreed that the variations in CL are associated primarily with the trace element concentrations and the Eh value of the original solutions and that they provide a picture of the changes in pore water chemistry during the process of formation (see, e.g., Meyers 1978, Frank *et al.* 1982). Sometimes this chemistry is constant for long periods; sometimes it undergoes many changes.

Although most authors associate the zonation of CL with chemistry or oxidation state variations, we should note that ten Have and Heijnen (1985) observed zonation in some of their synthetic calcite crystals and were able to associate this zonation with crystal growth rates.

One of the puzzles is the observation that the boundaries between CL zones in cements are usually so distinct. Meyers (1978) has mentioned an example of a gradational transition, but the author is unaware of other published examples which illustrate a smooth, graded transition from one CL intensity level to another in a cement. Does this mean that pore solution chemistry conditions always change quickly compared to the time scale of actual cement formation? Or is there some other mechanism which can produce this type of pattern? Or are there other observations of examples of continuous transitions from one CL intensity level to another? The answers to these questions may have important geological implications.

When CL observations lead to the conclusion that the amounts of Fe^{2+} and Mn^{2+} in the lattice vary, this also raises the question about what other trace elements may also be fluctuating in concentration. Mg and Sr concentrations have been measured in a few instances but, to the author's knowledge, little or no work has been published yet on the subject of variations in other trace element concentrations in regions of differing luminescence.

The many examples which have been published of multiple alternating luminescence zones also illustrate the more general problem of measuring the abundance of manganese (and probably also iron and other trace elements) in such samples. The distribution of these elements may be very nonuniform and measurements of bulk samples represent only a grand average.

Similarly, one may ask about the variations of stable isotope ratios in the different zones. With the development of techniques for sampling small regions (Prezbindowski 1980), and mass spectrometric techniques suitable for very small samples, answers to some of these questions are beginning to come forth. CL is used to identify the individual generation of cement, a photographic map is made, and then a small sample is taken from each cement generation of interest for mass spectrometric analysis (Dickson & Coleman 1980, Moldovanyi & Lohmann 1984, Meyers & Lohmann 1985).

The correlation in time of events separated geographically is a technique often used to solve geological problems. CL patterns (luminescence signatures) may offer another means to develop these correlations.

The paper by Meyers and Lohmann (1985) is an example of a comprehensive approach utilizing CL, trace element analysis for Mg, Fe, and Mn with the electron microprobe, and stable isotope (C, O) analysis of individual limestone components in Mississippian limestones from New Mexico. From this study they developed a model for the cement zones forming at different times within chemically distinct pore water systems. They were able to develop 'time-sequence reconstructions of the chemistries and flow directions of paleo-groundwater systems that precipitated four major regionally extensive calcite cement zones in Mississippian skeletal limestones of southern New Mexico' (p. 223). Seven compositional zones were distinguishable with CL. Four of these were volumetrically important and could be correlated regionally.

In an earlier study, Meyers (1978) found that four major cement zones, which were identified by CL and other techniques, could be correlated regionally over an area of approximately 30,000 km^2, indicating that each zone is an 'approximate "time-stratigraphic" unit' (p. 381).

Ebers and Kopp (1979) found that the gangue dolomite in the Mascot-Jefferson City zinc district exhibits CL zonation which could be correlated throughout the district (approximately 20 × 60 km). They also demonstrated that sulfides were formed at one particular stage of dolomite formation throughout the district. They suggest the more general use of CL as an exploration tool for some Mississippi Valley-type ore deposits. Gregg (1985) has also used correlations of dolomite cement CL microstratigraphy in a study of dolomitization in the Bonneterre Dolomite in southeastern Missouri.

Harwood (1983) used CL to obtain evidence of the relative ages of barite, calcite, and anhydrite crystal growth in the Lower Magnesian Limestone.

We should note that some investigators have mentioned the difficulty in making such correlations, even over restricted small regions. Part of the resolution of these diverging conclusions may be associated with the degree of the open or closed nature of the particular diagenetic environment.

In summary, then, CL has proved useful to tell if conditions at a given location have been constant. In the case of changing conditions, CL has sometimes proved useful as a stratigraphic indicator over large geographical areas.

7.3.6 *Studies of crystal growth and fabric*

CL can often be a useful tool to determine the nature of crystal growth. As Frank *et al.* (1982) have said, 'A boundary between two successive zones reveals the shape of the crystal at a particular time in its history of growth. Irregular boundaries record stages of growth interrupted by some degree of solution of earlier cement; that is, the irregular boundaries mark solution disconformities within the cement' (p. 634).

CL is particularly useful in situations such as completely filled voids where the pattern may not be visible with TrL and PoL (Miller 1983). If there is zoning of the type discussed in Section 7.3.5, the zonation will reveal the shape of the crystal at any stage during its growth. The example of Freeman in Plates 7a and b is an illustration of this.

Dickson (1983) has carried out an elaborate study of the detailed manner of crystal growth in carbonate cements. He used graphical modeling to develop crystal growth diagrams for various assumed conditions. He provided supporting examples taken from natural samples in which the growth pattern was nicely revealed by the presence of zones of bright luminescence alternating with dull or nonluminescent zones. One of his examples is shown in Plates 7e and f. The correlation of the luminescent zones from one crystal to the next enabled him to develop the time relationships of the growth of several crystals (up to eight in one example) in detail.

Reeder and Prosky (1986) have also used CL quite effectively in the study of compositional sector zoning in dolomite. They were able to observe luminescence differences between some sectors which correlated with easily measurable Fe compositional differences between the different growth sectors. Slight differences in Mn were also noted. Their paper has a detailed discussion of crystal growth in dolomite.

The CL technique is less helpful if the cement is nonluminescent or the CL is uniform. Sibley (1982) for example, tried to use CL to establish whether examples of sucrosic dolomite grew on detrital dolomite cores, but an attempt to find evidence for this was unsuccessful because the sucrosic dolomite had uniform luminescence.

7.3.7 *Applications to fossiliferous samples*

The CL technique is also useful in studying fossiliferous rock samples, often providing good resolution of fine details. In many cases, the outlines of fossils and their internal structure are quite clearly revealed by CL, often much better than with TrL or with PoL, and it would seem that there is a wealth of useful information to be derived from its use. However, to date, a surprisingly small amount of work in this area has been published by paleontologists. Of course many observations have been made as a by-product of other studies of carbonate rocks and their cements.

Smith and Stenstrom (1965) showed examples of fossil crinoids and indicated some possible uses of CL to distinguish original cell structures in samples which had undergone partial or complete recrystallization. They believed that crinoids exhibiting monotonous CL had been recrystallized and the minor element distribution was evened out in this process. They also suggested that CL might be a useful technique for the selection of fossils for isotopic or other studies where it might be important to know the extent of recrystallization.

Popp *et al.* (1983, 1986a, b) have discussed this question of sampling of fossils for isotopic analysis in more detail. They stressed that some apparently well preserved specimens are seen with CL to be altered. Therefore this type of evaluation is useful to insure that one is truly sampling the desired fossil component and not simply performing something more analogous to a whole rock analysis because of such an alteration.

Martin and Zeegers (1969) pointed out that bioclasts of small size are sometimes distinguishable from their matrix only with the use of CL.

Sommer (1972b) found both luminescing and nonluminescing aragonite in biogenic carbonate 'shells' and found that CL is a useful tool to assist in distinguishing calcite from aragonite in shells and distinguishing regenerative material from normal shell material.

Ferrigno and Walker (1973) have studied suites of bryozoans which ranged from unaltered to badly altered specimens. Either TrL or CL gave clear photos of unaltered specimens (Plates 9a & b). However, they found that the structures in badly altered specimens were better revealed by CL than by TrL (Plates 9c & d).

Miller and Clarkson (1980) studied the development of the trilobite eye. In all of their specimens, the primary structure had been partially or wholly obliterated by diagenesis. Etching of the sample helped their TrL observations but also produced partial dissolution of some features. They were able, however, to use CL to confirm the internal structure which was partially, but less effectively, revealed by etching. They surmised that the manganese ions in the lattice kept their relative concentration during neomorphism and therefore, even after extreme recrystallization, the eye structures could be revealed by CL (Plates 8e & f).

Faupl and Beran (1983) used CL to help reconstruct the diagenetic history of radiolaria capsules and sponge spicules. Nonluminescing dolomite and two stages of calcite, one with dull CL and one with bright orange CL, were observed.

James and Klappa (1983) presented a comprehensive discussion of components of reef limestones, and their paper included a well reproduced color plate showing the CL characteristics of components of the reef limestone such as sponges, echinoderms, etc. One of their examples showed how well the spar filling in an ooid is illustrated by CL.

Stow and Miller (1984) have used CL as one of their tools for examining samples from the Deep-Sea Drilling Project. They found many cases where bioclasts (foraminifers, echinoderm fragments) were much more apparent with CL than with TrL.

Another very important application is the detection of fossils in recrystallized rocks. Smith and Stenstrom (1965) and Sippel and Glover (1965) have pointed out that there are cases where fossils could be recognized with CL in recrystallized rocks, whereas they are not visible with PoL.

Similarly Friedman (1975) has noted in certain cores from the Red Sea that 'tests of *Globigerina* have been progressively micritized with depth and ultimately "vanish" (but are still recognizable as ghosts in thin section under an electron beam). . . .' (p. 392).

7.3.8 *Calcite containing magnesium*

Discussion has appeared in the literature on the possible influence of Mg content on the CL spectrum and presumably also on the CL intensity of magnesium-containing calcite. Sommer (Section 7.1.4) observed an increase in the wavelength of the peak of the CL emission spectrum from Mn-activated calcite with increasing Mg content.

Amieux (1982) suggests that the CL 'color' of calcite (as well as that of other carbonates) varies with the amount of Mg. He also discusses the possibility that Mn has a valence of more than two in some samples and that the CL color tends towards the red in this case. Both of these important suggestions will require much work, e.g., with synthetic samples, to elucidate further.

Schrank and Friedman (1975) have commented that in their sub-bottom Red Sea carbonate sediments, the 'high-magnesian calcite contains enough manganese to luminesce but does not glow as brightly as low-magnesian calcite' (p. 3).

Adams and Schofield (1983) reported that magnesian calcite cements containing between 6 and 12 mole percent Mg luminesced bright yellow.

The observations by the various authors of the effect of increasing Mg content on the CL spectrum are not consistent and further work would be of value.

7.3.9 *Marbles*

The amount of published information on CL of marbles is relatively small. It is logical to suspect that the CL of marbles will tend to be rather uniform because they have undergone metamorphism. For some examples this appears to be the case. A specimen of the very fine uniform white marble from Carrara, Italy, for example, shows a uniform dull reddish CL. But in other cases (Sippel & Glover 1965) details in marbles show up with CL, details which may be obscured with TrL or PoL, lending support to the point made in Section 7.3.4 that the impurity distribution does not necessarily homogenize during recrystallization.

Renfrew and Peacey (1968) studied the CL of 43 thin sections of various marbles, using an electron probe operated at 25 kV, in an attempt to establish the source quarries for marbles used in ancient sculptures. They noted various luminescence colors and also some samples with little or no CL. One group had a characteristic CL and the method was considered to be of promise. To the author's knowledge, this potential application of CL has not yet been followed up.

7.4 Summary and recommendations for further research

Numerous other papers concerned with particular problems of cements and diagenesis, and sedimentary petrography in general, are continuing to appear. The detailed nature of the applications cannot all be discussed in this introductory text on CL. The significant point to note is that the use of CL in carbonate studies often provides the following types of information which may be difficult or impossible to obtain otherwise.

(a) CL intensity is an expression of the possible iron and manganese content and may be a guide to deciding whether the sedimentary carbonate was formed under oxidizing or reducing conditions.
(b) CL can sometimes be used to determine the sequence of events and the correlation of parallel events.
(c) The time-relationship of fracturing in relation to other events can often be diagnosed.
(d) The fabric of cements, of ooids, and cavity fillings can be revealed.
(e) The uniformity or nonuniformity of conditions during carbonate deposition can be revealed.
(f) Better details of relict fossils and other features are often presented with CL.
(g) Rapid delineation of calcite and dolomite is often possible.

Despite the present usefulness of CL, there remains important work to be done. Some avenues which seem important to the author include:

(a) What is the rôle of magnesium in modifying the CL properties of calcite? Is the color or intensity of calcite CL affected appreciably by the range of Mg contents found in natural calcites?
(b) What are the other trace element variations in zoned cements? An instrument of high sensitivity like the ion microprobe will probably be needed to determine this.
(c) Does ferric iron ever play a rôle as a quencher?
(d) A more quantitative evaluation of CL intensity versus Mn and Fe content would be valuable. Theoretical work in this area is also needed.

There seems little doubt that the promise of even more important applications of CL to carbonate studies will lead investigators into these studies.

Chapter 8

Some further geological applications of cathodoluminescence

A. N. MARIANO

Cathodoluminescence has proved to be an extremely useful tool in the study of sedimentary rocks, and the three previous chapters have been devoted primarily to this subject. Unsatisfying though it may seem, much progress has been made with it in sedimentary petrology, and a significant number of publications have resulted, without a full and clear understanding of the reasons for the CL of all of the minerals involved. In contrast, the published information on the applications of CL in nonsedimentary situations is much less than that in sedimentary samples and many of the potentially useful applications are known only to a limited number of investigators. The application of CL to these latter situations generally requires a better understanding of the reasons for the activation and quenching of the CL of the many different minerals involved. The studies are also facilitated if the investigator has suitable instrumentation for recording the CL emission spectrum.

In this chapter, Dr Anthony N. Mariano sets forth many examples of what has been done to date in areas other than sedimentary petrology. For many years he has been involved with the exploration, evaluation, and development of beneficiation processes for economic mineral deposits throughout the world. CL has been one of the more important tools that he has used in this work. From the beginning, he recognized the powerful advantages offered by the capability of obtaining CL emission spectra and comparing these with the spectra obtained from doped synthetic minerals with known impurities.

D. J. Marshall

8.1 Introduction

Cathodoluminescence as an analytical tool in geology is becoming more widespread in use. There is a noticeable increase in the number of publications involving the use of CL in mineralogy, petrology, and geochemistry. While the visual observation of CL in minerals is in itself of value, the generation and interpretation of CL emission spectra from these minerals can lead to an understanding of the cause of the luminescence and can provide far more significant information than simple visual observations. The mineral apatite, for example, often shows luminescence colors in the yellow region of the spectrum, but different apatites exhibiting yellow CL can have different activators and significantly different emission spectra. Mn^{2+} activation in apatite gives a broad-band emission that peaks between approximately 562 and 580 nm, yielding a yellow-green CL. A very similar color can result from line spectra due to Dy^{3+}, with or without the presence of Mn^{2+}. Visual observation of yellow CL in apatite is not sufficient to identify the activators responsible for that luminescence. For the most part, apatite crystallizing from a particular magmatic or hydrothermal event will show a specific CL emission spectrum that is a reflection of the physicochemical conditions that characterized the event.

The presence of overgrowths and lanthanide clustering in apatite are brought out by visual observation of the CL in this mineral. These features are not evident in normal TrL petrographic observation because the impurities that are responsible for CL colors are present only in trace amounts. These features are also difficult to identify by chemical analysis because the lanthanide activators occur in the 0.0*x* weight percent level or less, below the detection limits for the electron microprobe, the most common instrument choice for this type of analysis.

The CL information and conclusions presented in this chapter are heavily supported by CL emission spectra as well as by color plates. The emission spectra were taken with the basic arrangement described by Mariano and Ring (1975). In this arrangement, a fiber optics light pipe has its input on the objective focal plane of an optical microscope with high light-gathering power. The particular microscope used has an objective lens of 2.5× magnification and a numerical aperture variable up to 0.25. The light pipe normally used has an input diameter of 1,200 μm. This diameter at the objective focal plane, for an objective magnification of 2.5×, corresponds to a diameter of 480 μm on the sample. The CL emission from this selected area passes through the light pipe and into a grating monochromator which can be used over the range of 400–1,150 nm. The detector range is between 380 and 760 nm. The acquisition of CL spectra at shorter wavelengths would require the use of quartz optics and other modifications.

8.2 Determination of LREE versus HREE dominance of an environment from CL spectra

The lanthanides are usually divided into two groups: the light lanthanides, also known as the cerium sub-group, and the heavy lanthanides, also known as the yttrium sub-group. The division between these two groups is somewhat arbitrary but an even division giving the LREE as La–Gd and the HREE as Tb–Lu is probably workable for purposes of CL studies because, in HREE minerals, Tb^{3+} is often the most efficient activator among the heavy lanthanides, whereas in LREE enriched minerals Tb^{3+} activation is either very weak or not present.

8.2.1 Apatite

Apatite ($Ca_5(PO_4)_3(F,Cl,OH)$) is probably the most ubiquitous accessory mineral in all rocks and therefore an understanding of its trace element chemistry is useful. Apatite is also an important synthetic phosphor host and can accept many different elements as activators. Extensive information on the CL of doped synthetic apatite is available in the literature of the phosphor industry. The CL spectra of these different activators in geological samples are distinctive and can provide immediate information pertinent to the petrology and geochemistry of the rock units hosting the apatite.

Phosphate minerals have a strong affinity for lanthanides and the mineral apatite is an effective getter for REE elements in a crystallizing system. The information obtained from studying apatite CL rapidly indicates whether lanthanides occur in any quantity in the rocks, and the LREE versus HREE determination can be quickly established. This type of information is otherwise usually obtained by elaborate instrumental analyses including neutron activation, spark source mass spectrometry, or X-ray fluorescence, each of which requires prior chemical manipulation procedures as well as isotope dilution or solvent extraction processes. In contrast, CL emission spectra quickly provide similar information on the presence and distribution of REE and the sample preparation is simply the fabrication of thin sections or even the direct examination of rock fragments.

A pure apatite crystal without activator impurities or defects exhibits no CL* but virtually all naturally occurring apatites exhibit CL. The luminescence is due to impurity elements which function as activators in the apatite host. The activators which the author has identified in natural apatites include Mn^{2+}, Sm^{3+}, Dy^{3+}, Tb^{3+}, Eu^{2+}, and, rarely, Eu^{3+}. The luminescence of apatite containing any of these activators is so efficient that even the most minute crystals, as small as 20 μm in diameter, may be vividly revealed. This characteristic of apatite is an asset to the petrographer because, even though apatite is one of the most prevalent accessory minerals in rocks, it frequently occurs as micrometer-size crystals whose recognition is often difficult owing to intimate association with other minerals (Plates 2e & f).

Synthetic apatite is the main host material for phosphors used in fluorescent lights and as a result extensive research has been carried out by the phosphor industry on the CL of apatite with variations in host composition and many different activators and co-activators. In addition, rare earth-activated fluorapatite has been synthesized for solid-state laser crystals. CL emission spectra for most of the lanthanides in apatite are well documented in the phosphor and laser industry literature. Interpretation of CL emission spectra from natural apatites has been facilitated by correlation with this extensive amount of information on synthetic apatites.

In nature, apatites are represented principally by the formula $Ca_5(PO_4)_3F$. The possibilities for chemical variation are great and include substitutions for Ca, P, and F. Minor replacement of Ca in natural apatites is accounted for mostly by Sr, Mn, and the rare earth elements. In some rare cases substantial amounts of Na^+ may accompany REE substitution for Ca^{2+}, thus satisfying the charge balance requirements. Generally, other elements replace Ca in rock-forming apatites in only trivial amounts. Fluorine is replaced mainly by Cl and OH. Replacement of F by O or vacancies is relatively restricted. Probably the most common substitution for P is Si with a coupled atomic substitution for Ca by trivalent rare earth ions. Smaller amounts of As, S, and V also replace P in natural apatites.

Fe^{2+} is an effective quencher of luminescence in most hosts, even in the 1.0 weight percent range, and presumably behaves similarly in apatite. The fact that virtually all natural apatites cathodoluminesce indicates that very limited amounts of Fe^{2+} substitute for Ca in the structure. Nonluminescing (CL) apatite in nature is rare since most apatites contain at least sufficient amounts of Mn for activation. A relatively pure apatite can show activation from 0.0*n* weight percent Mn. For example, a pegmatite apatite from Palermo, New Hampshire (see Fig. 8.2) exhibits an intense CL emission peak with only 240 ppmw Mn. Alpine-type apatite from Val Cristallina, Switzerland, gave similar luminescence with only 360 ppmw Mn.

MANGANESE

Cathodoluminescence emission spectra for Mn^{2+} in apatite hosts consist of a broad band located in the yellow-green part of the spectrum. The wavelength of emission of Mn^{2+} activation in apatite is determined by the various energy levels of the Mn^{2+} in the crystal field of the host. Consequently, the wavelength of the emitted light is considerably influenced by the atomic species and configuration around the manganese ion. Owing to variations in apatite

* Synthetic apatites with no added impurities were examined by CL and found to show no luminescence. They include: (a) hydroxyapatite synthesized by Walter Eysel of the Materials Research Laboratory, the Pennsylvania State University, University Park, PA, 16802 (received 1972), and (b) General Electric Lamp Division standard synthetic apatites, received from R. A. Young of Georgia Institute of Technology, Atlanta, GA, 30332. The samples are fluorapatite: $Ca_{10}(PO_4)_6F_2$ and chlorapatite: $Ca_{10}(PO_4)_6Cl_2$.

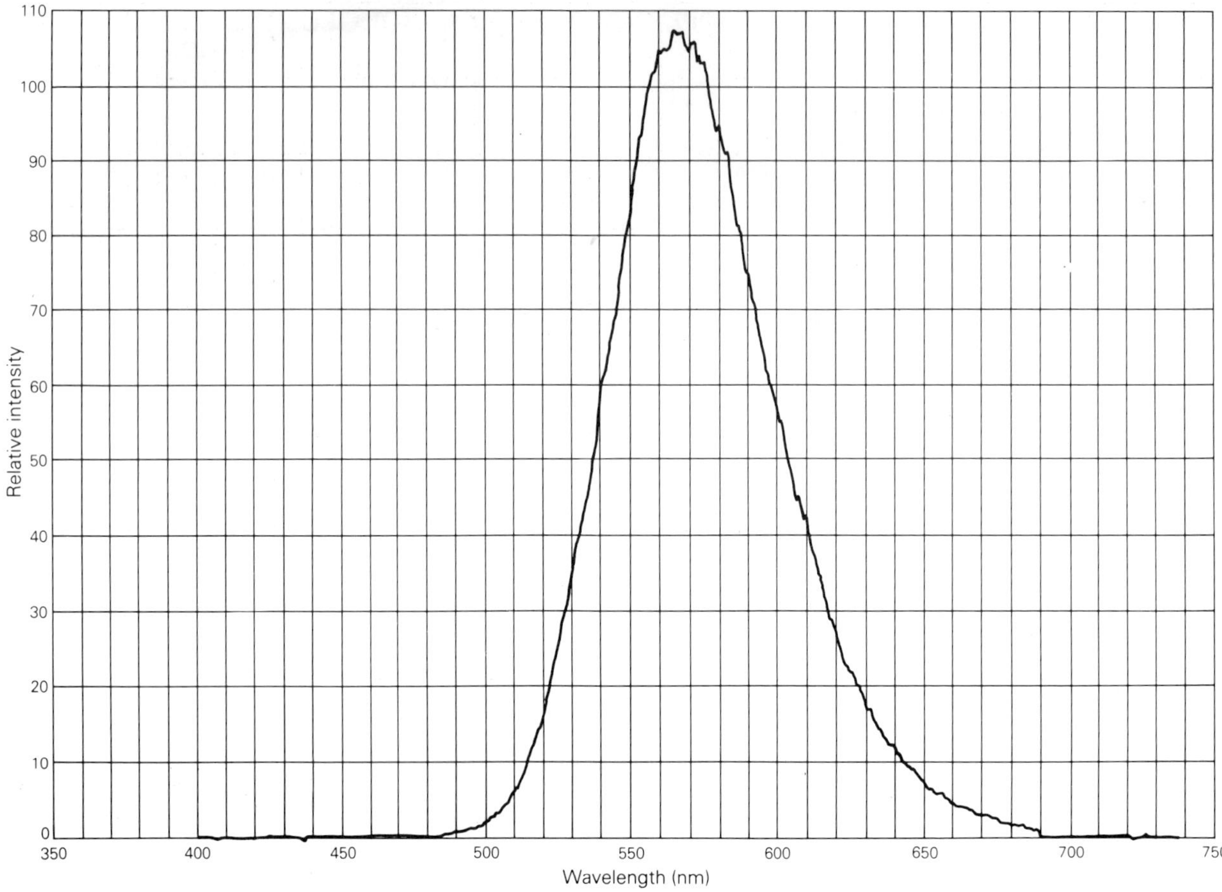

Figure 8.1 Synthetic Mn^{2+}-activated fluorapatite with 1.09 weight percent Mn (sample furnished by T. L. Barry, GTE Sylvania). Peak emission at 565 nm. CL is yellow. CL conditions: 7 kV, 0.15 mA.

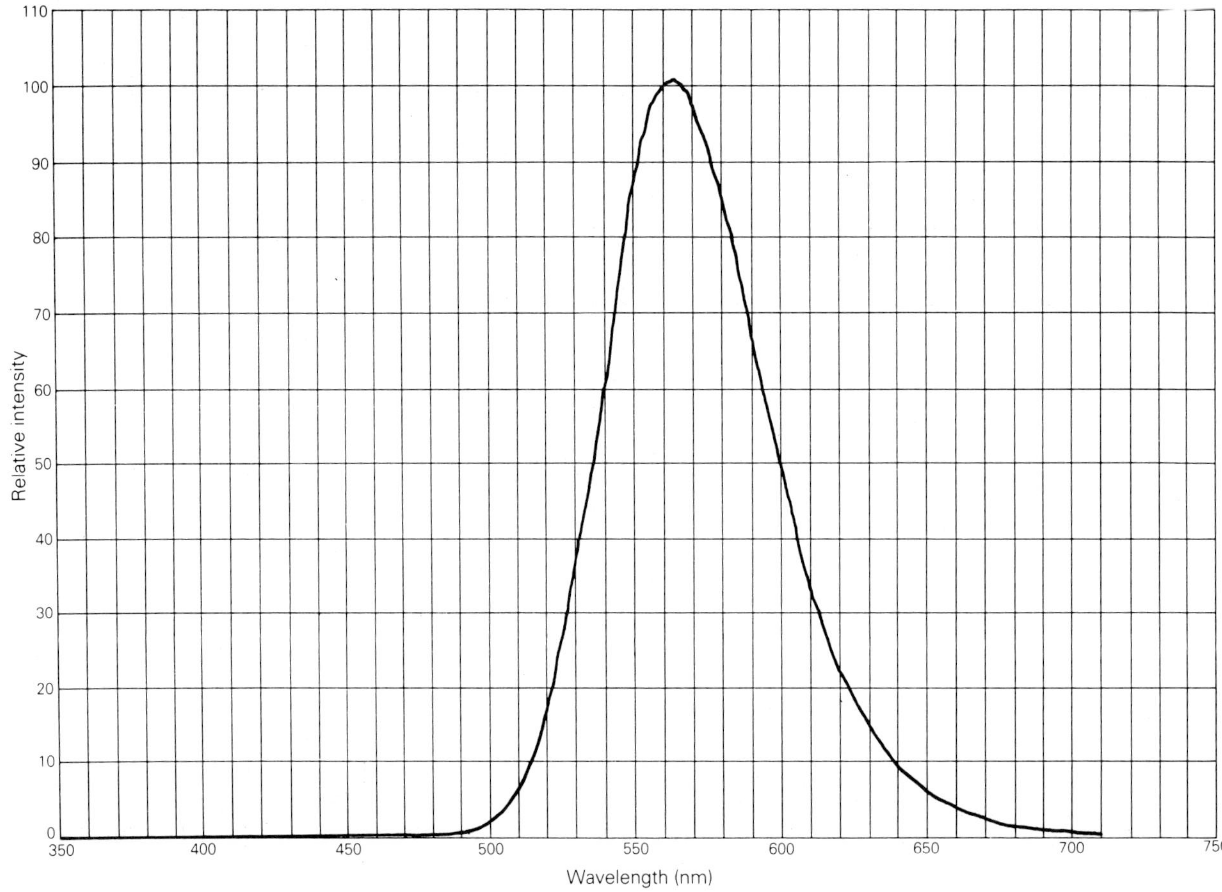

Figure 8.2 Mn^{2+}-activated pegmatite apatite from Palermo, New Hampshire. Mn content is 240 ppmw. Peak emission at 565 nm. CL is yellow. CL conditions: 13 kV, 0.10 mA.

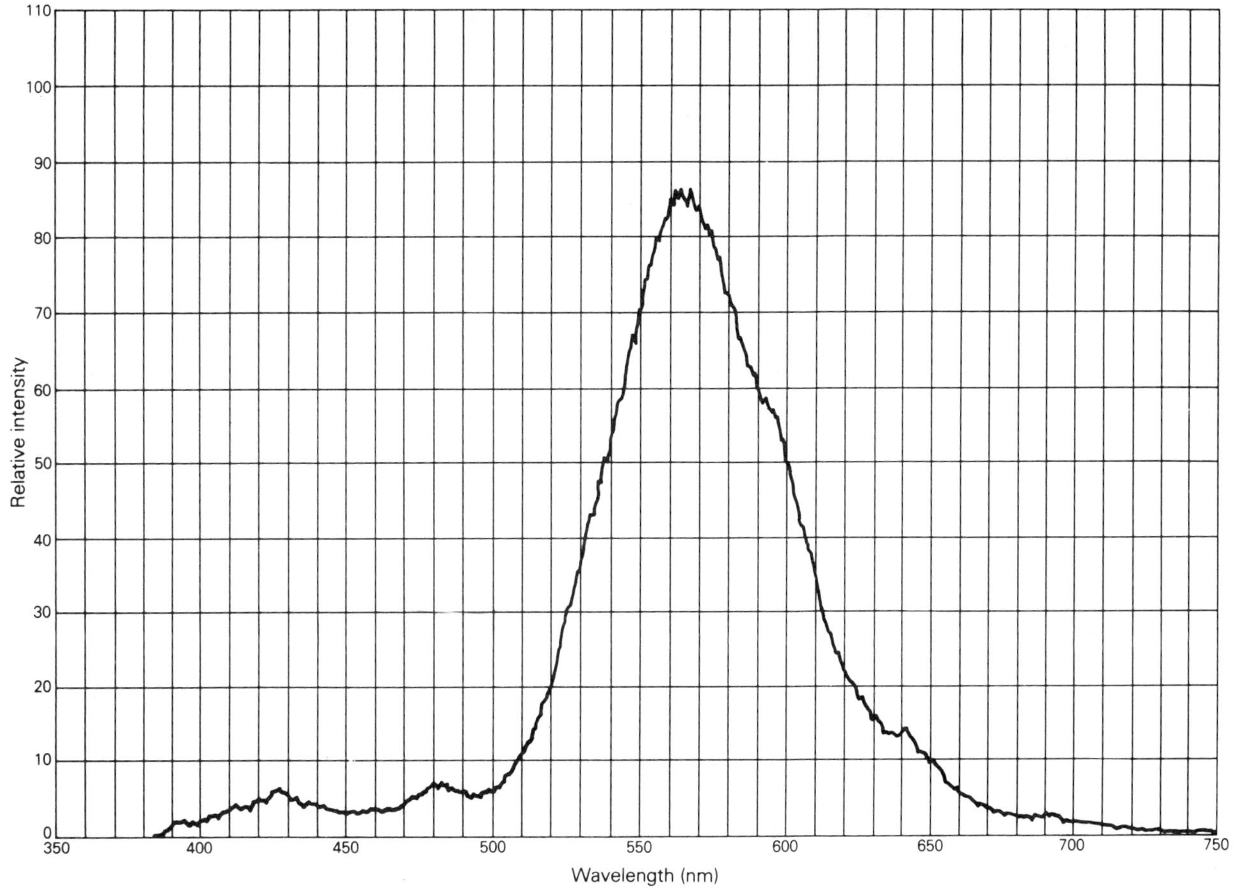

Figure 8.3 Hydrothermal apatite, Mn^{2+}-activated, surrounding molybdenite from Alice Arm, British Columbia. Peak emission at 570 nm. CL is yellow. Both Sm^{3+} and Dy^{3+} peaks are observable, superimposed on the Mn^{2+} spectrum. CL conditions: 9 kV, 0.3 mA.

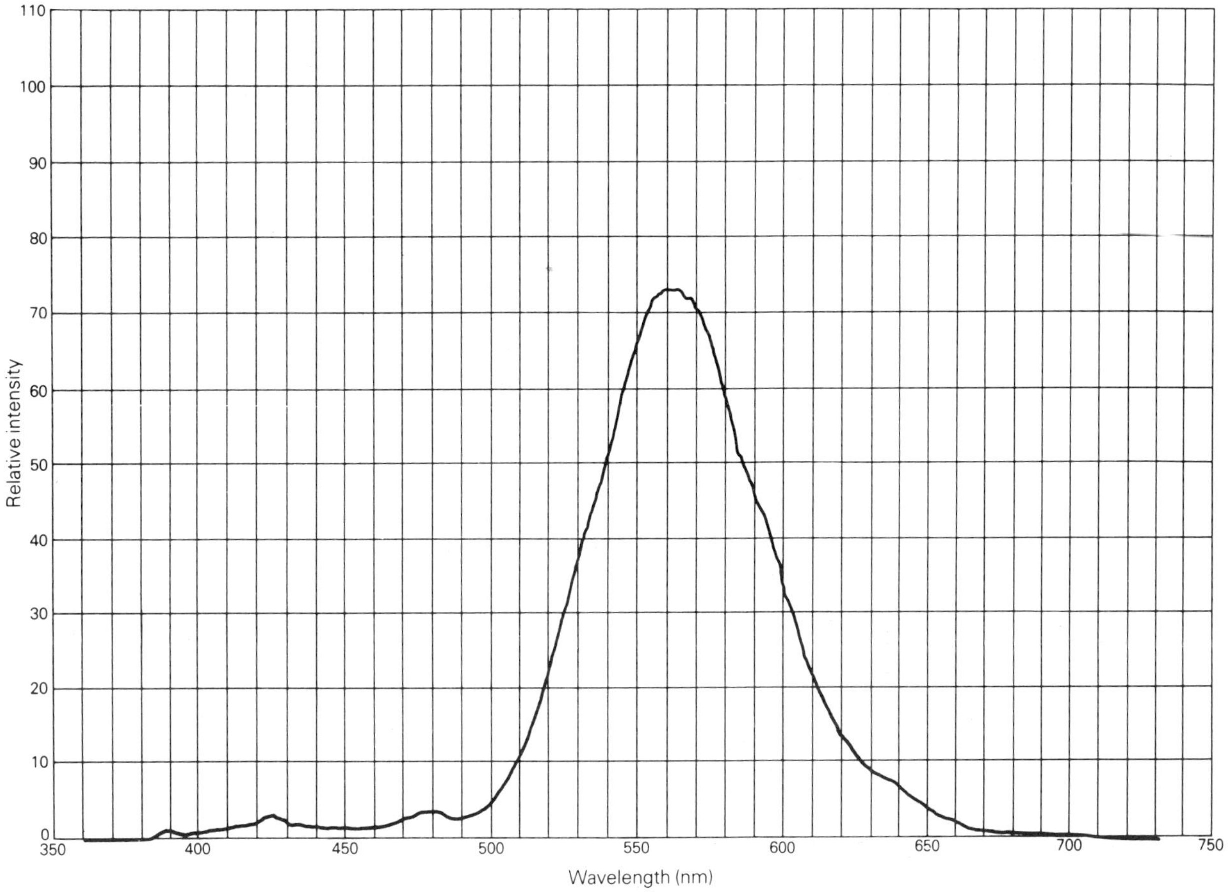

Figure 8.4 Metamorphic apatite from a kyanite quartzite, Ogilby, Imperial County, California. The principal activator is Mn^{2+}. Peak emission at 562 nm. CL is yellow. CL conditions: 15 kV, 0.8 mA.

chemistry, the peak emission from manganese in this mineral varies between approximately 562 and 580 nm, with the average at 570 nm. The CL emission spectrum for synthetic fluorapatite with 1.0 weight percent Mn is shown in Figure 8.1 and can be compared with the spectra for Mn^{2+}-activated apatite from a pegmatite (Fig. 8.2), hydrothermal mineralization (Fig. 8.3), and a metamorphic rock (Fig. 8.4). All four show nearly identical spectra with predominant Mn^{2+} activation and peak emissions in the 560–570 nm range.

LANTHANIDES

There are extensive data in the literature concerning the lanthanides as PhL and CL activators in apatite (see, e.g., Morozov *et al.* 1970). Divalent Eu produces a broad-band CL spectrum with a variable emission wavelength because the interaction with the crystal field of the host mineral is important (Section 1.3.3). In contrast, the trivalent REE activators each emit several narrow bands whose wavelength is only slightly affected by the crystal field of the host. This is true when they substitute not only in apatite but also in other minerals including fluorite, calcite, strontianite, anhydrite, scheelite, zircon, etc. The trivalent lanthanides Sm^{3+}, Tb^{3+}, and Dy^{3+} each have a unique set of line emissions. The wavelengths and relative intensities of the lines for Tb and Dy are essentially the same, even in different host lattices. The relative intensities for the Sm^{3+} lines at approximately 600 and 645 nm vary in some hosts. As a result, the identification of these elements, when they are activators in different minerals, is unambiguous. Trivalent europium spectra from different hosts show some variation but in this work Eu^{2+} and Eu^{3+} emission in the mineral apatite is correlated with synthetic apatite containing these activators.

Samarium and dysprosium

In apatite Sm^{3+} produces narrow bands at 600, 645, and 710 nm, with the relative intensities being $600 > 645 > 710$. Another band from Sm^{3+} occurs at 560 nm but it is frequently masked by the major Dy^{3+} line at 575 nm. The narrow bands from Dy^{3+} in apatite are at 480 and 575 nm. The intensity of the 575 band is about twice that of the 480 band (Figs. 8.5 & 8.6).

In general, the peak heights in lanthanide CL emission spectra are not directly related to relative concentrations of the elements because the luminescence efficiency varies for each element and other impurities can either enhance or suppress the luminescence. However, in a study by Buchanan *et al.* (1967) on rare earth-activated yttrium oxide with 0.3 weight percent Sm^{3+} and 0.3 weight percent Dy^{3+}, the efficiency of emission for these two activators was found to be approximately the same. In the case of apatite, the CL intensities of the strongest lines of Sm^{3+} and Dy^{3+} can be used as an indication of their relative concentration since they occupy equivalent crystallographic sites.

Terbium

The CL emission spectrum of the Tb^{3+} ion is a series of very narrow bands. The strongest narrow band at 545 nm is considerably more intense than the others. The Tb content of most naturally occurring apatites is quite low but the efficiency for luminescence of the terbium ion is very high. As a result, the strongest emission peak at 545 nm is frequently observed in apatite. For example, an apatite from Mineville, New York, with 70 ppmw Tb (Fig. 8.5) and an apatite from the Kiglapait Intrusion, Labrador, with 38.4 ppmw Tb (Fig. 8.6) both show a well resolved Tb^{3+} emission at approximately 545 nm because the Mn content is low and the Tb^{3+} line is well resolved from the Dy^{3+} and Sm^{3+} lines. The successful identification of Tb^{3+} from CL spectra in lanthanide-activated apatite is dependent both on the concentration of the terbium and the presence of relatively low quantities of Mn^{2+}. When Mn^{2+} is present in high quantities its CL emission masks the Tb^{3+} emission.

Europium

Europium shows two types of activation in the mineral apatite, one from Eu^{2+} and the other from Eu^{3+}. Eu^{2+} produces a broad-band emission that has a variable emission maximum wavelength in the blue part of the spectrum. The wavelength of the emission maximum is strongly affected by the composition of the apatite. Eu^{3+} is infrequently observed in apatite but when present it produces narrow-band emissions in the red part of the spectrum with wavelengths at 590, 615, and 695 nm. The 615 peak is usually the most intense.

Blue CL in apatite (e.g., Plate 10b), possibly due to Eu^{2+} activation, is most frequently encountered in carbonatites and some alkaline igneous rocks. It is usually characteristic of primary apatite mineralization. Often in carbonatites several generations of apatite are present in the same rock. These apatites may sometimes be differentiated by morphology or mineral association, but they are more effectively differentiated by CL. The CL emission spectra are specific for each generation of apatite.

Trivalent europium activation is less common in apatite but it has been observed in the carbonatites of Tapira, Minas Gerais, Brasil; Oka, Quebec; Magnet Cove, Arkansas; and Blue River, British Columbia (Mariano 1978).

Apatite from Llallagua, Bolivia (Smithsonian #114,175) shows Eu^{2+}, Eu^{3+}, and Mn^{2+} emission. Figure 8.7 is the spectrum from Llallagua apatite and this can be compared with a synthetic apatite that has been doped with Eu^{2+} and Eu^{3+} impurities (Fig. 8.8). Figure 8.9 is also from a Llallagua apatite when it is excited with UV instead of CL. In this instance, the Mn^{2+} emission band which appeared with CL is not present, indicating that the europium is not functioning as a sensitizer for Mn^{2+} luminescence (see Section 1.3.3).

The appearance of Eu^{2+} and Eu^{3+} in the Llallagua apatite establishes that europium occurred in two valence states in the source solutions, and the presence of europium as the only lanthanide activator suggests that a positive europium anomaly occurs in the apatite lanthanide distribution. This is contrary to the more frequently obtained REE distribution analyses for apatite which usually show negative Eu anomalies (Roeder *et al.* 1987).

Emission spectra from apatite from varied geological environments have been published by Portnov and Gorobets (1969; PhL), Ovcharenko and Yur'ev (1971; probably PhL), Mariano and Ring (1975; CL), and Mariano (1978;

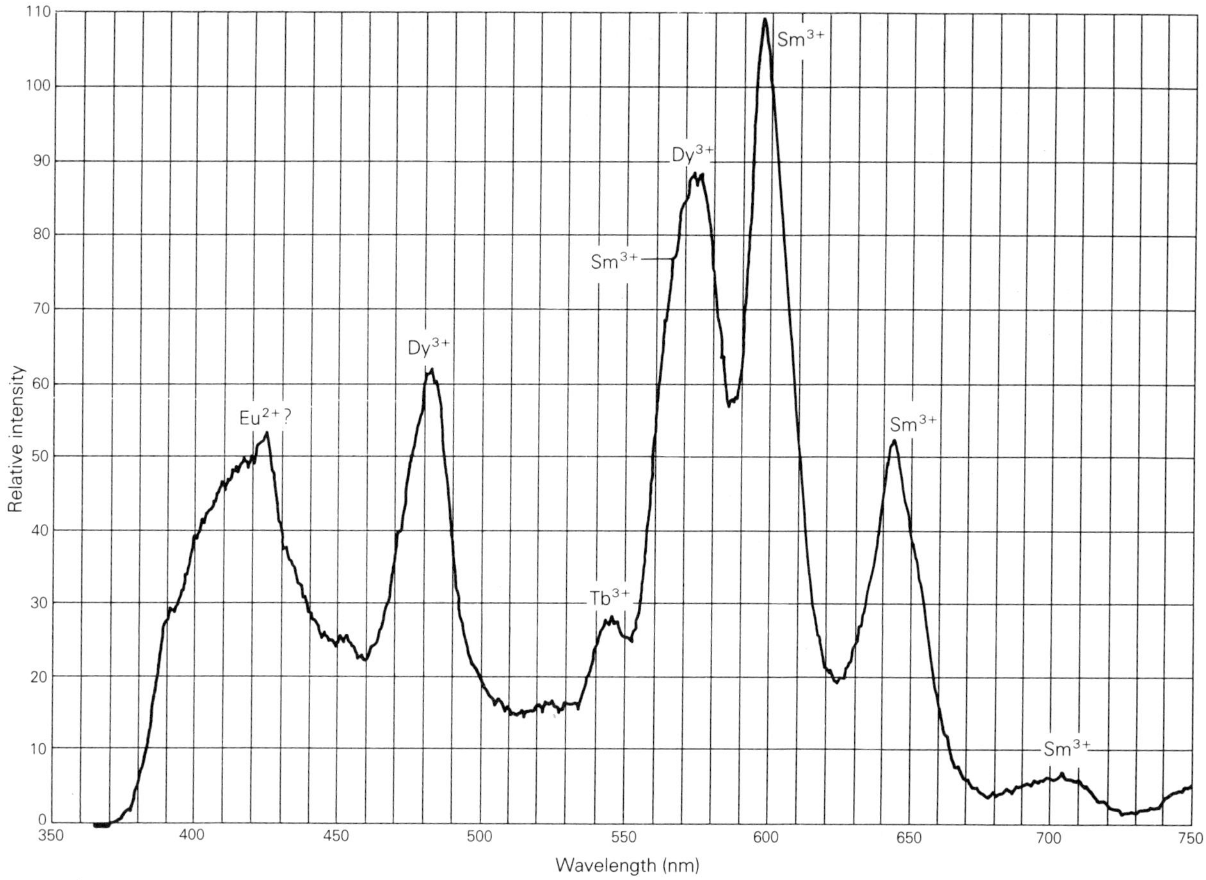

Figure 8.5 Apatite from Mineville, New York. Sm = 500 ppmw, Eu = 50 ppmw, Tb = 70 ppmw, Dy = 440 ppmw (analyst: L. Miller Arnold). CL is violet. CL conditions: 17 kV, 0.15 mA.

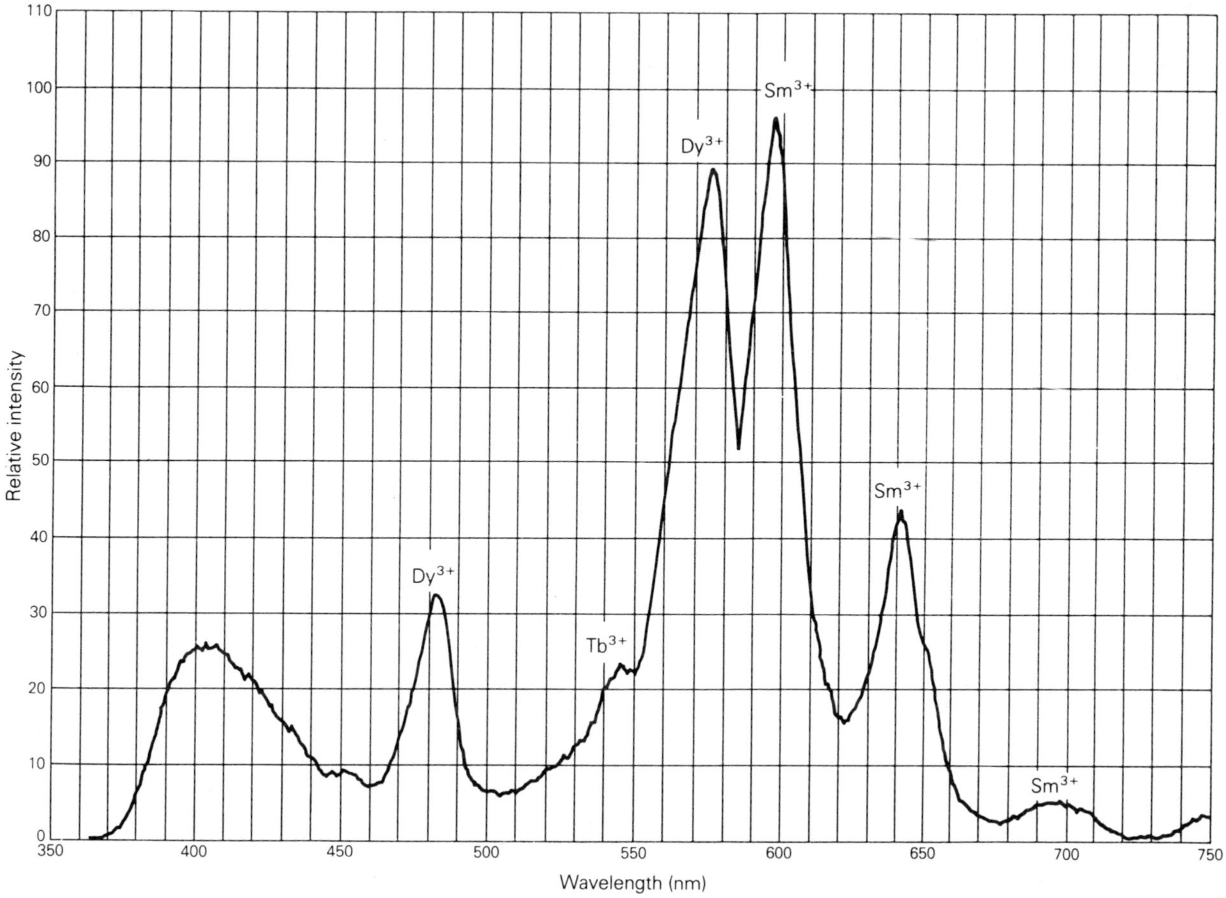

Figure 8.6 Apatite from Kiglapait Intrusion, Labrador. Sm = 291 ppmw, Eu = 57.7 ppmw, Tb = 38.4 ppmw. CL is light violet. CL conditions: 12 kV, 0.45 mA. (Sample # 3,384, Morse & Nolan 1985.)

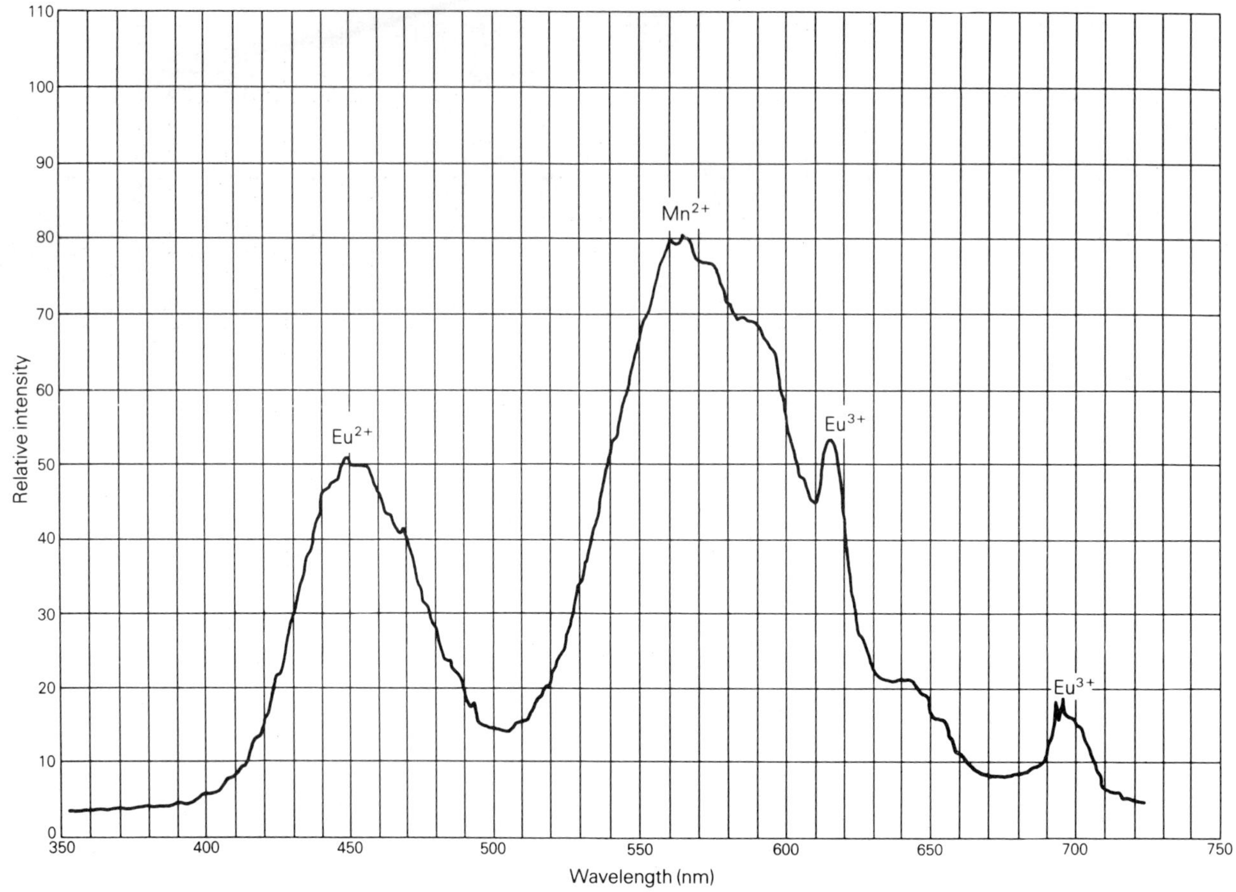

Figure 8.7 Apatite from Llallagua, Bolivia (Smithsonian # 114,175). The Eu^{2+} and Eu^{3+} emission are shown, with the major Eu^{3+} line superimposed on the Mn^{2+} broad band. To the eye, the CL emission is blue with some yellow areas. CL conditions: 6.5 kV, 0.7 mA.

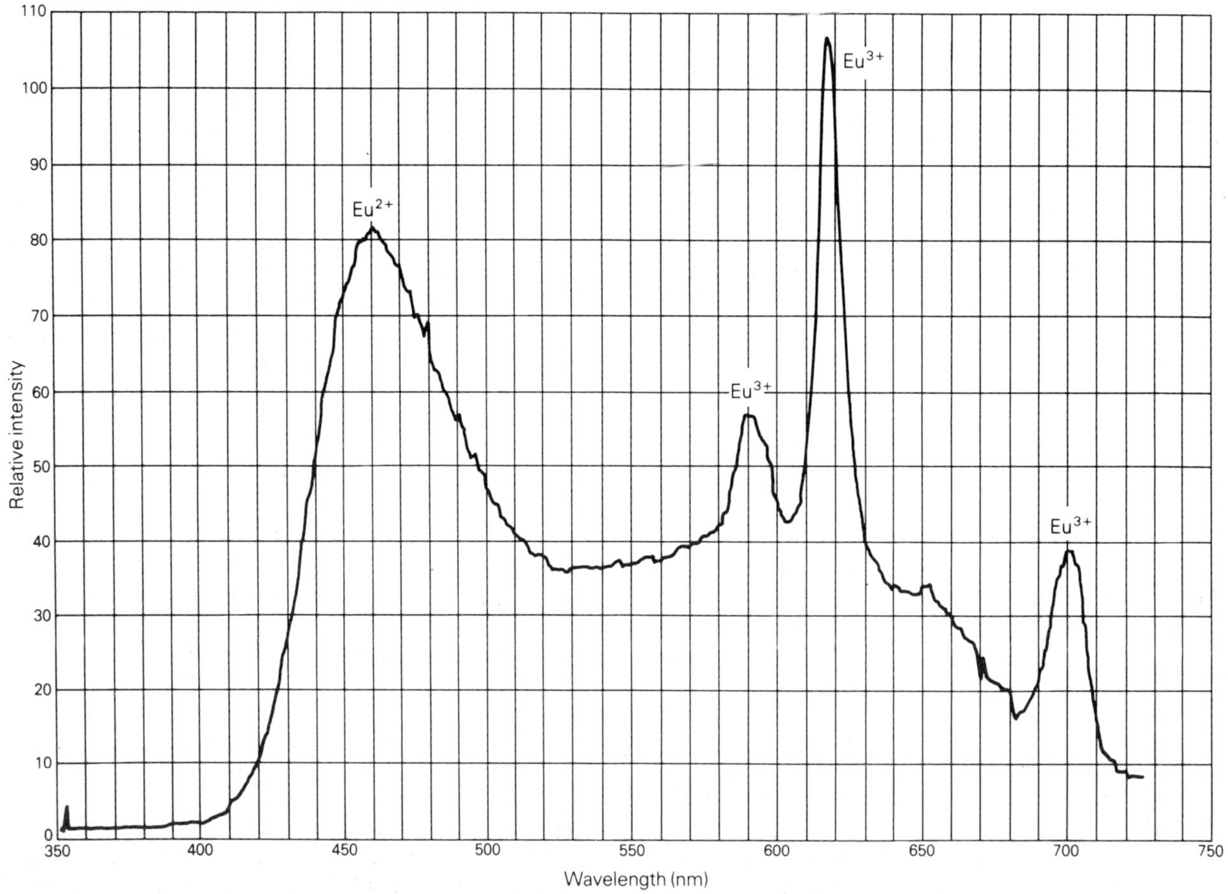

Figure 8.8 Synthetic apatite, Sylvania FS 53–91X1. $Ca_{4.95}Eu_{0.05}(PO_4)_3F_{0.95}Cl_{0.05}$. The CL is violet. This sample shows the characteristic emission from both Eu^{2+} and Eu^{3+}. (Sample furnished by T. L. Barry, GTE Sylvania.) CL conditions: 6 kV, 0.6 mA.

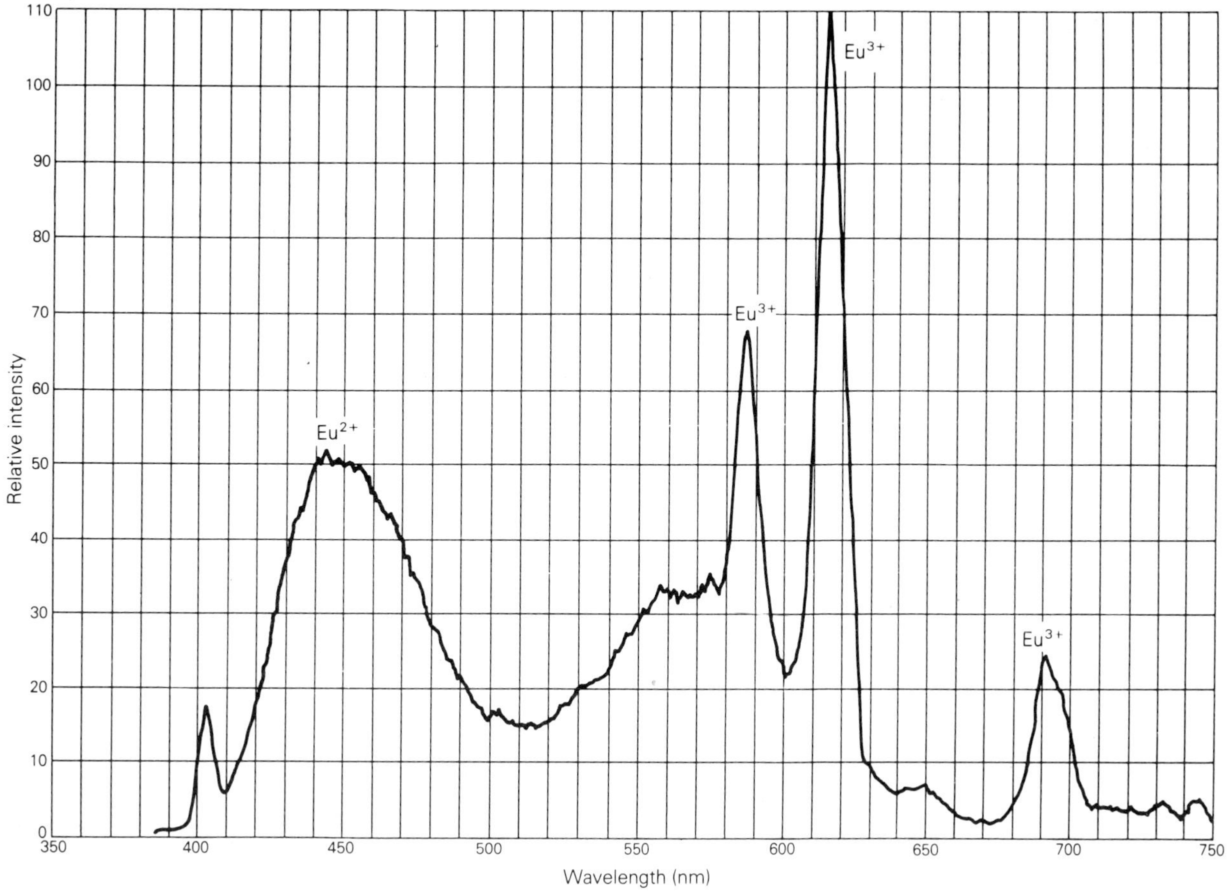

Figure 8.9 Apatite from Llallagua, Bolivia (Smithsonian # 114,175). UV excitation (2,537 Å). Only the europium lines are excited by this source, even though the apatite contains both Mn and Eu.

CL). Portnov and Gorobets (1969) report Ce^{3+}, Eu^{2+}, Sm^{3+}, Dy^{3+}, and Mn^{2+} activation. They do not report Eu^{3+} activation in apatite. Ovcharenko and Yur'ev (1971) have examined the luminescence (probably PhL) of Ukrainian apatite and made interpretations based on Sm^{3+} and Dy^{3+} activation.

The studies performed by all the above investigators are in general agreement concerning the interpretation of the CL spectra of apatite and the geological implications.

CL, combined with the capability for spectral analysis, is an extremely useful tool for quickly identifying the trace element constitution of apatite. The determination can be done on a standard thin section or even on a slab or rough chip in most cases.

Knowledge of the activators and their relative abundance in apatite serves as a petrogenetic indicator in the following manner.

(a) Blue CL is most frequently encountered in primary apatite from carbonatites and their genetically related igneous rocks. The color of CL may vary from blue to violet, depending on the distribution of the REE activators Eu^{2+}, Sm^{3+}, and Dy^{3+}. Sm^{3+} in apatite from these rocks is most often more pronounced than Dy^{3+}. Minor Mn^{2+} activation may also be present in carbonatite apatites, in which case it can be recognized by a loss in resolution between the principal line spectra of Sm^{3+} and Dy^{3+}.

Although blue luminescence in the mineral apatite has been largely attributed to Eu^{2+} activation (Portnov & Gorobets 1969, Mariano & Ring 1975, Mariano 1978), in a recent study by Roeder *et al.* (1987) it is shown that a blue-luminescing apatite probably does not contain sufficient Eu to dominate the color of luminescence. Furthermore, the peak wavelength for the blue CL band in carbonatite apatites occurs at about 410 nm, whereas Eu^{2+} in synthetic apatites and Llallagua apatite (Fig. 8.7) peaks at about 450 nm. The strong blue emission that is so frequently observed in carbonatite apatites may be due mainly to a structural defect produced by anomalous amounts of Sigma REE substituting for Ca^{2+} in the structure.

(b) In alkaline igneous rocks and kimberlites, Sm^{3+} and Dy^{3+} are the dominant activators with the concentration of Sm^{3+} greater than that of Dy^{3+}. Europium and Mn^{2+} activation may occur in subordinate amounts.

(c) Apatites from granites usually show dominant Mn^{2+} activation with superimposed weaker line spectra from Sm^{3+} and Dy^{3+}. Dysprosium emission is always stronger than that of Sm^{3+}, reflecting an HREE dominance.

(d) If apatite is present in alkali granite plutons the lanthanides are the major activators and the REE distribution can be variable depending on the total mineralogy. In this case apatite CL spectra cannot be used as an indicator for the overall REE distribution in the alkali granites.

(e) Secondary apatite co-crystallized with porphyry copper, stock work molybdenite, Pb–Zn ores, etc., generally shows only Mn^{2+} activation, implying relatively low amounts of REE.

(f) Apatite from pegmatites, metamorphic rocks, and alpine veins usually shows only Mn^{2+} activation, although in some cases, low-intensity CL emission peaks from the lanthanides may be superimposed on the Mn^{2+} broad band.

(g) Apatite crystallizing under supergene conditions in laterites derived from carbonatites, kimberlites, and alkaline igneous rocks generally shows only Sm^{3+} and Dy^{3+} activation. The narrow emission bands of Sm^{3+} and Dy^{3+} are easily distinguished due to the absence of Mn^{2+} activation. Divalent europium activation is rarely present but well resolved Eu^{3+} lines are often observed. The concentration of Sm^{3+} is usually higher than that of Dy^{3+}, but in some cases, where the lanthanides have been mobilized, variations in the migratory capacity of the LREE versus HREE can produce supergene minerals with strong HREE dominant activation.

(h) An example of nonluminescing apatite was observed in supergene mineralization from the wall of an exploration adit in the porphyry copper deposit of Safford, Arizona. Stout euhedral hexagonal prisms of colorless apatite appeared intimately crystallized with chrysocolla, neotocite, and other supergene minerals of Cu and Mn. Spark source mass spectrographic analyses on these nonluminescing apatite crystals showed only 25 ppmw Mn and less than 200 ppmw total REE. It is assumed that during this apatite crystallization Mn was oxidized to a high valence state (as indicated by the presence of oxide zone minerals) and therefore was not able to substitute in the apatite structure. With the added condition of low REE in the system, no activators were available to substitute in the crystallizing phases, resulting in the formation of nonluminescent apatite.

Some of the above conclusions are in contrast to those summarized in Marfunin (1979, p. 211).

In summary, the examination of CL emission spectra in apatite gives direct evidence of the chemistry of the crystallizing solutions with respect to the presence of REE and their distribution. CL emission spectra in apatites are very sensitive to geochemical changes in the crystallizing solutions. Overgrowths of apatite from different generations can be very effectively observed with CL. Lanthanide clustering in apatites is frequently observed by CL but the oscillatory zoning that is revealed so effectively with CL for some minerals is only rarely encountered in apatite.

For the most part, the conditions listed above are consistent for apatites examined from many occurrences on a world level. When apatite from a specific environment shows CL spectra that are contradictory to these conditions, it reflects an anomalous situation that is most often explained and supported by other mineralogical or geochemical characteristics of the system.

8.2.2 Fluorite

The CL characteristics of fluorite are quite variable and may be attributed to both activator impurities and other more complex processes, including vacancies that trap electrons, color centers in ionic crystals, and other intrinsic defects.

Examination of synthetic CaF_2 with known impurities shows that luminescence in the blue part of the visible spectrum may be associated with several different impurities but also appears in high purity CaF_2. According to Raman (1966), blue photoluminescence in the mineral fluorite from Amba Dongar, Baroda, India, has its cause from crystal imperfections rather than impurity activation.

Table 8.1 is a compilation of CL data for 22 different synthetic fluorites that have been examined by the author. Some of the fluorites were synthesized as impurity-free crystals. The data suggest that certain impurities enhance intrinsic luminescence in the blue part of the visible spectrum. This is particularly true of Eu^{2+} in synthetic fluorite which shows a strong blue CL with relatively low conditions of excitation voltage and current density. However, similar, but less intense, CL spectra can be obtained from impurity-free CaF_2.

In environments that are enriched in LREE, fluorite usually shows a strong blue CL which peaks between 405 and 430 nm. Since europium is enriched in LREE-dominant environments, it is likely that Eu^{2+} is at least partly responsible for the strong blue CL. Infrequently in these same environments, the blue CL band may be accompanied by line spectra from Sm^{3+} and Dy^{3+}, in which case the Sm^{3+} peak is almost invariably greater in intensity than the Dy^{3+} peak. In contrast, in HREE environments, the blue CL is usually absent in fluorites and the major activator is Dy^{3+}, with a discernible peak from Tb^{3+} at 545 nm, and the color of luminescence is usually a light blue-grey or lilac.

Examples for HREE-dominant CL activation are given in Figures 8.10, 8.11, and 8.12 respectively for fluorites from Rafin Gabis near the Afu Hills in Nigeria, fluorite from the Muzo Emerald mine, Colombia, and yttrofluorite from Hundholmen, Norway. The Rafin Gabis fluorite spectrum is complex, consisting of at least ten sharp lines which are attributed to lanthanide activation. The major activator is Dy^{3+} but Ho^{3+}, Er^{3+}, Tm^{3+}, and Tb^{3+} are also suspected to be present. The Afu Hill granites contain many high charge density elements including yttrium and the HREE. An analysis by atomic absorption of the Rafin Gabis fluorite gave 0.10 weight percent Y.

Fluorite from the Muzo Emerald mine in Colombia is often intimately crystallized with emeralds and parisite (Mariano 1974, Escobar & Mariano 1976). Trace element analyses of Muzo parisite shows an enrichment in heavy lanthanides, as compared to parisite from LREE environments, and the CL spectra of the Muzo fluorite show activation from Dy^{3+} and Tb^{3+}, confirming the HREE dominance of the environment.

Yttrofluorite from Hundholmen, Norway, is known to be enriched in heavy lanthanides (see Fleischer 1969, Table 1, analysis No. 10 and Fig. 8.12). The Sm:Dy ratio in atomic percent is 0.4. The CL emission spectra reveal major activation from Dy^{3+} and a very strongly defined peak for the major Tb^{3+} line at 544 nm.

Yttrofluorite from the White Cloud pegmatite in Platte County, Colorado, gives a CL emission spectrum that is identical to that of the Hundholmen yttrofluorite (sample furnished by D. M. Wayne).

A CL emission spectrum that is typical of fluorite crystallized in an LREE-enriched environment is shown in Figure 8.13 for fluorite in carbonatite from Panda Hill, Tanzania.

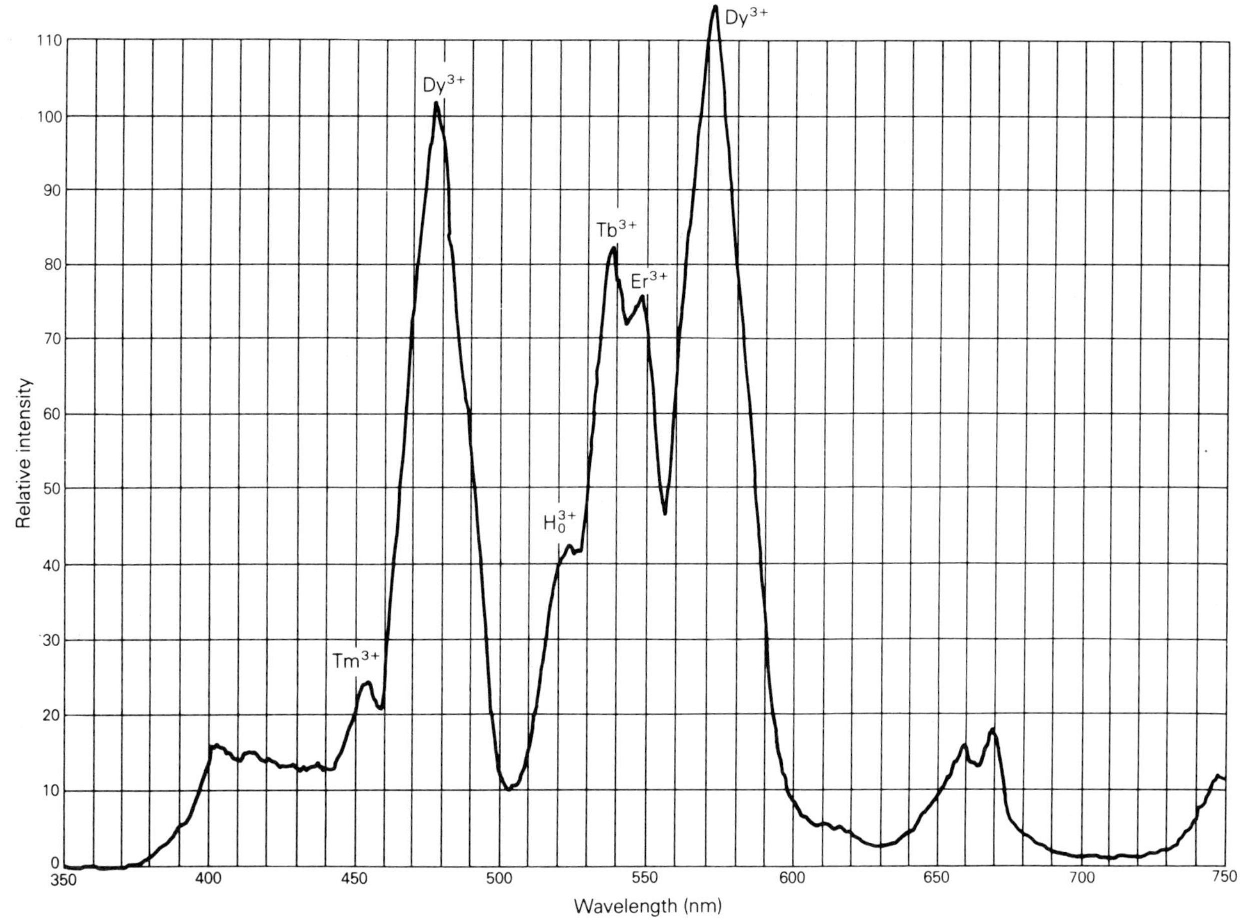

Figure 8.10 CL emission spectrum for fluorite, Rafin Gabis, Benue Province, Nigeria. The CL is a pale straw yellow. The strongest line emissions in this spectrum, at 476 and 572 nm, are due to Dy^{3+} activation. The major line for Tm^{3+} peaks at 452 nm; the major line for Ho^{3+} occurs at 537 nm; and the major line for Er^{3+} peaks at 549 nm. The major line for Tb^{3+} at 540 nm is probably present but would be masked by the Ho^{3+} and Er^{3+} major lines. This fluorite is associated with base metal mineralization, and the appearance of HREE dominance in the CL emission spectrum suggests that it may have been influenced by the Younger Granite mineralization in the nearby Afu Hills. Fluorite from albitized granites from nearby Agwalande in Nigeria give CL emission spectra that are identical to the Rafin Gabis fluorite. CL conditions: 8 kV, 0.8 mA.

Table 8.1 CL characteristics of synthetic fluorites*.

Impurity†	Luminescence color and wavelength (nm)	Excitation conditions: excitation voltage (kV) excitation current (mA)	Color of crystal in natural lighting
(1) Co	light blue, very broad band sweeping across visible spectrum with slight rise at 430 nm	7.5 kV 0.60 mA	violet
(2) Fe	light blue broad band at 425 nm	10.5 kV 0.40 mA	colorless
(3) U 0.04 wt %	blue-green, broad band at 500 nm	12 kV 0.70 mA	red
(4) Yb^{2+}	medium blue broad band at 500 nm	18 kV 0.75 mA	dark reddish-brown
(5) U^{4+}	blue-green broad band at 500 nm	14 kV 0.70 mA	light reddish-brown
(6) Eu^{2+}†	deep blue, relatively narrow band at 430 nm	5.5 kV 0.5 mA	violet-red
(7) U^{3+} 0.01 at %	light blue broad band at 410 nm	11 kV 0.7 mA	colorless
(8) impurity-free	light blue broad band at 410 nm	8 kV 0.2 mA	colorless
(9) impurity-free (powder)	blue and green, mottled; two broad bands at 407 & 490 nm	8.5 kV 0.3 mA	white (powder)

Table 8.1 CL characteristics of synthetic fluorites* *(continued)*.

Impurity‡	Luminescence color and wavelength (nm)	Excitation conditions: excitation voltage (kV) excitation current (mA)	Color of crystal in natural lighting
(10) impurity-free	relatively narrow band, 405 nm	8 kV 0.40 mA	colorless
(11) Ir 0.1 wt %	blue, relatively narrow band at 405 nm	8.5 kV 0.8 mA	colorless
(12) PbF_2 2 wt %	blue-green, two broad bands at 405 & 490 nm	9.5 kV 0.6 mA	colorless
(13) Cr^{3+} 0.2 wt %	blue, relatively narrow band at 405 nm	9.5 kV 0.7 mA	colorless
(14) YF_3 0.01 wt %	deep blue, relatively narrow band at 407 nm	9 kV 0.65 mA	colorless
(15) Pr^{3+} 0.1 wt %	Pink with narrow spikes at: nm / intensity: 405 70; 428 46; 488 100; 525 24; 537 18; 606 60; 642 40	11 kV 0.1 mA	amber-brown
(16) Nd^{3+}	lilac to blue with narrow spikes at: 512 100; and broad bands at: 450 36	11.5 kV 0.8 mA	light lilac

(continued)

Table 8.1 CL characteristics of synthetic fluorites* *(continued)*.

Impurity‡	Luminescence color and wavelength (nm)		Excitation conditions: excitation voltage (kV) excitation current (mA)	Color of crystal in natural lighting
	525	20		
	585	30		
	640	25		
	655	30		
	687	15		
(17) Gd^{3+}	blue with		9.5kV	colorless
0.1 wt %	broad band at:		0.6mA	
	413	100		
	and spikes at:			
	435	80		
	543	40		
(18) Sm^{3+}	yellow orange with		8.5kV	green
0.5 wt %	narrow spikes at:		0.7mA	
	564	80		
	584	25		
	602	100		
	650	30		
(19) Tb^{3+}	light blue-white with		7.25kV	colorless
0.1 wt %	narrow spikes at:		0.35mA	
	385	12		
	416	51		
	436	70		
	470	17		
	487	35		
	542	100		
	581	18		
	619	13		
(20) Ho^{3+}	green with single		5.5kV	colorless
0.1 wt %	narrow spike at 537nm		0.65mA	
(21) Er^{3+}	green with narrow		6.5kV	colorless
0.1 wt %	spikes at:		0.65mA	
	404	28		
	472	18		
	526	28		
	540	65		
	549	100		
	559	77		
	668	13		
(22) Tm^{3+}	bright blue with		7.5kV	light
0.1 wt %	narrow spike at:		0.50mA	blue
	453	100		
	and relatively narrow band at:			
	405	10		
	515	15		
	660	17		

* The above results were obtained by the author.
† Gives same spectrum with shortwave (2,537Å) UV excitation.
‡ Valence states are indicated only where they are relatively certain.

The peak emission is about 405nm, whereas synthetic $CaF_2:Eu^{2+}$ has a peak emission at 430nm (Mariano & Ring 1975). This amount of shift can be due to crystal field effects on the divalent Eu^{2+}. It is assumed that Eu^{2+} is responsible for the strong blue luminescence of fluorites in LREE-enriched occurrences. It is also possible that Ce^{3+} can contribute to CL in this region, according to some of the phosphor industry literature.

In several alkali granite complexes that are well mineralized with high charge density elements including the HREE, Y, and Zr, fluorite with light lilac-colored luminescence is conspicuous and gives a complex spectrum from REE activators that can serve as a diagnostic indicator for exploration geologists.

One such example of unusual lanthanide activation in fluorite is shown for this mineral from the Bayan Obo REE–Fe deposit in Inner Mongolia, People's Republic of China, which is reported to be genetically related to alkali granosyenites (see Vlasov 1966, pp. 566–7, Wang 1981, and Bai & Yuan 1985). The CL emission spectrum for Bayan Obo fluorite (Fig. 8.14) shows a broad band at 415nm that I interpret to be due to Eu^{2+} and Ce^{3+}. The strongest lines for Dy^{3+} appear at 475 and 575nm. The strongest line for Tb^{3+} peaks at 540nm and the two strongest lines for Sm^{3+} appear at 560 and 600nm. The activators responsible for the small sharp spikes at 670 and 755nm have not been identified, but these lines frequently appear in CL emission spectra of fluorites that crystallize in alkali granite-type environments that are well mineralized with high charge density elements. Notice that these spikes also appear in the fluorite from Rafin Gabis, Nigeria (Fig. 8.10).

Fleischer (1969) has shown that the lanthanide distribution in fluorite can often be used as an indicator of the environment, the HREE being concentrated in granite pegmatites and sialic rocks, while in alkalic rocks fluorite concentrates the LREE. CL emission spectra of fluorites can very often provide this same information about the presence of lanthanide activators and their relative concentrations.

8.2.3 *Strontianite*

REE ACTIVATION IN STRONTIANITE

Lanthanide activation is frequently observed in strontianite when the mineral is found crystallizing in an environment that is enriched in REE.

When strontianite is encountered in carbonatite environments, CL is invariably due to activation by LREE impurities. In some cases the luminescence is a deep blue from a broad-band emission that shows a maximum at approximately 410nm (Fig. 4.5a). This has been interpreted as Eu^{2+} activation (Mariano & Ring 1975) and has been observed in carbonatites from Kangankunde, Malawi; Mountain Pass, California; Bonga, Angola; Kerimasi and Panda Hill, Tanzania; Araxá, Brasil; and others.

In carbonatite environments strontianite can be found in different rock units including fenites, late crystallization in the carbonatite rocks or associated alkaline igneous rocks, and as hydrothermal mineralization. Within a given carbonatite complex the strontianite from the different rock units usually shows different CL that is discernible by visual observation and in emission spectra. For example, at Mountain Pass, California, the deep blue-luminescing strontianite is usually found in the fenites, while late hydrothermal strontianite veins may show multiple activation from Eu^{2+}, Sm^{3+}, and Dy^{3+} (with the intensity of the major Sm^{3+} line being greater than that of the major Dy^{3+} line), yielding a light blue-violet luminescence (Fig. 8.15).

Another case of multiple LREE activation can be found in strontianite occurring as late hydrothermal veins in søvite from the Bonga carbonatite in Quilengues, Angola (Fig. 8.16). In this case the CL spectrum shows a broad band from Eu^{2+} at 415nm and line emission from Sm^{3+} activation.

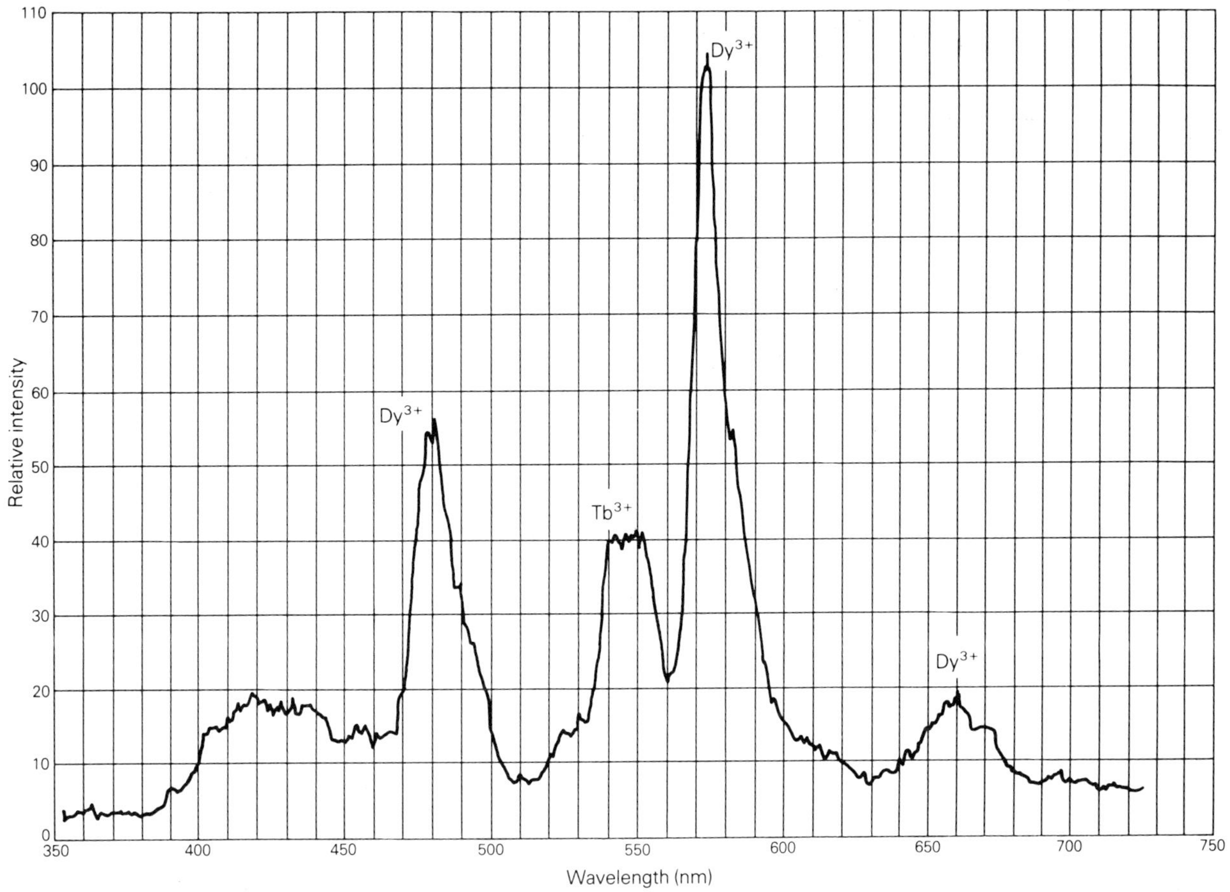

Figure 8.11 CL emission spectrum for fluorite from the Muzo emerald mine, Colombia. Included mass spectrographic analysis for the Muzo fluorite shows a relatively high Dy value which is confirmed by the CL spectrum showing line emissions at 480, 574, and 659 nm. Mass spectrographic analysis (all in ppmw) are as follows:

La	70	Ce	105	Pr	70
Nd	145	Sm	25	Eu	10
Gd	80	Tb	15	Dy	105
Ho	15	Er	40	Tm	4
Yb	25	Lu	5		

The Tb^{3+} major emission at 546 nm is also prominent, despite the fact that the mass spectrographic analysis shows only 15 ppmw Tb (± a factor of 3). The HREE dominance of the Colombian emerald environment is also reflected in the mineral parisite as revealed in lanthanide distribution analyses relative to parisite from carbonatite environments (LREE dominance). The CL for the Muzo fluorite is a whitish-violet. CL conditions: 8 kV, 0.7 mA.

In some carbonatites, late hydrothermal strontianite with only Sm^{3+} activation can be encountered in open vugs where it forms euhedral crystals often associated with lanthanide minerals. An example of this is given for strontianite that is intimately crystallized with hydrothermal ancylite and carbocernaite in the Bear Lodge carbonatite rocks of Wyoming (Fig. 8.17). Identical spectra are obtained for late hydrothermal strontianite in vugs in carbonatite from Jacupiranga in Brasil (Fig. 8.18).

Unlike the case for many of the common carbonate minerals, activation by Mn^{2+} has not been observed in strontianite. The crystal chemistry of strontianite is similar to that of aragonite and it is assumed that both of these minerals usually discriminate against sufficient manganese substitution for activation.

In some environments strontianite occurs with different morphologies and CL spectra. For example, strontianite from the Faylor quarry in Windfield, Pennsylvania, occurs both as acicular prisms and short stubby crystals. The two morphologies show significantly different CL emission spectra, reflecting changes in the conditions of crystallization. The spectrum from the Windfield strontianite is complex and the cause of the luminescence has not been determined.

Strontianite from the classical locality of Argyllshire, Scotland (Smithsonian #48,904) also shows lanthanide activation with Dy^{3+} as the dominant activator and weaker activation from Sm^{3+} and Tb^{3+} (see Fig. 4.5b). An almost identical spectrum was also obtained from a strontianite from Carlton County, Ontario (Smithsonian #80,299). This is shown in Figure 8.19.

Examination of strontianites from varied localities showed mostly REE activation with either LREE or HREE dominance. The significance can only be ascertained by studying each environment. However, it is already apparent that strontianite CL spectra show the variations in lanthanide content and this can be used in geochemical and petrological studies.

8.2.4 *Calcite*

REE ACTIVATION IN CALCITE

The activator that is responsible for CL in calcite is almost invariably Mn^{2+} but in rare cases calcite may show REE

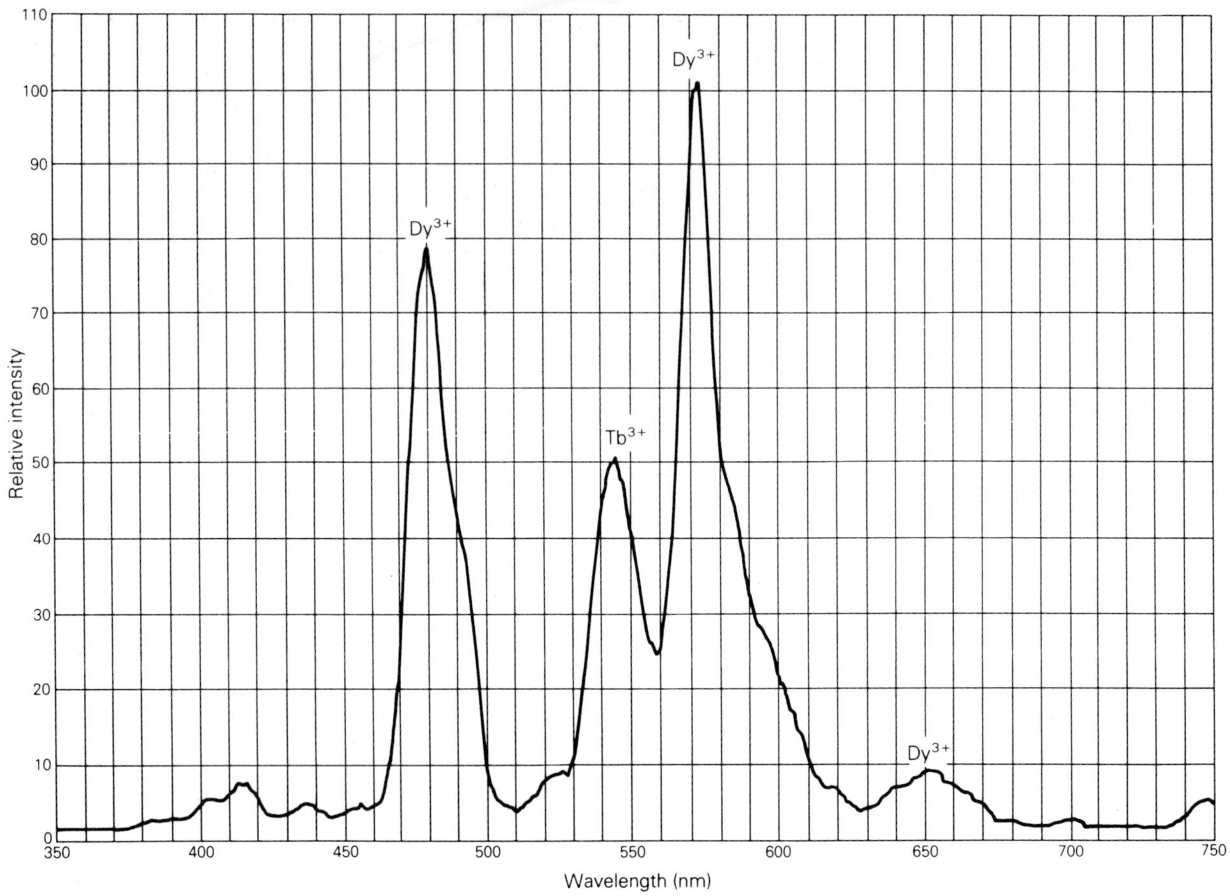

Figure 8.12 CL emission spectrum for yttrofluorite, Hundholmen, Norway (American Museum of Natural History, #17,654). CL is light grayish-white. The emission spectrum for Hundholmen yttrofluorite is identical to that for yttrofluorite from the White Cloud pegmatite of Platte County, Colorado. The Dy^{3+} lines occur at 479, 572, and 653 nm. The line at 544 nm is the major emission of Tb^{3+}. Excitation conditions: 7.5 kV, 0.2 mA.

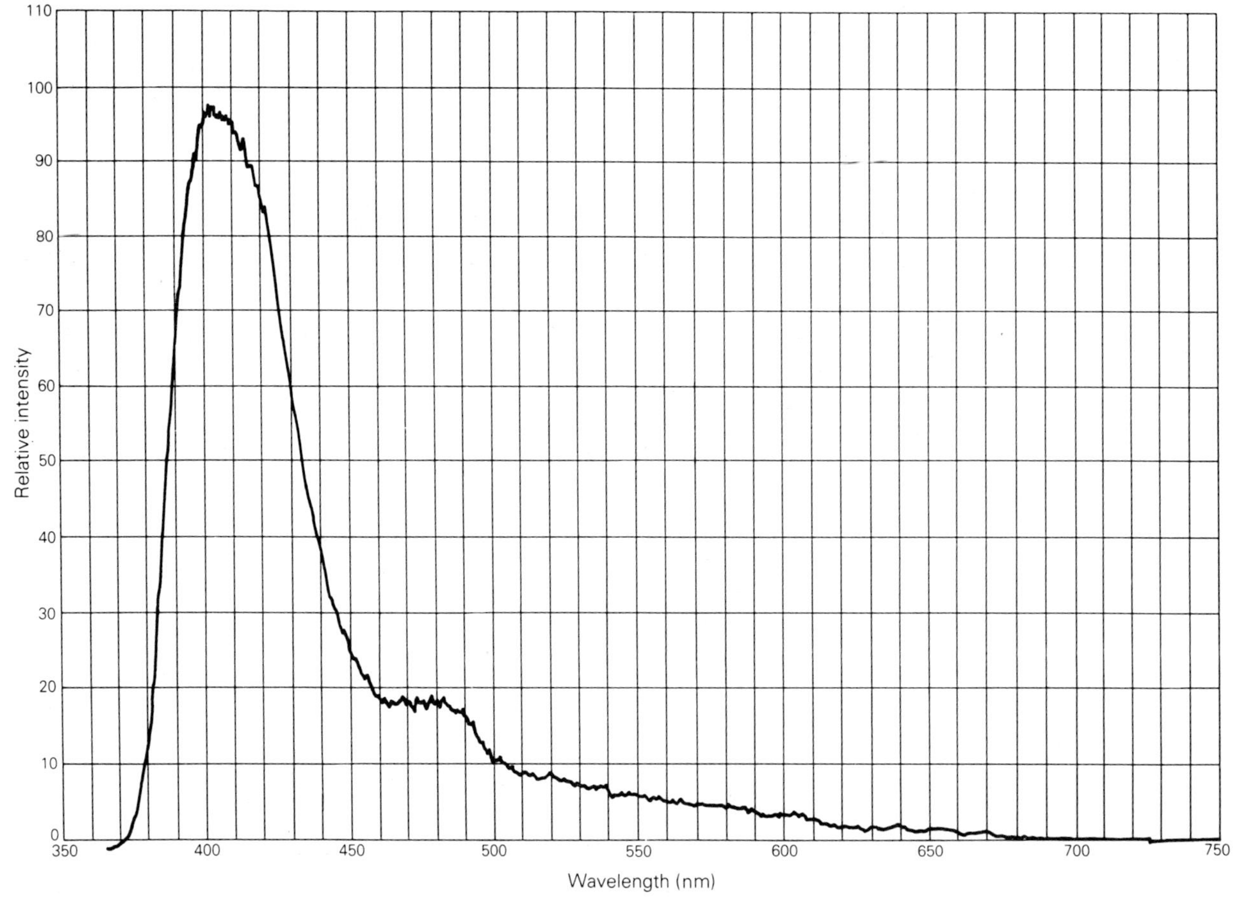

Figure 8.13 CL emission spectrum for fluorite from carbonatite, Panda Hill, Tanzania. The CL is a deep blue. This spectrum is typical for fluorite in carbonatites and alkaline igneous rock environments. The dominant peak in the blue region of the visible spectrum is attributed to Eu^{2+} activation and possible contributions from Ce^{3+}. Excitation conditions: 8.75 kV, 0.8 mA.

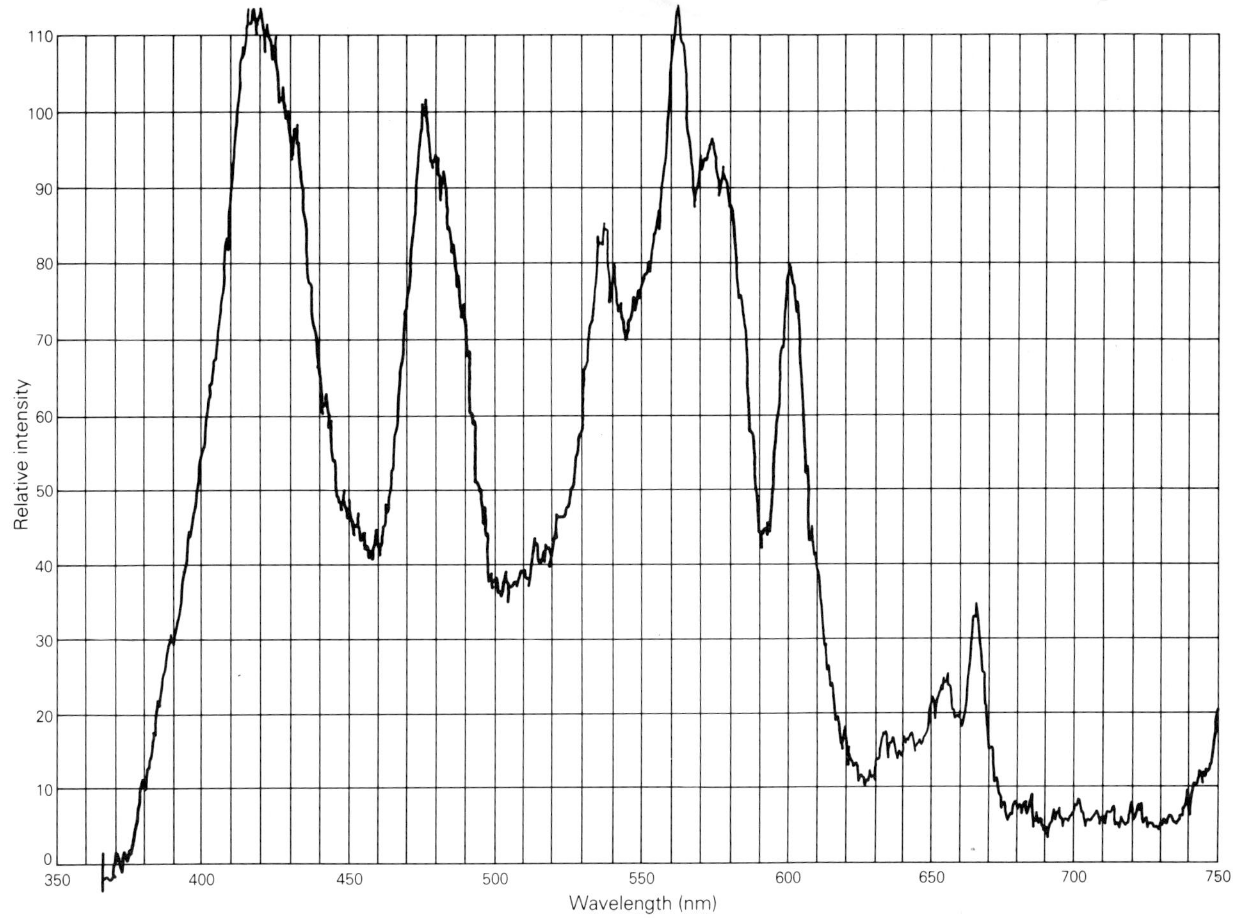

Figure 8.14 CL emission spectrum for fluorite, Bayan Obo REE–Fe deposit, Inner Mongolia, People's Republic of China. The luminescence is a light blue. This spectrum is an example of complex rare earth activation and an exact identification of the peaks has not yet been made. This same spectrum has been observed in fluorites crystallized in several alkali granite environments enriched with high charge density elements. CL conditions: 11.5 kV, 0.7 mA (sample from D. G. Bryant).

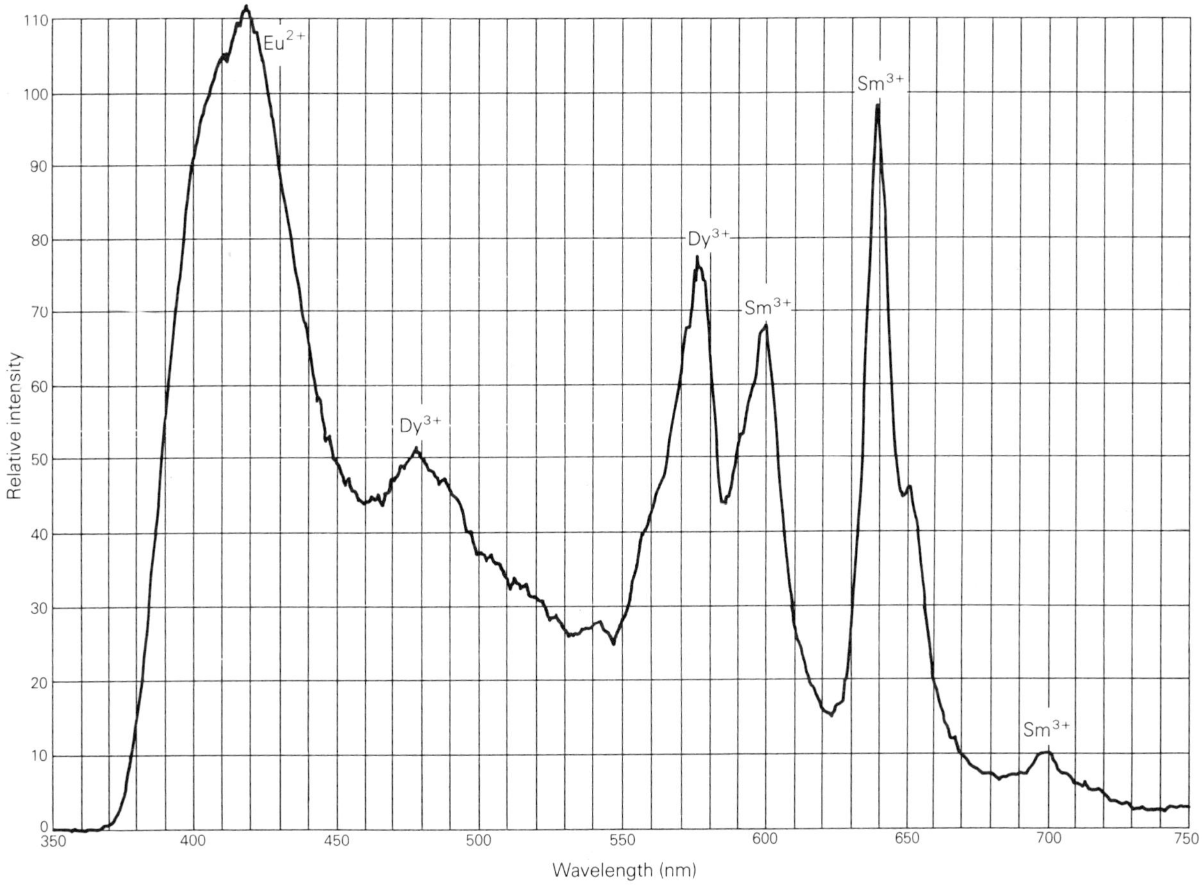

Figure 8.15 CL emission spectrum for hydrothermal strontianite in carbonatite from Mountain Pass, California. The luminescence is light blue-violet. Activation in this spectrum is as follows: Eu^{2+} 417 nm; Sm^{3+} 600 nm, 640 nm, 700 nm; Dy^{3+} 477 nm, 575 nm. Another Sm^{3+} emission at 560 nm is partly obscured by the major emission of Dy^{3+} at 575 nm. Excitation conditions: 18 kV, 0.5 mA.

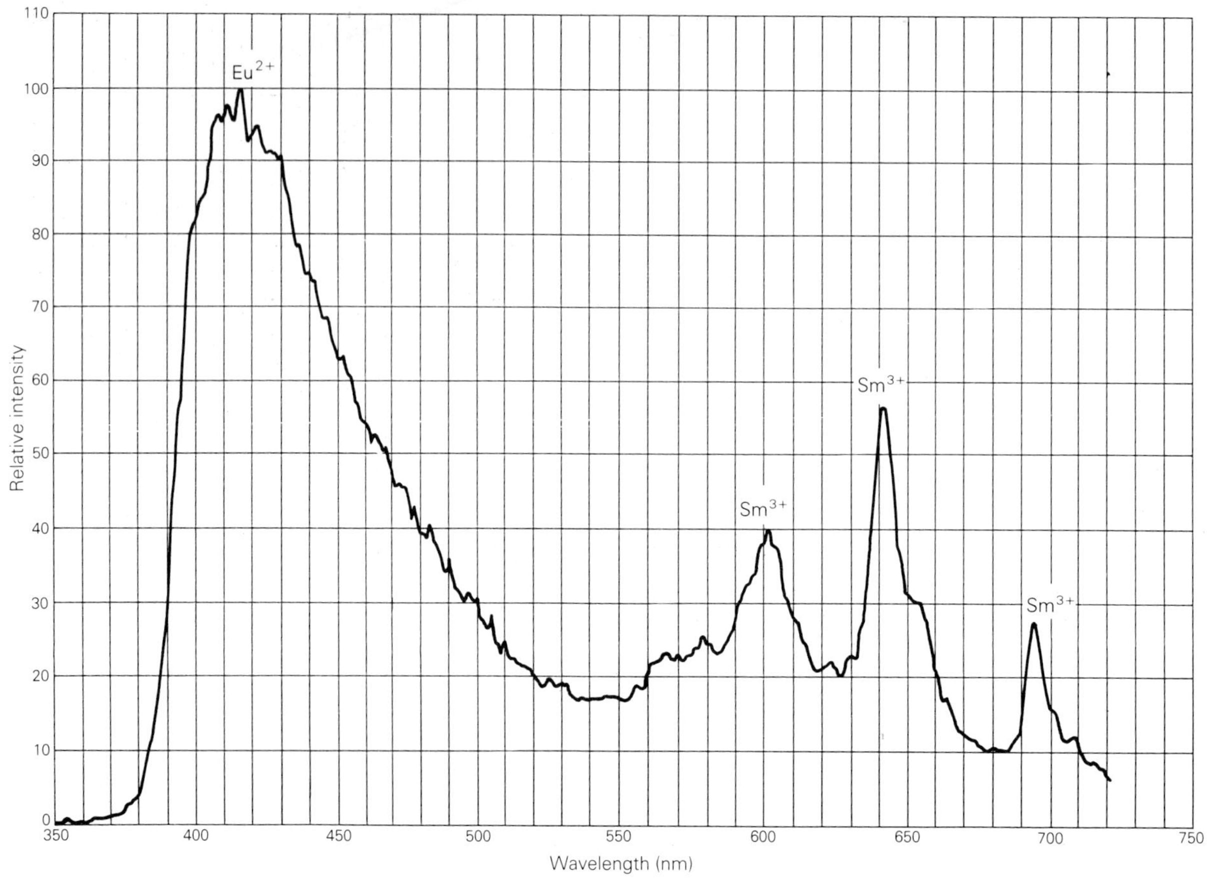

Figure 8.16 CL emission spectrum for strontianite from an apatite søvite, Bonga carbonatite, Quilengues, Angola. The broad band peaking at 415 nm is interpreted to be from Eu^{2+} activation. The narrow-band emissions at 601, 641, and 695 nm are from Sm^{3+} activation. CL conditions: 10 kV, 0.6 mA.

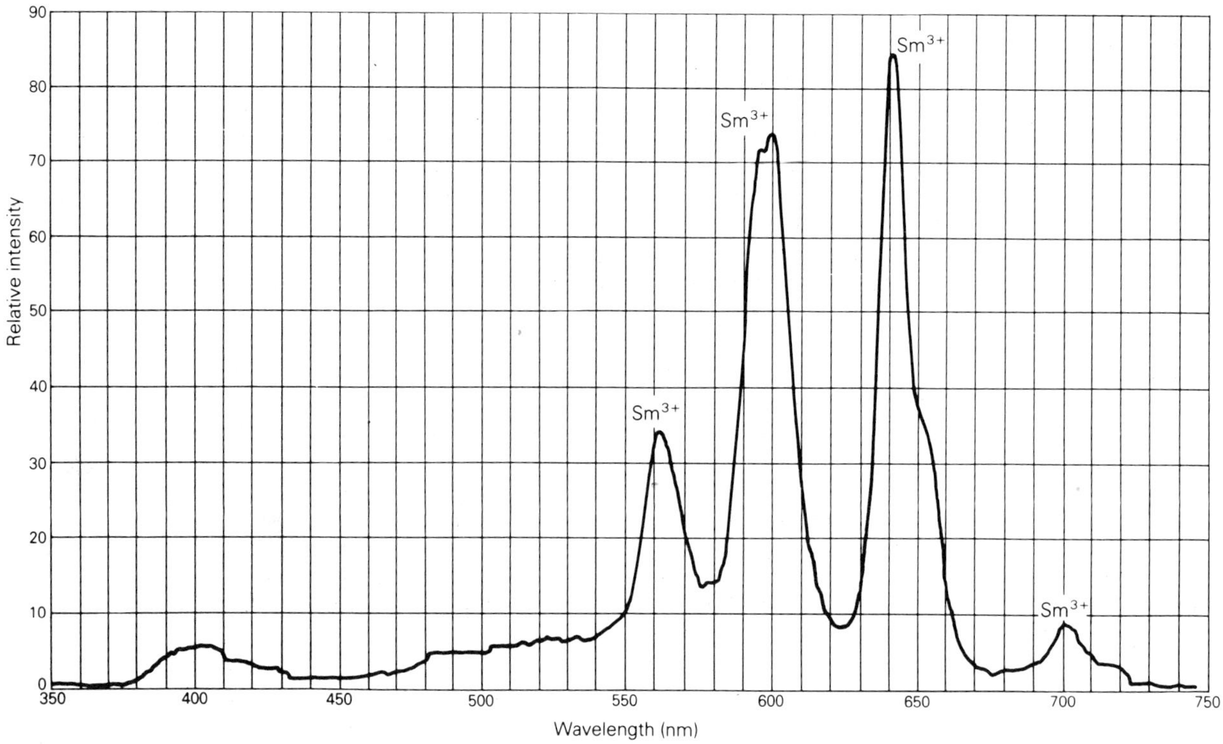

Figure 8.17 CL emission spectrum from hydrothermal strontianite in søvite, Bear Lodge Mountains, Wyoming. The sharp lines in this spectrum are exclusively from Sm^{3+} activation. The wavelengths of the lines and their relative intensities compare favorably with Sm^{3+}-doped synthetic $CaWO_4$ and Sm^{3+}-doped synthetic labradorite. The spectrum also compares in wavelength with Sm^{3+}-doped CaF_2 and $SmPO_4$ but in the latter two materials the 600 nm peak is more intense than the 641 nm peak. The wavelengths of the peaks for the Bear Lodge strontianite occur at the following values (in nm): 561, 600, 641, 701. This spectrum is virtually identical in wavelength and relative intensity to hydrothermal strontianite from the Jacupiranga carbonatite of São Paulo State in Brasil (see Fig. 8.18). The CL to the eye is a light pink-salmon. CL conditions: 5 kV, 0.5 mA; beam diameter approx. 2 mm.

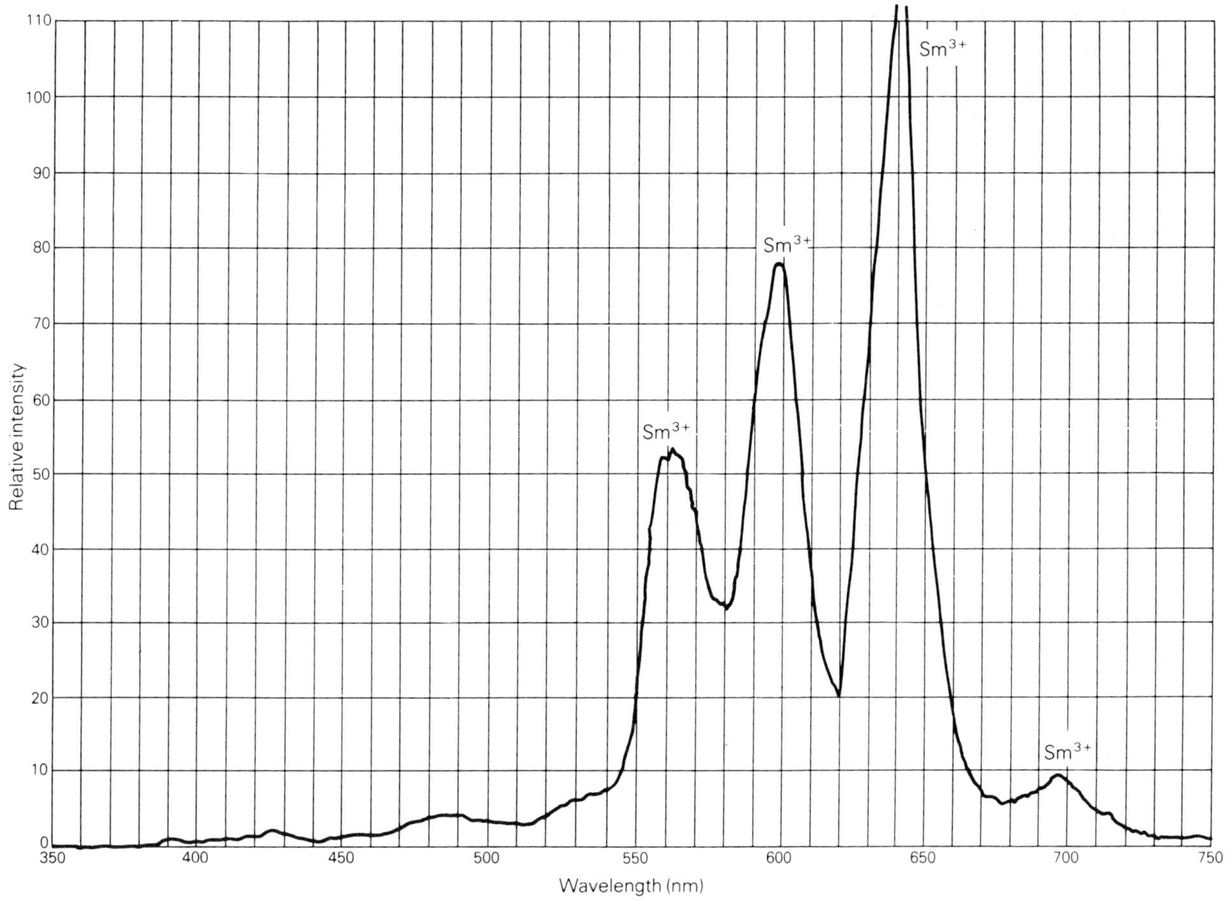

Figure 8.18 CL emission spectrum for hydrothermal strontianite, Jacupiranga carbonatite, São Paulo State, Brasil. Except for small shifts in wavelength, the Sm^{3+} activation peaks in this spectrum are almost identical in wavelength and relative intensity to the hydrothermal strontianite from the Bear Lodge carbonatite (Fig. 8.17). CL is a light orange-violet. CL conditions: 11 kV, 0.3 mA.

activation. In order to show REE activation, it is not sufficient for calcite to be crystallizing in a high REE environment because the lanthanides substitute in the calcite structure only in limited amounts (less than 0.2 weight percent Sigma REE). These amounts would be enough for activation of CL in a calcite that is otherwise free of impurities, but Mn^{2+} is always present in natural calcites and therefore its spectrum is almost invariably the only one observed. However, in carbonatite rocks calcite contains anomalous amounts of Sigma REE as substitutional impurities. This is a characteristic of carbonatite calcite that differentiates it from limestone and marble calcite. The latter rarely contain more than 200 ppmw Sigma REE whereas the carbonatite calcites most often contain 0.*n* weight percent Sigma REE. However, carbonatite calcites also contain substantial amounts of substitutional Mn^{2+} and other impurities including Sr and Ba. As a result they always show Mn^{2+} activation with a broad band of CL occurring at approximately 615 nm. Even in carbonatite magmas where REE occur in major amounts such as Mountain Pass, California, Kangankunde, Malawi, and Bayan Obo, People's Republic of China, calcite that is in intimate crystallization with REE minerals shows only Mn^{2+} activation.

HREE ACTIVATION IN CALCITE

The first reported occurrence of REE activation of CL in calcite is from the Snowbird fluorite mine of Mineral County, Montana (Mariano 1978). This is a very unusual hydrothermal carbonate deposit consisting mostly of calcite with subordinate ankerite, fluorite, and quartz. Parisite occurs in accessory quantities as prisms that are often 5 cm in length. The bulk of REE content in the system is substitutional in calcite and fluorite. Both of these minerals show HREE enrichment based on chemical analyses and CL emission spectra. The calcite CL spectrum (Fig. 8.20) consists of a broad-band emission from Mn^{2+} and line spectra from Dy^{3+} and Sm^{3+} with Dy notably more intense than Sm. Chemical analysis revealed a dysprosium content of 96 ppm, samarium content of 13 ppm, and manganese content of approximately 200 ppm.

The unusual conditions for lanthanide activation in the Snowbird calcite include relatively high HREE with concomitant low Mn and unusually low P*. Xenotime (YPO_4) has been observed at Snowbird (Metz 1971, Metz *et al.* 1985) but only in trace amounts. Abnormal amounts of lanthanides have partitioned into carbonate and fluorite phases. Because of the low Mn content of the system, the HREE-dominant activation appears in the calcite as $Dy^{3+} > Sm^{3+}$ line spectra, superimposed on the Mn^{2+} broad band (see Fig. 8.20).

Through CL examination of the Snowbird rocks it became apparent that the REE not only partitioned into accessory independent lanthanide minerals but also substituted in minerals that constitute a major proportion of the Snowbird deposit.

* When P occurs in solutions in the presence of REE, they have a very strong affinity and immediately combine to form an REE phosphate mineral.

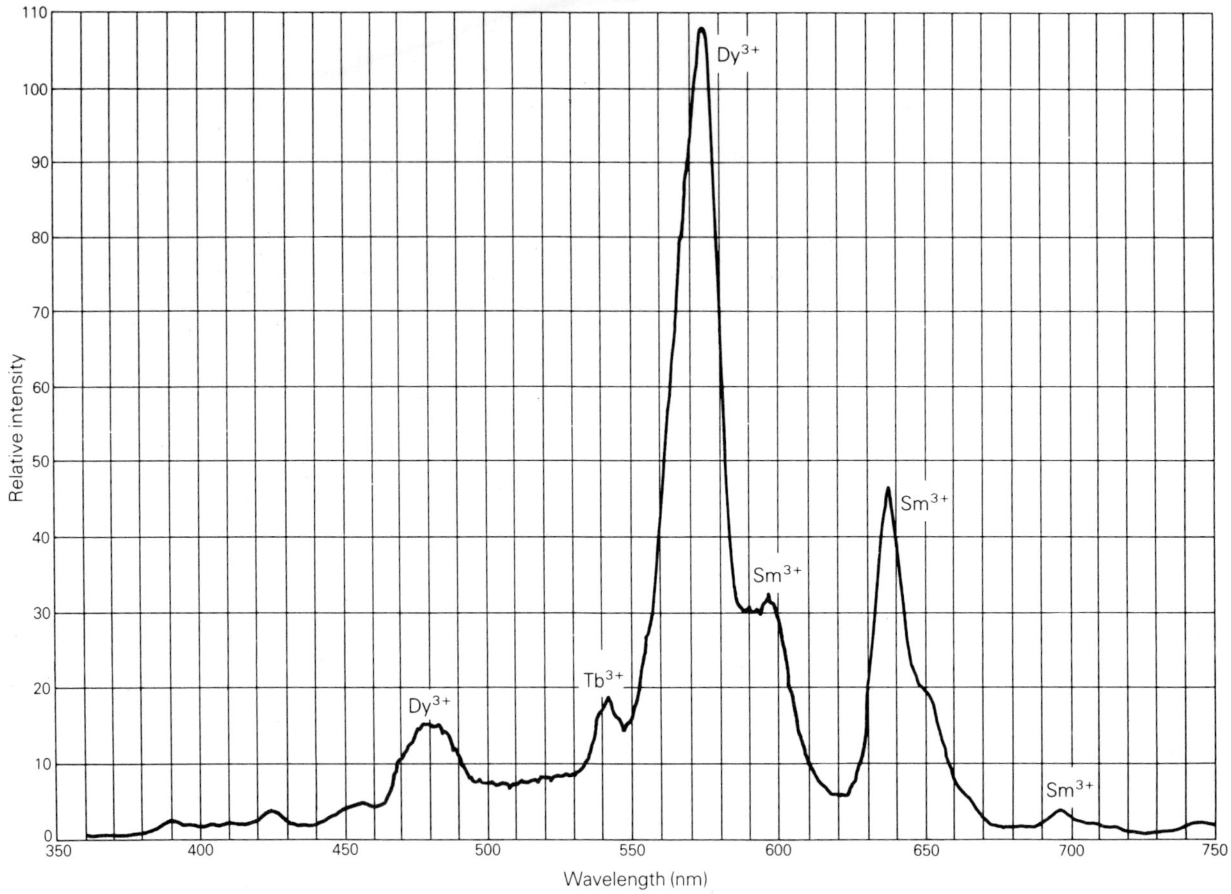

Figure 8.19 CL emission spectrum for strontianite from Carlton County, Ontario (Smithsonian # 80,299). This strontianite occurs in Ordovician limestone but unlike the carbonatite-crystallized strontianite it shows HREE-dominant activation. The major activator is Dy^{3+} with emissions at 480, 575, and 640 nm. Subordinate activation from Sm^{3+} occurs at 597, 639, and 697 nm. The small hump resolved at 542 nm is interpreted to be the strongest emission from Tb^{3+} activation. The CL is light pink. This spectrum is identical to the classical locality strontianite from Argyllshire (Strathclyde), Scotland (Fig. 4.5b). CL conditions: 9 kV, 0.8 mA.

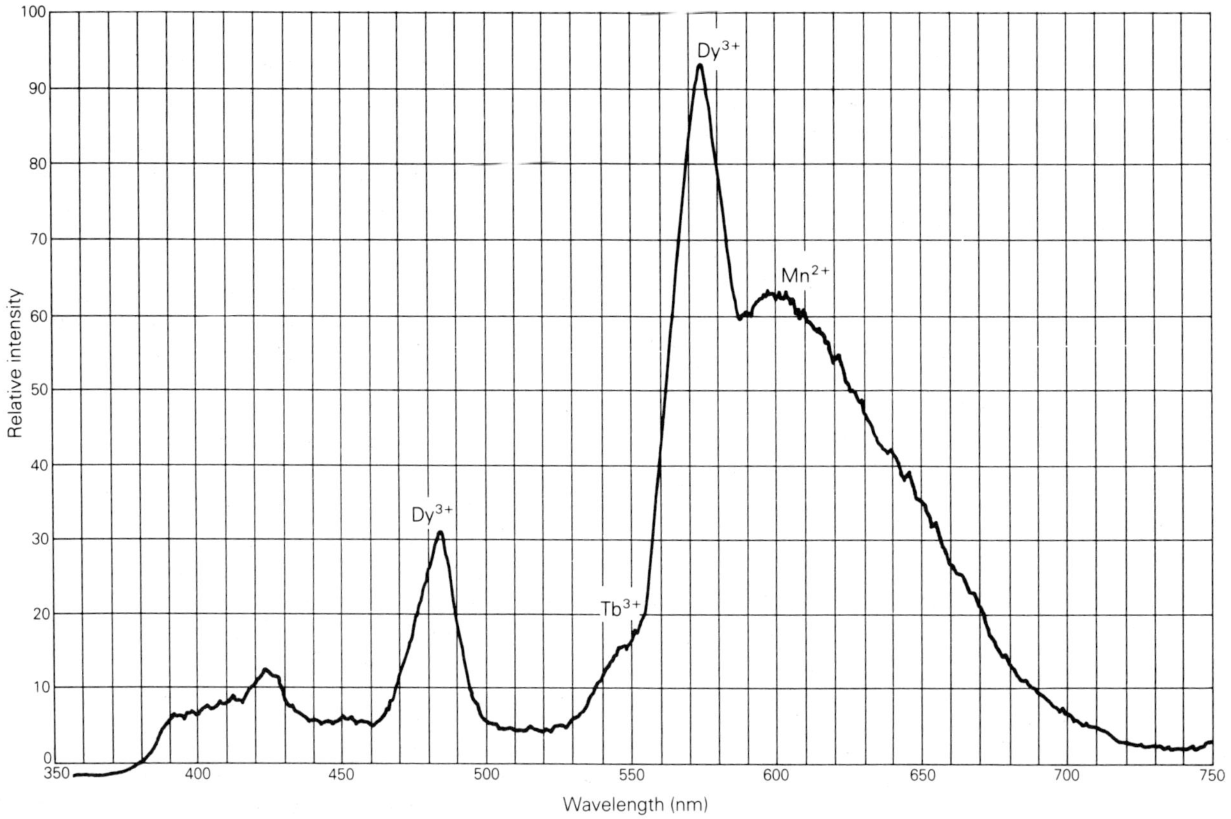

Figure 8.20 CL emission spectrum from calcite in the Snowbird fluorite deposit, Mineral County, Montana. The CL is light orange-yellow. This spectrum indicates dominant activation by Dy^{3+}, producing lines at 480 and 575 nm. The broad band with the peak wavelength at 605 nm is from Mn^{2+} activation. The main emission from Tb^{3+} is suggested at 550 nm. Particular attention was paid, when this spectrum was obtained, to insure that the CL was coming from the calcite and that no mineral inclusions, extraneous to the calcite structure, were present to contribute to the CL. CL conditions: 12 kV, 0.7 mA.

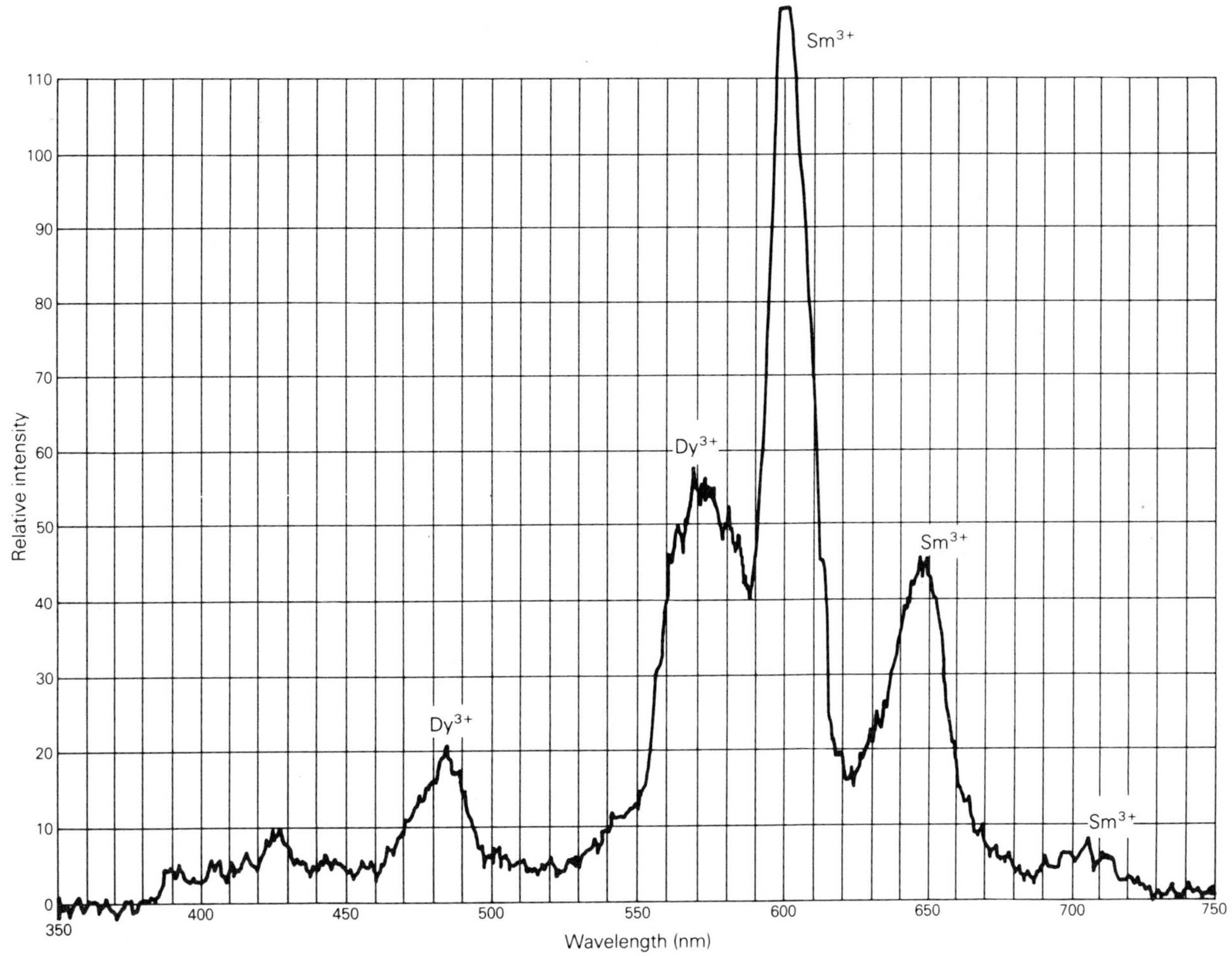

Figure 8.21 LREE-activated calcite from laterite overlying carbonatite in Western Australia. The CL is light yellow-orange to pink. Both Sm and Dy are activators in this calcite and Mn activation is not present. CL conditions: 10.5 kV, 0.15 mA.

LREE ACTIVATION IN CALCITE

Calcite is usually unstable in the supergene horizon of tropical zone carbonatites. However, in isolated cases, after the total leaching of carbonatite, late low temperature aqueous solutions traveling through the laterite may deposit minerals including quartz, feldspar, barite, apatite, celestite, and especially supergene REE phosphates and carbonates. In rare cases calcite may be crystallized in isolated areas that remain free from subsequent leaching. Because of the crystallization under conditions of strong oxidation these calcites are free of Mn^{2+}. They may, however, contain anomalous concentrations of REE that are contained in the aqueous solutions which permeate the carbonatite-derived laterite. Calcite found in these special conditions shows only LREE activation and the narrow bands of the trivalent lanthanides are very well resolved. Divalent manganese and divalent europium activation are absent. An example of LREE-activated calcite is given in Figure 8.21.

It is interesting to note that despite the low coordination number* of Ca in calcite of six, the REE distribution is variable, reflecting the geochemistry of the crystallizing solutions.

In addition to establishing the presence of REE activation in calcite that is found in laterites, the absence of Mn^{2+} activation establishes that it is of supergene origin (i.e., crystallized under conditions of strong oxidation). When supergene calcite with LREE activation has been found in laterites it is accompanied by manganese oxide minerals such as psilomelane and pyrolusite, proving that Mn was available in the system but could not substitute in the calcite because of the high valence state brought on by the strong oxidizing conditions.

In summary it should be emphasized that despite the account of these unusual occurrences, REE activation of CL in carbonates of the calcite group is indeed rare in nature.

* Low coordination number sites in minerals are reported to concentrate the HREE preferentially (Semenov 1958).

8.3 Cathodoluminescence in feldspars

8.3.1 Ferric ion activation

The action of ferric iron as an activator for CL in synthetic feldspars has been established by Geake *et al.* (1973) and Mariano *et al.* (1973). The wavelength of emission for Fe^{3+} for the mineral feldspar is variable, ranging within the red part of the visible and IR spectrum. The CL emission spectrum for a synthetic labradorite with Fe^{3+} activation is shown in Figure 5.4.

Iron in optical quality orthoclase from Itrongay, Madagascar, is reported to occur mainly in the ferric state, based on chemical analyses (Deer *et al.* 1971, p. 45) and optical absorption spectroscopy (Faye 1969). A sample of Itrongay

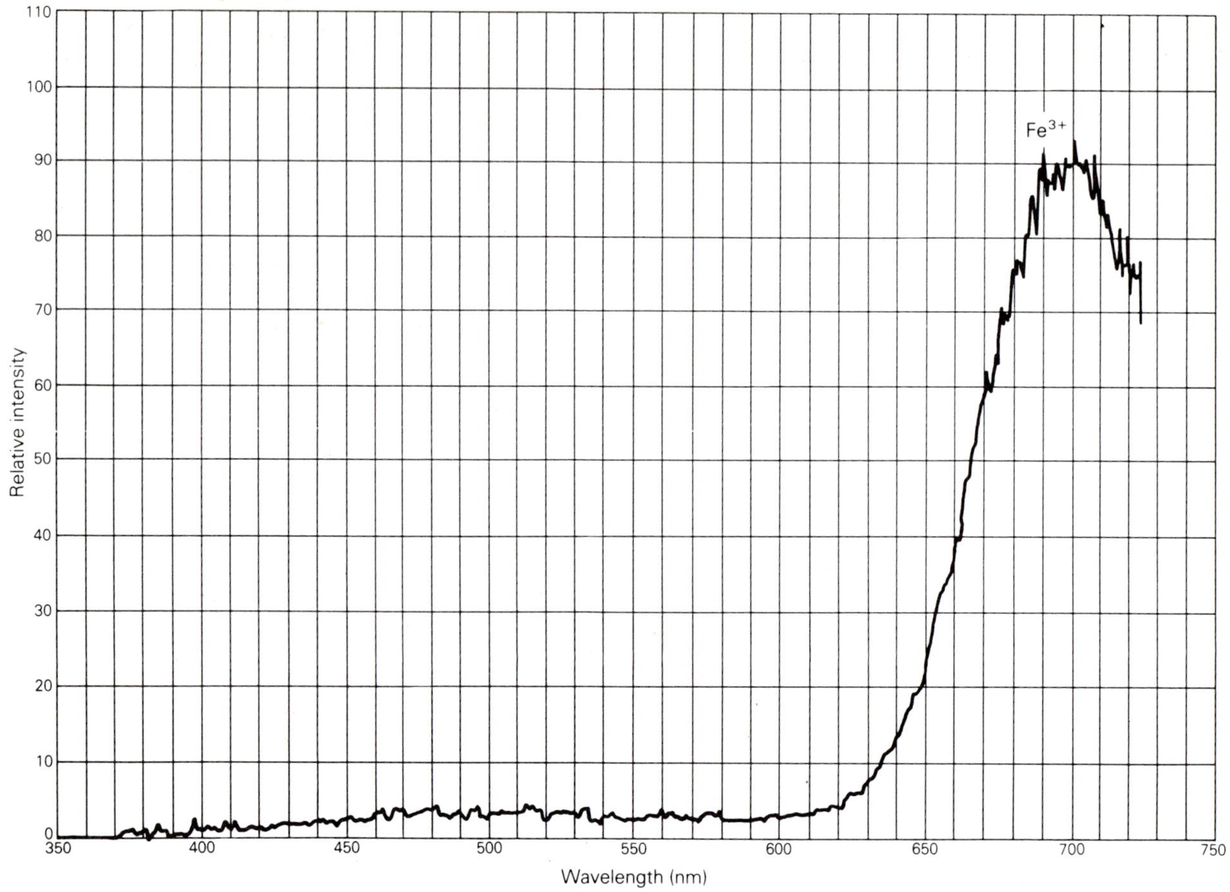

Figure 8.22 The CL emission spectrum from an orthoclase from the Itrongay region, Madagascar. The CL of this optical quality colorless crystal is red to the unaided eye and is due to the Fe^{3+} activator, confirmed by chemical analyses and optical absorption data (see text). This particular sample contains 0.23 weight percent of Fe (analysis by Jun Ito). Sample from Ward's. CL conditions: 8.5 kV and 0.5 mA.

orthoclase, obtained from Ward's Natural Science Establishment, was analyzed by Jun Ito and found to contain 0.29 weight percent Fe. A CL emission spectrum for this sample, shown in Figure 8.22, has a broad band at approximately 700 nm which is attributed to Fe^{3+} activation.

The luminescence color from Fe^{3+} activation alone is red but in many cases other activators compete with Fe^{3+} and the feldspar may exhibit a blue, green, or yellow CL. In such cases the activators responsible for the color can only be identified by use of a spectroradiometer. Examples of multi-activation in feldspars will be given later in this chapter.

8.3.2 Cathodoluminescence as a tool for fenite and alkaline igneous rocks petrology

The brilliant red CL that is observed in feldspars which crystallize under conditions of high alkalinity in peralkaline igneous rocks, including both agpaitic and miaskitic nepheline syenites and alkali granites, is due to Fe^{3+} activation. This same condition also appears in all feldspars that crystallize from the metasomatic effects of fenitization*. The red-luminescing behavior of feldspars which crystallize under conditions of high alkalinity and moderate to high temperatures can be explained by the experimental work of Carmichael and Nicholls (1967). They found that in silicate liquids that are held under constant temperature and constant oxygen fugacity, an increase in alkalis stabilizes iron in the ferric state. They labeled this the alkali-ferric-iron effect (AFE). Experiments with basalt and silicate liquids that gave similar results were performed by Roeder (1974), Waff (1977), and Thornber *et al.* (1980).

The stabilization of Fe^{3+} under conditions of high alkalinity is reflected in the mineralogy and trace element chemistry of the crystallizing mineral phases. Amphiboles and pyroxenes are peralkaline. Magnetite is not stable and Fe^{2+}-bearing minerals, particularly in the last stages of crystallization, are not formed. The presence of peralkaline amphiboles and pyroxenes and feldspathoids is usually an indication of crystallization under conditions of high alkalinity, although feldspathoids may be crystallized in conjunction with magnetite under conditions of low silica activity, whereas in strongly alkali solutions magnetite is not stable. For this reason it is well known to exploration geologists in carbonatite programs that fenite rock units are magnetic

* Fenitization is alkali metasomatism resulting from the late residual fluids of an alkali magma that act upon previously crystallized rocks. The process should involve the replacement of earlier minerals of the invaded rock by the introduction of extraneous alkali cations. This in itself is manifest mainly by the introduction of Na and K ions, either being dominant depending on the geochemical characteristics of the parent alkali magma. The term 'fenitization' was first introduced by Brøgger in his 'Memoir on the Fen' (Brøgger 1921, p. 156).

lows. In the classical Fen area of Norway, Saether (1957, p. 9) reported that 'The fenite is very poor in magnetite, and in harmony with this it gives negative magnetic anomalies.'

The identification of strongly alkaline igneous rocks or fenitization is usually made by the presence of peralkaline amphiboles and pyroxenes and/or strong turbidity in alkali feldspar-rich rocks which is due to disseminated ferric iron oxides. However, these features are not always present and furthermore turbidity in feldspar is not confined to conditions of strong alkalinity. The most subtle indication of strong alkalinity in igneous rocks and fenites is the presence of red luminescence in feldspars due to substitutional Fe^{3+} in the structure. The quantity of Fe^{3+} may vary from 0.0*n* to greater than 0.*n* weight percent but the CL is always intense red and the conditions for excitation require only low electron beam voltage and current. The use of CL in feldspars for the identification of fenites as an exploration guide in carbonatites was first proposed by Mariano (1978).

8.3.3 *Classification of fenitization by CL*

INCIPIENT FENITIZATION

Incipient fenitization is developed by fenite solutions invading adjacent country rock or a pre-existing igneous rock in an alkaline complex. Incipient fenitization appears as veinlets, often on the micro scale, that consist predominantly of alkali feldspar. The fenite feldspar invasion may be accompanied by carbonate introduction, alkali pyroxenes and amphiboles, melanite, ferric iron oxides, and apatite with a dominant LREE activation as shown in CL emission spectra. In some cases, the invading fenite solutions develop predominantly peralkaline amphiboles and pyroxenes but in most cases they are almost exclusively alkali feldspar, the recognition of which is often not possible without resort to CL. Incipient fenitization may be observed in carbonatite complexes on a world level and usually appears at the contact with the carbonatite rock units; but it may also appear in country rock well removed from the carbonatite or in isolated igneous bodies that are genetically related to the carbonatite. An example of incipient fenitization is given for quartz diorite country rock, adjacent to the Ste. Honoré Carbonatite of Chicoutimi, Quebec, in Plate 11b, which shows a vein of red-luminescing K-feldspar and a central string of light violet-blue-luminescing apatite. This vein is transecting a mottled blue- and green-luminescing microperthite which is the major mineral of the country rock. Large anhedral grains of yellow-luminescing country rock apatite are in sharp contrast to the violet-blue apatite of the fenite vein. Orange-luminescing calcite appears in clusters, replacing earlier minerals, and also as microscopic grains enveloping the fenite vein.

In Figure 8.23, CL emission spectra for the apatite and feldspar from the fenite vein shown in Plate 11b are compared to the primary apatite and feldspar of the Chicoutimi country rock quartz diorite. The fenite apatite (Fig. 8.23c) shows the typical LREE-dominant activation, whereas the country rock apatite (Fig. 8.23d) shows major activation from Mn^{2+} and superimposed line spectra of REE with Dy^{3+} greater than Sm^{3+}, indicating HREE dominance. The vein feldspar (Fig. 8.23a) shows the broad-band spectrum for Fe^{3+} with a peak wavelength at approximately 710 nm. The country rock microperthite (Fig. 8.23b) shows the very broad Ti^{4+} band at approximately 475 nm and a discernible hump at 550 nm which is interpreted to be from Fe^{2+} activation in plagioclase.

Other CL examples illustrating incipient fenitization veinlets intruding country rock in the carbonatite environment of Kangankunde, Malawi, and Cerro Impacto of Estado Bolivar in Venezuela are shown by Mariano (1978, p. 50). It should be emphasized that this type of fenitization is so subtle that it may often totally escape recognition without the use of CL.

MASSIVE FENITIZATION

In massive fenitization, pre-existing rocks, both country rock or igneous rocks of the carbonatite complex, are totally replaced on a massive scale by fenite mineralization. Usually remnant texture of the pre-existing rock is preserved. However, when carried out to the extreme case, the fenite rock may assume an igneous texture (Haggerty & Mariano 1983). An example of massive fenitization where remnant country rock feldspar is almost totally engulfed and replaced by fenite feldspar and calcite is shown in Plate 11c for the fenitized country rock of the Fen from the Filadelfia road cut in Ulefoss, Norway. It exemplifies incipient fenitization on an advanced massive scale where the country rock is almost totally replaced.

INTERNAL FENITIZATION

As crystallization proceeds in an alkaline magma the late stage fluids become increasingly enriched in alkalis, and iron in the system becomes stabilized in the ferric state. These solutions react with previously formed primary minerals along their edges and within their microfractures. They especially act on feldspars, replacing them with a ferric iron-substituted feldspar. The alkali solutions also act on primary amphiboles or pyroxenes, mantling them with peralkaline amphibole or pyroxene mineralization. Although this type of mineral alteration operates within a closed system without the introduction of extraneous solutions, it is a replacement phenomenon similar to deuteric effects. The term 'internal fenitization' is used to describe this process by some because the alteration products are identical to those found in other forms of fenitization. Internal fenitization is not specific to carbonatite-associated alkaline igneous rocks. It may appear in nepheline syenite complexes, in syenite rocks not related to carbonatites, and especially in alkali granites.

In many cases the alteration minerals that are formed during internal fenitization can be recognized during normal petrographic examination but more subtle alterations can only be observed by CL. In Plate 11e an example of internal fenitization of a hornblende syenite from the McClure Mountains of Colorado is demonstrated by CL. The illustration consists predominantly of primary perthitic feldspar with blue Ti^{4+}-induced luminescence. The feldspar grains have red-luminescing outlines due to peripheral replacement by a secondary feldspar that is activated by Fe^{3+} impurities. In this case the alkali build-up is very subtle and is indistinguishable without resort to CL. Specks of

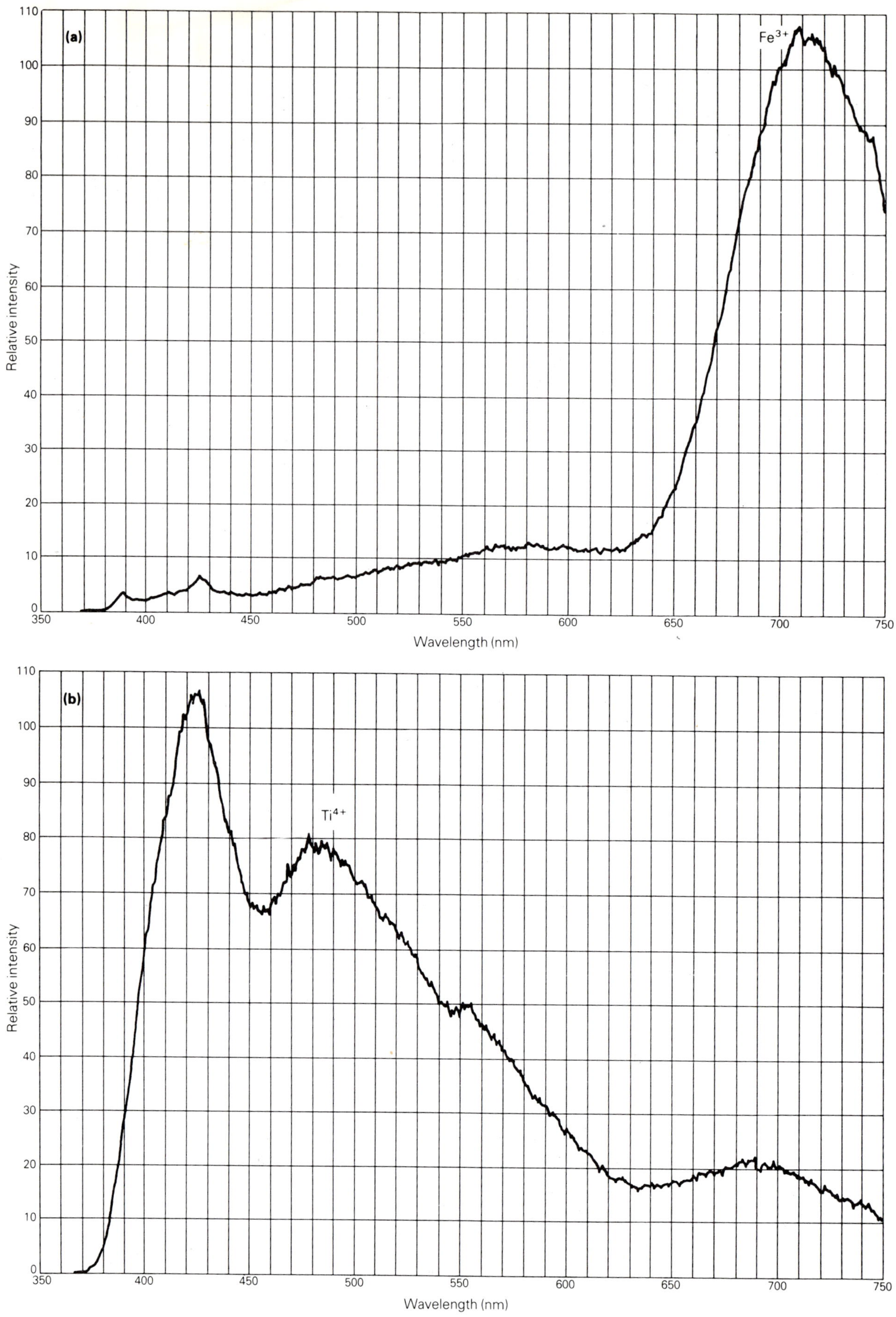

Figure 8.23 CL emission spectra for a quartz diorite from Ste. Honoré Carbonatite, Chicoutimi, Quebec, showing incipient fenitization of country rock (Plate 11b): (a) CL emission spectrum from fenite feldspar – the CL is red (CL conditions: 15 kV, 0.16 mA); (b) CL emission spectrum from feldspar in host rock – the CL is blue (CL conditions: 15 kV, 0.5 mA); (c) CL emission spectrum from apatite in fenite vein – the CL is blue (CL

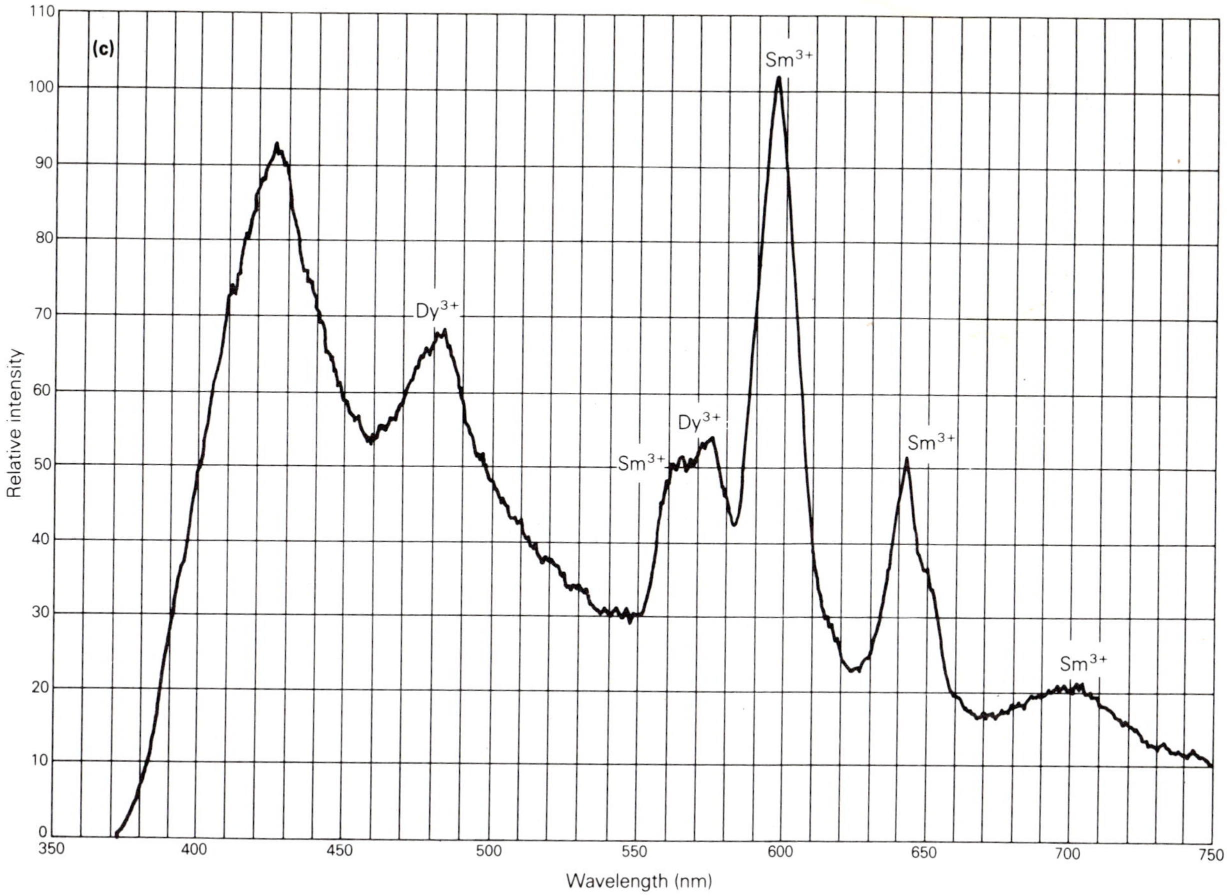

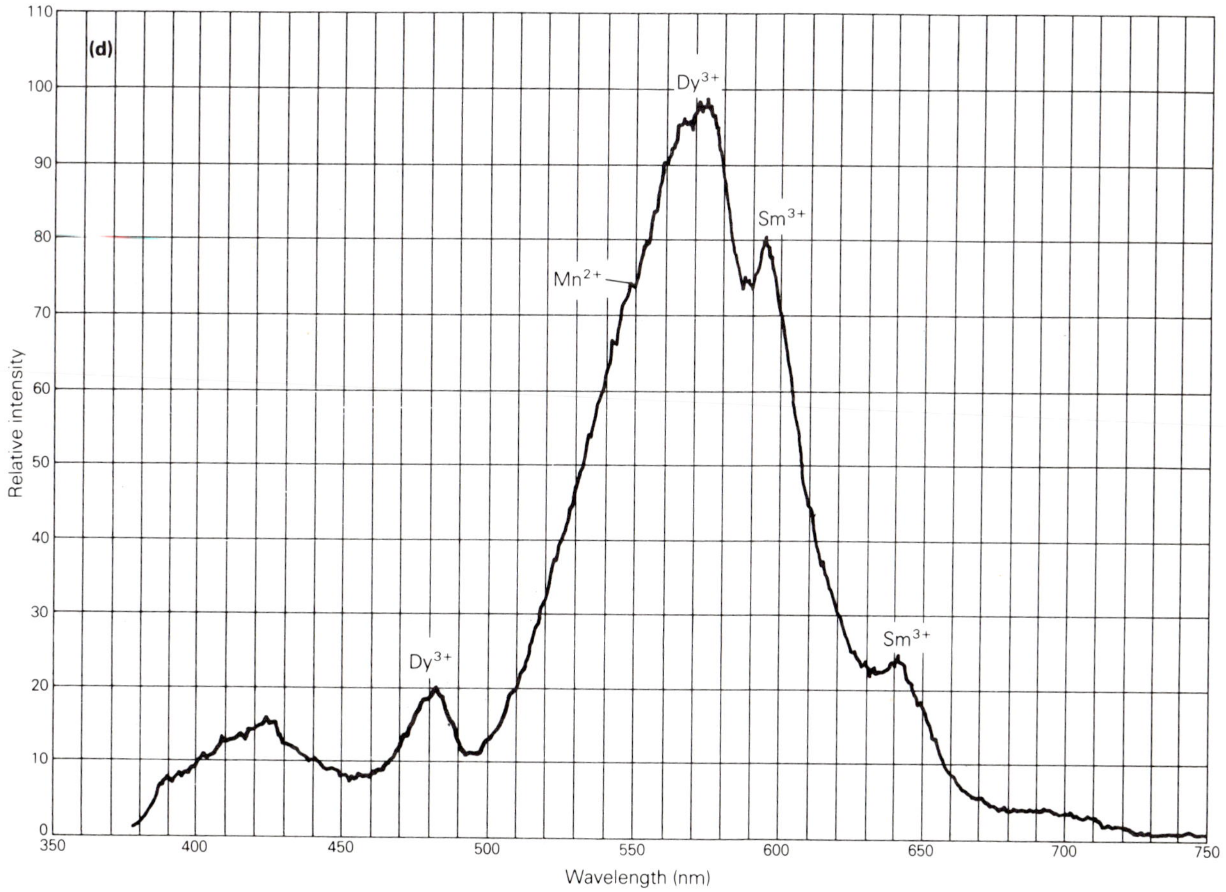

conditions: 15 kV, 0.8 mA); (d) CL emission spectrum from a manganese-activated apatite in host rock – CL is yellow (CL conditions: 10 kV, 0.6 mA). For further discussion of this example, see text and also the comments in Section 8.6 on peak assignment in the blue region.

calcite are also observed with orange CL. They tend to be localized along the feldspar grain boundaries, in association with the red-luminescing bands, suggesting co-crystallization.

Red CL in feldspars resulting from Fe^{3+} activation may be a primary igneous feature and is indicative of conditions of high alkalinity or strong oxidation. For example, in the Devil's Tower of Wyoming the early crystallizing feldspar phenocrysts show blue luminescence, and their CL emission spectra give no indication of Fe^{3+} activation. The outer edges of the phenocrysts and microfractures show brilliant red CL from late stage Fe^{3+}-activated feldspar crystallization.

No feldspathoid phenocrysts appear in the Devil's Tower rocks but pyroxene phenocrysts consisting of aegirine-augite, with a thin outer mantle of aegirine, also indicate a late stage increase of alkalis and Fe^{3+}. The aphanitic groundmass is composed predominantly of microscopic orthoclase laths in subparallel position, intergrown with needles of aegirine and disseminated grains of sodalite. The groundmass orthoclase gives a strong red CL, establishing that, during the crystallization of the aphanitic components, the alkalinity was sufficiently high to stabilize iron in the ferric state. The sodalite is easily identified by an orange luminescence. It comprises about 15 volume percent, and is the dominant feldspathoid of the Devil's Tower rocks. The average grain size of the sodalite is less than 50 μm and it escaped definitive recognition by Pirsson (1894) and Johannesen (1907). The Devil's Tower rocks are an excellent example of how routine use of CL can aid the petrographer in mineral identification and recognition of alkali build-up with progressive crystallization.

An interesting example of contrasting CL in feldspars is shown in Plate 11d for a brecciated rock from the Fen Carbonatite of Norway. Feldspar fragments show both red and blue CL. The red fragments are from fenites while the blue are from the nonfenitized parts of the country rock.

8.3.4 Ferrous ion activation in feldspars

Ferrous iron is a well known quencher for luminescence in many phosphors. However, there is strong evidence, based on controlled synthesis experiments and on mineral samples from several world localities, that supports Fe^{2+} as an activator for CL in plagioclase feldspar.

In a feldspar synthesis program conducted by the late Jun Ito, anorthite containing ferrous iron was found to give a strong green CL with an emission peak at approximately 550 nm (Fig. 5.3). The material was crystallized by heating hydrothermally for 20 hours a mixture of ferrous oxalate and chemically prepared amorphous hydrated calcium alumino silicate with the composition of $Fe_{0.1}Ca_{0.9}Al_2Si_2O_8{*}nH_2O$ at 800 °C under total P_{H_2O} of 1 kbar. The crystals produced from this synthesis were examined by X-ray diffraction and found to give a pattern for anorthite. They were further examined by scanning electron microscopy and energy dispersive X-ray spectroscopy (EDS) which established that the crystallized product was one homogeneous, crystalline phase with an EDS spectrum, appropriate for anorthite, that shows a noticeable Fe (K-alpha) peak at 6.403 keV.

Several areas were examined by EDS and each point gave the same spectrum with constant peak heights for all of the elements including Al, Si, Ca, and Fe, indicating that the Fe is substitutional in the anorthite phase. An EDS spectrum and secondary electron image for the synthetic anorthite is shown in Plate 13.

A Mössbauer spectrum for the anorthite is given in Figure 8.24. It shows mainly Fe^{2+} as indicated by the two main absorption peaks. The presence of Fe^{3+} cannot be resolved from the spectrum. The position for one of the Fe^{3+} peaks is coincident with the larger Fe^{2+} peak and the other is buried in between the two peaks of Fe^{2+}. One may, however, estimate the relative fraction of Fe^{3+}/Fe^{2+} as approximately 0.2 (with a large error), from the area difference of the two large peaks.

Trace element analyses performed by Jun Ito on the synthetic anorthite showed only Ag as an impurity above the 0.00n weight percent level. It is unlikely that this small amount of Ag impurity is the cause of the green luminescence, particularly since flux grown labradorites doped with Ag, also synthesized by Jun Ito, failed to show any CL.

Despite the fact that Fe^{2+} is an effective quencher for luminescence in many host crystals, the results on the Fe^{2+}-doped synthetic anorthite give strong evidence that ferrous iron is functioning as an activator in plagioclase. Furthermore, many analyzed natural plagioclases, both terrestrial and lunar, contain as much as 1 weight percent Fe, and for lunar feldspars the iron is mostly in the ferrous state. These feldspars exhibit luminescence indicating that the iron is not functioning as a quencher at or near the 1 weight percent level. For many terrestrial plagioclases that show green CL with a peak emission at about 555 nm, EDS spectra reveal the presence of Fe in at least the 1 weight percent level. When no emission occurs in the red part of the visible spectrum, it is safe to assume that the iron is not occurring in the ferric state. Finally, the difference in peak emission wavelength between Mn^{2+}- and Fe^{2+}-activated synthetic plagioclase is sufficient for their differentiation. Hopefully, more concentrated research will be carried out in this area by synthesis of plagioclase with controlled impurities and combined CL, X-ray diffraction, EPR, and Mössbauer studies to establish definitively whether or not Fe^{2+} is indeed an activator in plagioclase.

An example of Fe^{2+} activation in a natural feldspar is shown in Plate 11f for a quartz andesite from Safford, Arizona (Robinson & Cook 1966). The green patches of mineralization in this illustration are secondary oligoclase that is co-crystallized with yellow-luminescing apatite and zoned brown- and green-luminescing anhydrite. The spectrum from the green-luminescing patches is shown in Figure 8.25 and compares favorably with the synthetic Fe^{2+}-activated anorthite of Figure 5.3.

Green CL in secondary plagioclase has been observed in several occurrences and is remarkably different from the red CL seen in fenite-derived feldspars of other environments.

8.3.5 Ti^{4+} activation in feldspars

The most frequently oberved color of luminescence in feldspars is blue. From luminescence studies on phosphors (Leverenz 1968) it is known that SiO_2 and several silicates exhibit CL emission spectra as broad bands that extend from

Figure 8.24 Mössbauer spectrum of synthetic anorthite ($Fe_{0.1}Ca_{0.9}Al_2Si_2O_8$). The doublet present is due to ferrous iron. Small quantities of ferric iron may be present also but cannot be resolved due to interference from the dominant ferrous iron doublet. (Spectrum furnished by Prof. J. H. Chen.)

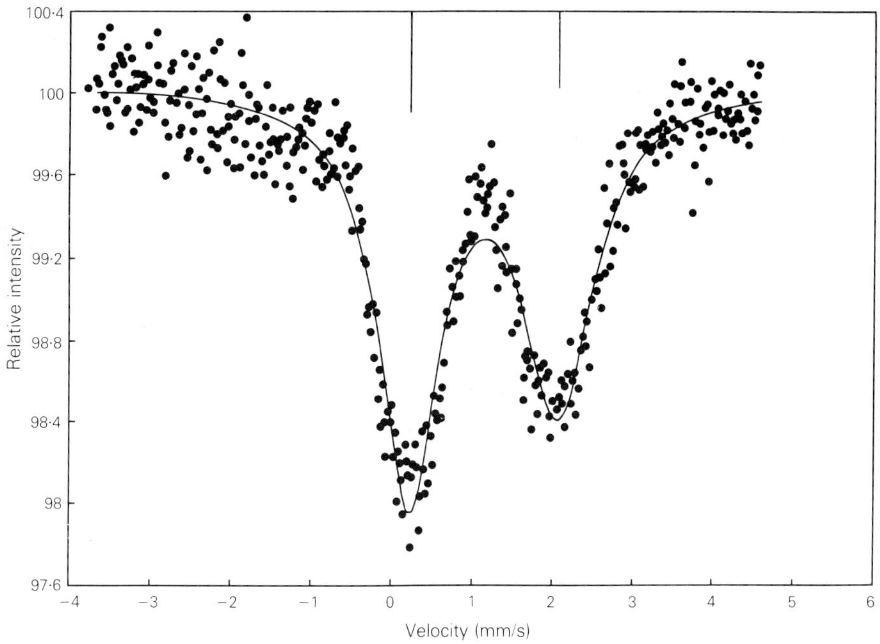

the UV into the blue part of the visible spectrum. The emission is considered to be associated with the SiO_4 group or the Si—O bond. It has been further established that the introduction of titanium impurities intensifies the otherwise weak intrinsic CL emission band of these silicates (see Leverenz 1968, pp. 192, 220–1). The intensification action of Ti as an activator in silicate phosphors is attributed to the anomalously high electronic polarization of an oxygen ion, O^{2-}, in the field of an adjacent titanium ion, Ti^{4+} (Shockley 1948). In addition, according to Leverenz (1968, p. 193) 'Ti prefers to be 6-coordinated with O, as compared with the normal 4 coordination of Si with O, so traces of Ti produce a structural perturbation which may upset the normal host-crystal selection rules in the immediate vicinity of a Ti impurity atom (ion).'

An example of a blue-luminescing feldspar is provided in Plate 11a for a secondary K-feldspar vein transecting other minerals in a quartz andesite from Safford, Arizona. This example of secondary K-feldspar contrasts vividly with the bright red CL of the fenite K-feldspar (e.g., Plate 11b).

A spectrum from a natural blue-luminescing feldspar is shown in Figure 8.26.

8.3.6 Multi-activation in feldspars

Examples where feldspars show more than one CL band are given in Figures 8.26 and 8.27. Figure 8.26 is for an orthoclase phenocryst from the Mt Fubilan quartz latite porphyry of Ok Tedi, New Guinea (Bamford 1972). Orthoclase in this rock occurs as large phenocrysts together with smaller plagioclase phenocrysts. The orthoclase phenocrysts show a light blue CL and the emission spectrum includes a broad band with a peak wavelength of about 450 nm. This band is attributed to Ti^{4+} impurities that act to intensify the weak UV and blue luminescence that appears in many silicates. A higher intensity band also appears at approximately 700 nm which is the usual emission for Fe^{3+}. Without resort to CL emission spectra, the red band would not have been assumed to be present. The smaller plagioclase phenocrysts of the quartz latite porphyry show red luminescence and their CL emission spectra contain only the red band with a peak maximum at about 710 nm. This type of luminescence in feldspars from porphyry copper deposits is unusual and it suggests increasing oxygen activity with progressive crystallization of feldspar phenocrysts.

Figure 8.27 is another example of a feldspar CL spectrum showing two emission bands. The feldspar is an oligoclase phenocryst from a chalcopyrite-mineralized quartz andesite from Safford, Arizona. The color of luminescence for the phenocryst is green but the spectrum contains a broad band at approximately 700 nm from Fe^{3+} activation. The band at 555 nm which is responsible for the green luminescence is assumed to be due to Fe^{2+} activation since Mn^{2+} emission occurs at about 570 nm. The occurrence of more than one activator in plagioclase is not unusual. It is possible that the emission at 555 nm may be a combination of both Fe^{2+}- and Mn^{2+}-activation but if the band is exclusively Fe^{2+} it suggests the possibility of using CL spectra in some plagioclases for Fe^{2+}/Fe^{3+} determination.

8.4 Cathodoluminescence zoning in minerals

Several minerals exhibit rhythmic zoning with CL that is not evident in normal petrography. The cause of the luminescence in the different minerals may be due to subtle defects or to impurities that occur at the ppm level, impurities which are often not detectable with an electron microprobe or even proton excitation (Blank *et al.* 1984). Some of the minerals that may exhibit CL zoning include:

anhydrite	diamond	lepidolite	topaz
apatite	dolomite	nepheline	tremolite
baddeleyite	emerald	plagioclase	wöhlerite
calcite	fluorite	polylithionite	zircon
cassiterite	garnet	quartz	
corundum	isokite	sphalerite	

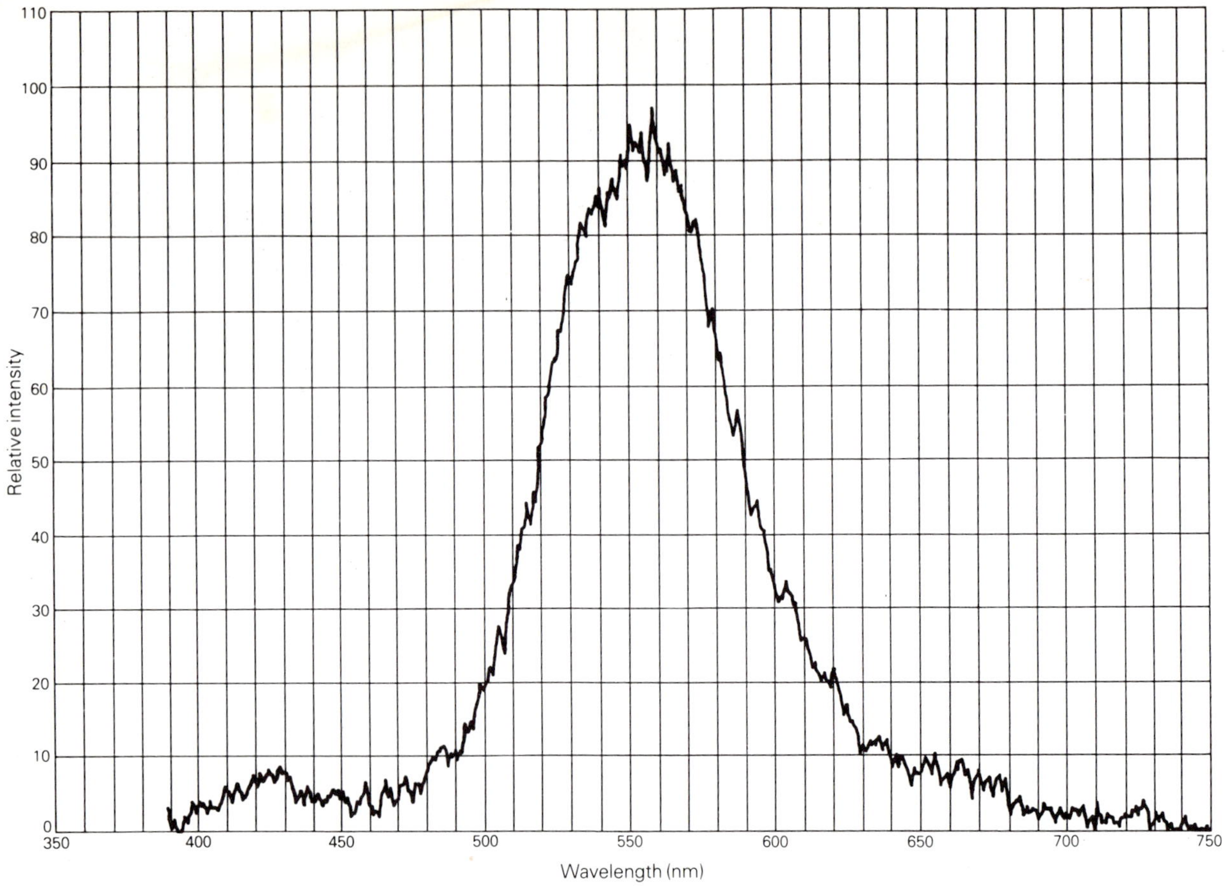

Figure 8.25 CL emission spectrum of Fe^{2+}-activated oligoclase from quartz andesite, Safford, Arizona. The peak wavelength at 550 nm is more favorable for Fe^{2+} activation than for Mn^{2+} activation.

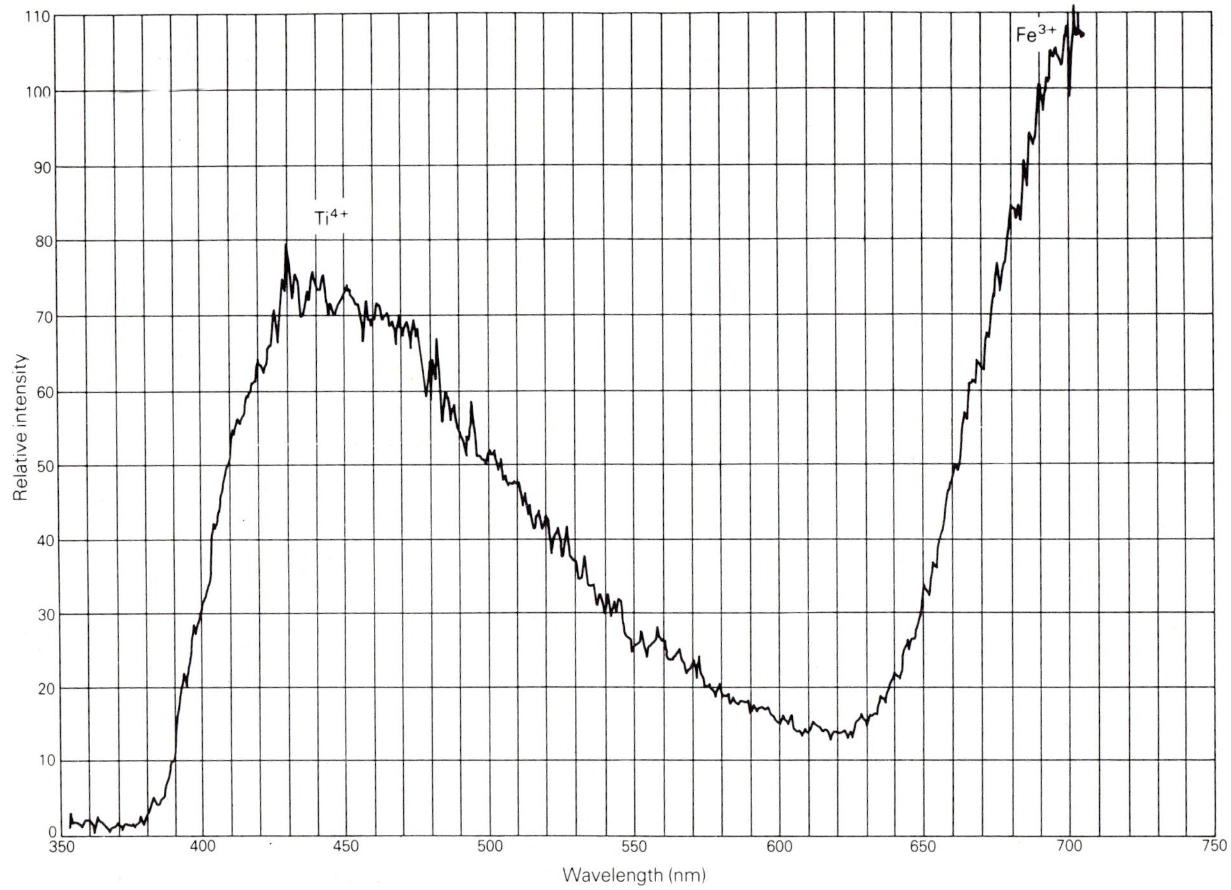

Figure 8.26 CL emission spectrum from an orthoclase phenocryst showing two emission bands. This phenocryst is from the Mt. Fubilan quartz latite porphyry of Ok Tedi, New Guinea (Bamford 1972). The luminescence of this crystal is light blue. Without resort to CL emission spectra, the presence of the red band could not have been predicted. The broad band peaking at about 440 nm is attributed to lattice defects produced by substitutional Ti^{4+}. The band which peaks at approximately 700 nm is typical of Fe^{3+} activation. CL conditions: 9.5 kV, 0.6 mA.

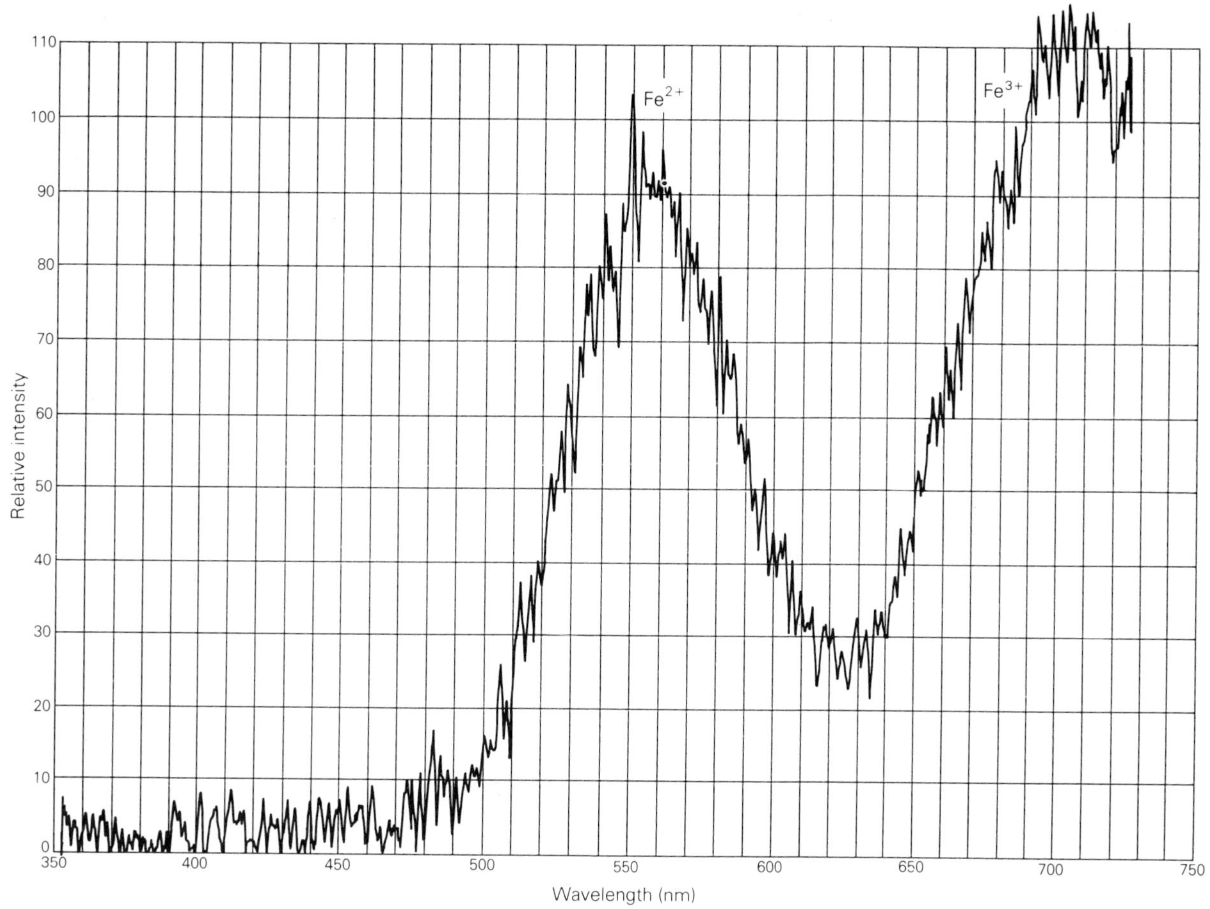

Figure 8.27 The CL emission spectrum from a plagioclase from Safford, Arizona. The CL of this sample is green to the unaided eye. Both ferrous and ferric iron peaks are present in the spectrum. CL conditions: 11 kV, 0.4 mA.

In some cases the CL is due to specific activators which have been determined by examination of CL emission spectra. In other cases the cause of the CL can only be inferred.

Examples of CL zoning in minerals are illustrated for apatite in Plate 12a, for fluorite in Plate 12b, for baddeleyite in Plate 12f, for anhydrite in Plate 12d, for anhydrite in Plate 12e, and for calcite in Plate 12c.

8.4.1 *Anhydrite*

Striking examples of rhythmic zoning have been observed in anhydrite occurring as a hydrothermal alteration mineral in porphyry copper deposits. The anhydrite is found in veins or adjacent to them, where it is intimately crystallized with chalcopyrite, apatite, calcite, secondary K-feldspar, and secondary plagioclase.

In the porphyry copper deposit of Safford, Arizona, hydrothermal anhydrite in quartz andesites displays rhythmic zoning with luminescence colors of green and various shades of tan (Plates 12d & e).

The most frequent cause of CL in anhydrite is the trivalent lanthanides Sm^{3+} and Dy^{3+} (Mariano & Ring 1975). Since these activators emit sharp line spectra, whose wavelength is largely unaffected by the crystal field of the host crystal, their identification is unambiguous. Spark source mass spectrographic analyses performed on anhydrite from the Faraday Mine, Bancroft, Ontario (American Museum #36,014) showed Sm and Dy values of 3 ppmw ± a factor of 3. The same analysis of Safford anhydrite failed to reveal these elements at a detection limit of 1 ppmw; yet in both cases the CL spectrum had well resolved line spectra for Sm^{3+} and Dy^{3+}.

Emission spectra for Sm^{3+}- and Dy^{3+}-activated anhydrites from Safford, Arizona, Ajo, Arizona, and the Faraday Mine of Bancroft, Ontario, are shown in Figures 8.28, 8.29, and 8.30. In Figure 8.31 the CL emission spectrum for Mn^{2+}-activated anhydrite from Safford, Arizona is also shown.

Green CL that appears in the Safford anhydrite is assumed to be from Mn^{2+} activation. It has a broad band which peaks at approximately 520 nm and shows an asymmetrical tail on the long wavelength side of the peak. Activation by Mn^{2+} in anhydrite has been reported by Medlin (1963a) for ThL and also for X-ray excitation. He reports a similar asymmetric broad band of green luminescence near 5,000 Å.

Thermoluminescence spectra for anhydrite showing Dy^{3+} and Sm^{3+} activation are also reported by Tarashchan (Marfunin 1979, p. 236).

In general, the mineral anhydrite is relatively pure with only limited amounts of substitutional impurities. This characteristic must be partly instrumental in producing the excellent resolution and effective CL from activators such as Sm^{3+} and Dy^{3+} that occur only on the few ppm level.

Cathodoluminescence data for anhydrites from varied geological localities is given in Table 8.2. In some cases no CL is observed in anhydrites. This is assumed to be due to lack of sufficient activator impurities because anhydrite does not easily accept iron or other possible quencher

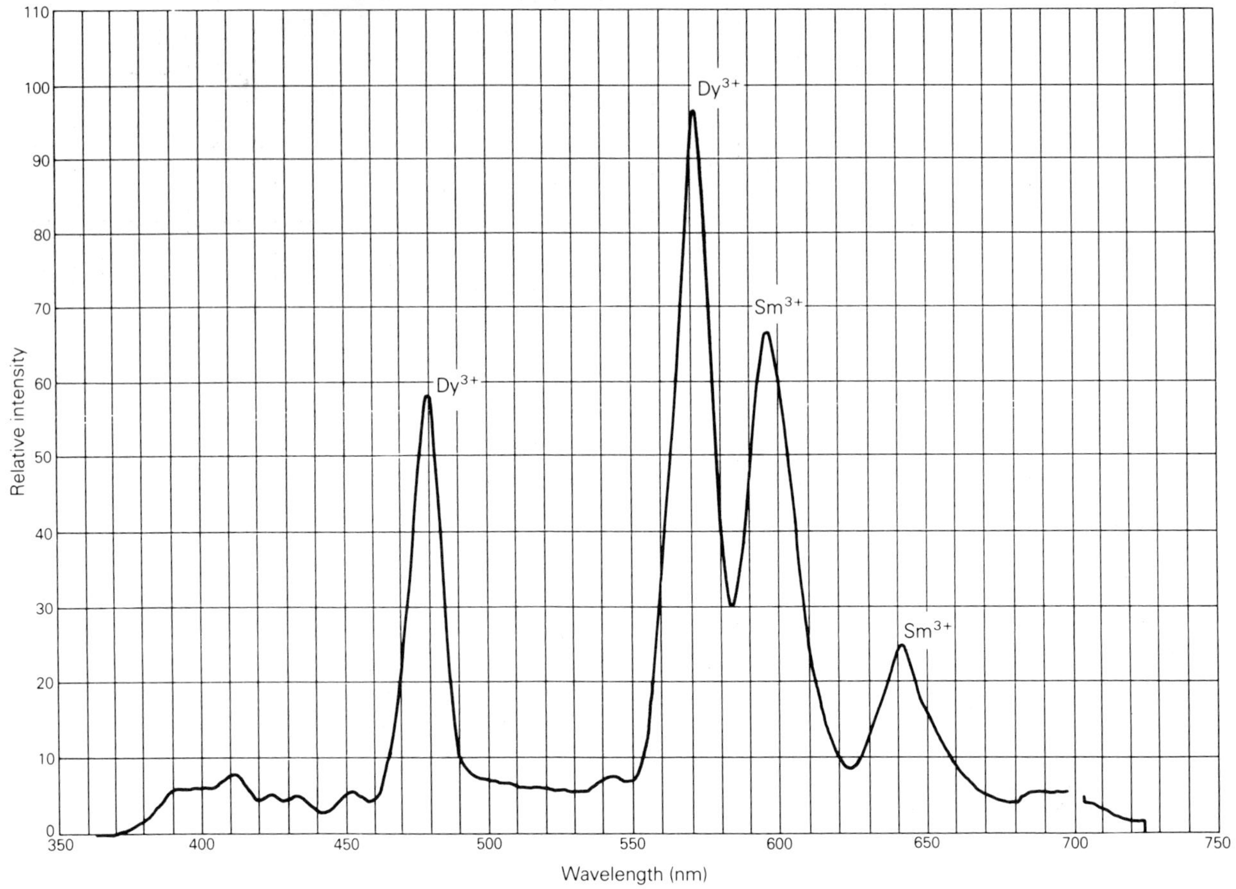

Figure 8.28 CL emission spectrum from anhydrite, Safford, Arizona. The CL is tan. CL conditions: 16 kV, 0.6 mA.

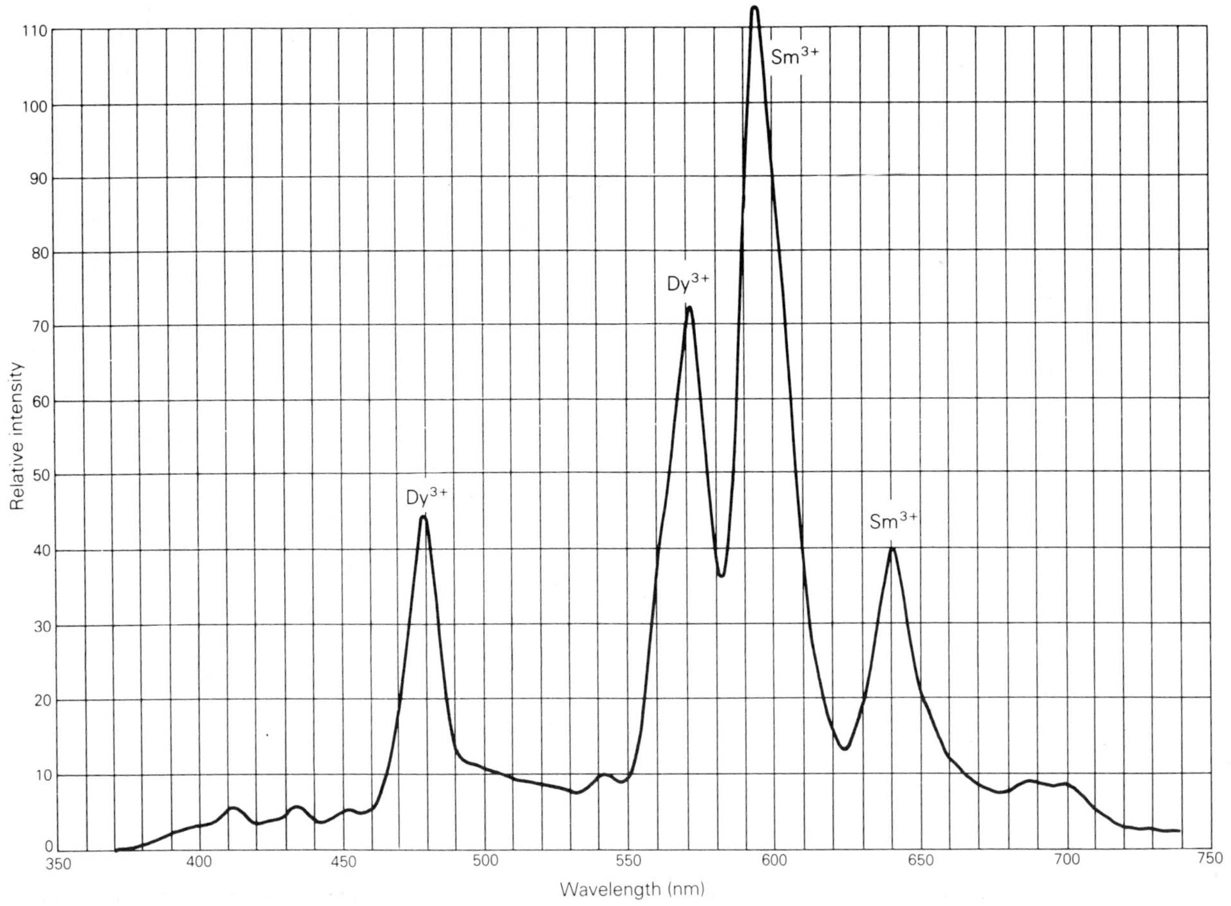

Figure 8.29 CL emission spectrum from anhydrite, Ajo, Arizona. CL color is salmon. CL conditions: 14.5 kV, 0.3 mA.

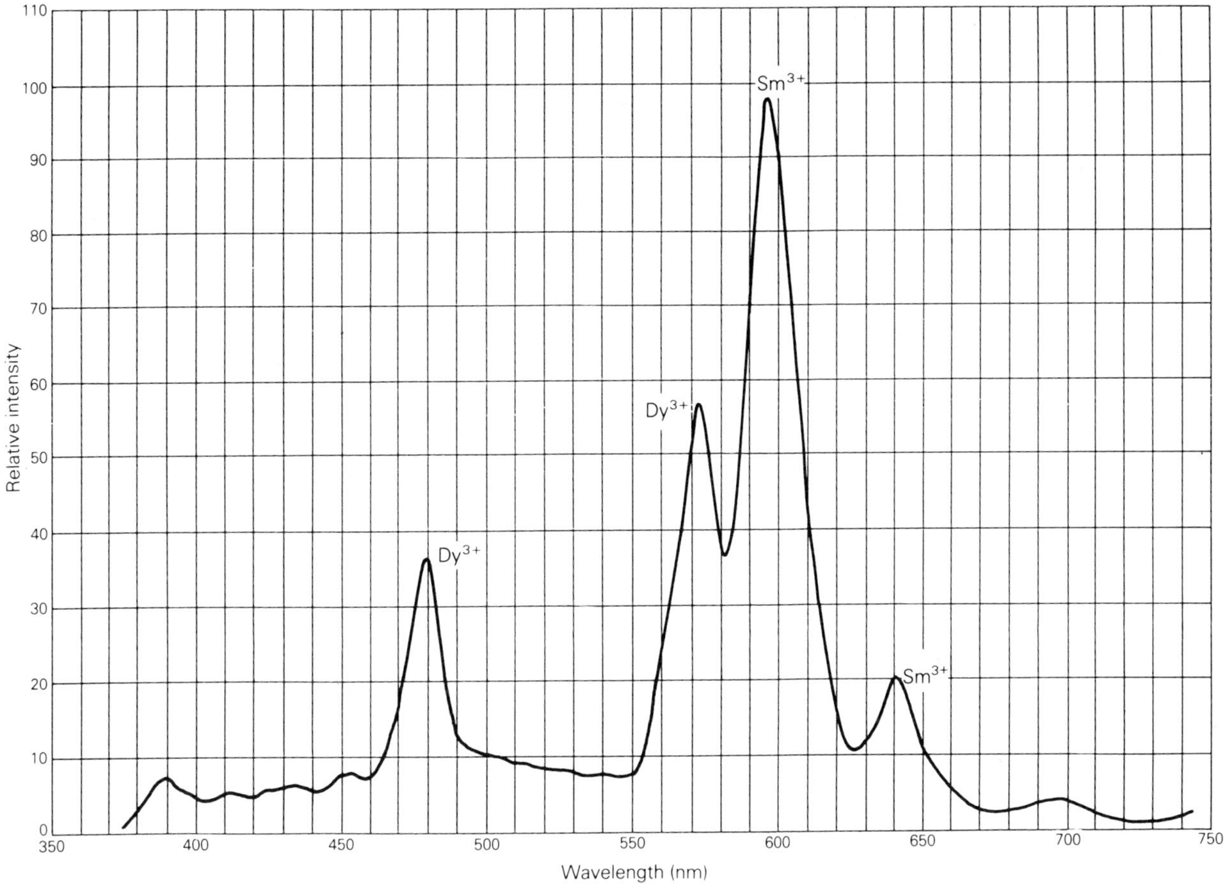

Figure 8.30 CL emission spectrum from anhydrite, Faraday mine, Bancroft, Ontario. CL is tan. CL conditions: 11.5 kV, 0.7 mA.

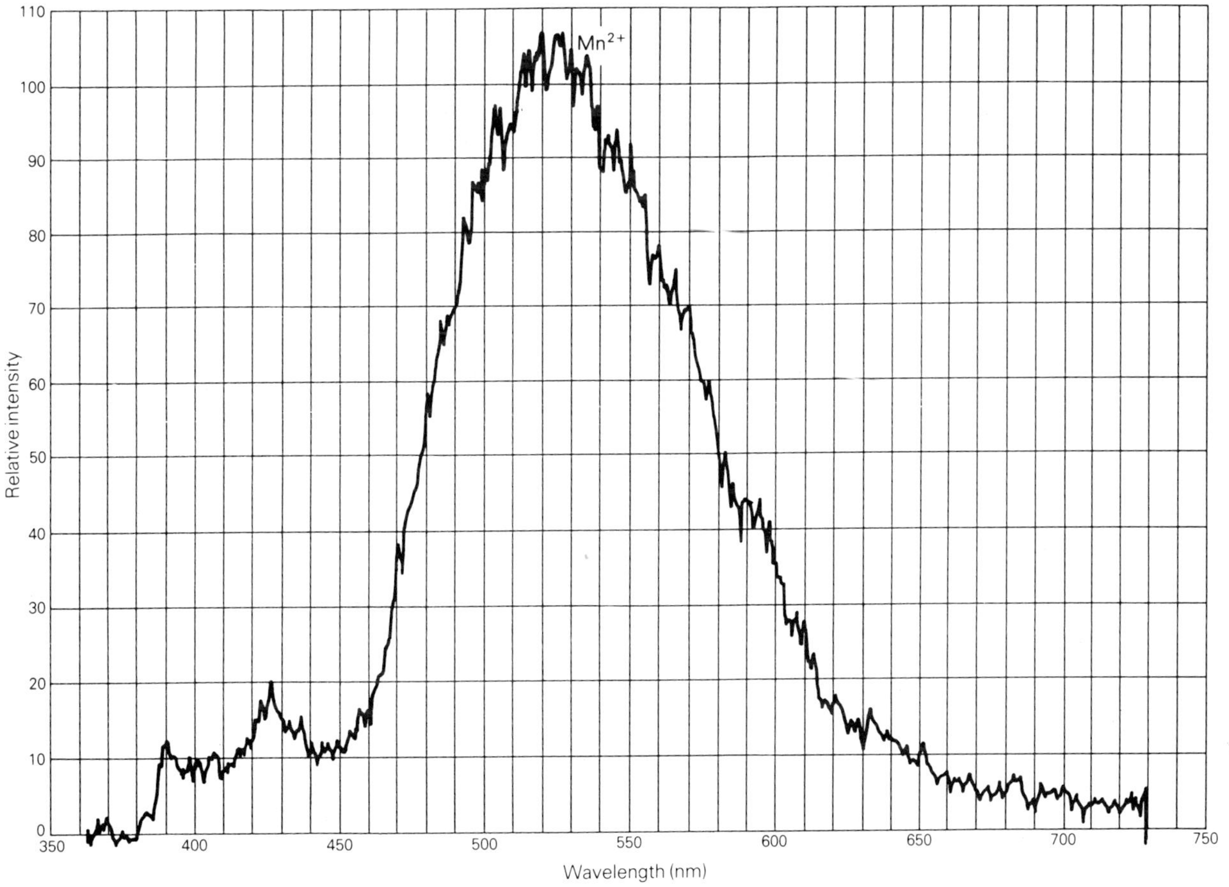

Figure 8.31 CL emission spectrum from anhydrite from Safford, Arizona. The CL is a soft green. CL conditions: 9.5 kV, 0.2 mA.

Table 8.2 CL data on anhydrites from various geological localities.

Locality	Environment	CL color	Activators
Safford, Arizona	hydrothermal in porphyry copper deposit	tan and green, zones	Mn^{2+}, Sm^{3+}, Dy^{3+}
Ajo, Arizona	hydrothermal in porphyry copper deposit	tan and green, zones	Mn^{2+}, Sm^{3+}, Dy^{3+}
Heleco, Puerto Rico	hydrothermal in porphyry copper deposit	tan and green, zones	Mn^{2+}, Sm^{3+}, Dy^{3+}
Braden, Chile	hydrothermal in porphyry copper deposit	tan and green, zones	Mn^{2+}, Dy^{3+} $>Sm^{3+}$
Faraday Mine, Bancroft, Ontario	gangue mineral in Faraday uranium mine	tan	$Sm^{3+}>Dy^{3+}$
Sam Goosly Lake, British Columbia	hydrothermal replacement in Ag–Cu sulfide ore deposit	tan	Mn^{2+}, $Sm^{3+}>Dy^{3+}$
Taraskan Pyrenees, France, Smithsonian R11,402	rock salt deposits	tan	$Sm^{3+}>Dy^{3+}$
Edwards, New York, Am. Mus. # 27,161	Balmat–Edwards hypogene gangue mineral in metalliferous veins	light orange	$Sm^{3+}>Dy^{3+}$
Stassfurt, Germany, Smithsonian #R6,016	rock-salt deposits	green	Mn^{2+}, $Sm^{3+}>Dy^{3+}$
Patterson, New Jersey, Smithsonian #124,349	low temperature hydrothermal in trap rock	no CL	
Zipaquira', Cundinamanca, Colombia, Smithsonian #122,432	low temperature hydrothermal in salt deposit	no CL	
Naica, Chihuahua, Mexico, Smithsonian #120,716	hypogene gangue mineral in Pb, Zn, Ag, (Au, Cu) veins	no CL	

elements. It may be largely temperature controlled. Anhydrite crystallizing under low temperature conditions may be relatively free of impurities.

Rare earth element distribution patterns for hydrothermal anhydrites from the porphyry copper deposits Sierrita Pit, Arizona, and Cerro Colorado, Panama, have been reported by Morgan and Wandless (1980). They report neutron activation analyses for Sm in anhydrite from the two localities as 23.6 and 13.2 ppmw respectively. These quantitative analyses further establish that only a few ppmw of Sm and Dy are found substitutional in hydrothermal anhydrites.

The rhythmic zoning that appears in hydrothermal anhydrites suggests that each zone with different CL coloration is reflecting a change in the composition of the source fluids. In some veins the anhydrite grains show a homogeneous tan luminescence, exclusively from Sm^{3+} and Dy^{3+} activation, while other grains nearby show only green CL from Mn^{2+} activation. In contrast some veins show a mixture of grains with green CL, tan CL, and include grains with rhythmic zoning in tan and green. In studies involving the distribution of lanthanides in coexisting hydrothermal minerals CL can be a useful tool for detecting REE activation in minerals and for discerning the existence of incomplete reaction between zoned crystals and compositional changes in the residual hydrothermal fluids.

8.4.2 *Calcite*

Zoning in calcite has been mostly observed as overgrowths in hydrothermal mineralization. In 1973 the writer extracted a sample of rock outcropping on the top rim of the Kerimasi carbonatite volcano in the Gregory Rift Valley of Tanzania. This rock contains phenocrysts of calcite in an aphanitic matrix of calcite. Examination of the rock by CL showed luminescent zones in the calcite phenocrysts (Plate 12c). The phenocrysts also exhibited flow texture in the nonluminescing aphanitic calcite matrix (Mariano 1978, pp. 55–6). This was interpreted by Mariano (1978) and Mariano and Roeder (1983) as a lava flow unit, mostly because of the textural characteristics and the CL zoning. According to Hay (1983) most of the calcite from the Kerimasi volcano is a pseudomorphic replacement after nyerereite ($Na_2Ca(CO_3)_2$). Dawson (1964), earlier, had suggested that the rocks were calcite replacements of leached sodium carbonate flows. If it is assumed that sodium carbonate flows were replaced by calcite, several questions are apparent. What is the source of the calcite? Since these rocks outcrop on the top rim of the volcano, the source would have to be from depth. This requires that a later extrusive pulse of igneous carbonate has replaced the sodium carbonate flow. If so, how is the zoning in the calcite phenocrysts explained, and why are the phenocrysts exhibiting CL while the matrix calcite is quenched?

The Kerimasi calcite phenocrysts contain numerous vestiges of spherical inclusions that may have originally contained trapped liquids and gases. Within these spherical forms solid phases have been detected by electron microprobe analysis (P. L. Roeder, pers. comm.) and by SEM-EDS (A. N. Mariano). They include Na carbonate, Na–Ca carbonate and K–Ca carbonate. A silicate phase that may have originally been a silicate liquid was also observed by P. L. Roeder.

Further unpublished work by P. L. Roeder and the writer has revealed the presence of monticellite inclusions in the calcite phenocrysts. Monticellite is a high-temperature mineral and is unlikely to form under conditions of surface alteration. Monticellite also appears as an accessory mineral in søvite which outcrops in the crater floor of Kerimasi (Mariano & Roeder 1983).

The alkali carbonate inclusions identified by P. L. Roeder are similar to the minerals from the flows of the adjacent active carbonatite volcano, Oldonyo Lengai. If the calcite phenocrysts were due to replacement of a leached sodium carbonate flow, how is the presence of the alkali carbonate inclusions explained? Many of the calcite phenocrysts have been fractured during flow, and sections of the phenocrysts have been displaced or faulted. However, the zones revealed by CL examination indicate that the zoning was preflow. If the displaced crystal fragments are rearranged, the zone bands can be seen to have continuity across the displaced fragments. This is good evidence that the zoning was a primary feature and that the calcite crystals are not products of replacement.

Ion microprobe analysis for Sr isotope measurements was conducted by D. E. Scatena-Wachel on a portion of the same sample used in the studies by Mariano (1978) and Mariano and Roeder (1983). The $^{87}Sr:^{86}Sr$ ratios for 17 analyzed areas including phenocrysts and matrix were found to be the same, with an average value 0.7038 $\pm$0.0004 (2σ error) (Scatena-Wachel *et al.* 1986). This Sr value is indicative of upper mantle rocks (average = 0.7037 for oceanic-island basalts). The average value for crustal material is 0.719 (Faure & Powell 1972).

In addition to the CL zonation, the strontium isotopic data combined with the presence of monticellite inclusions and Na and K carbonate phases within fluid inclusion vestiges, offers further evidence that the calcite phenocrysts in the Kerimasi volcanic rocks are features of primary crystallization.

8.5 Clustering in apatite

The substitution of lanthanides in the apatite structure requires a valency compensation, since the predominantly trivalent lanthanides displace divalent calcium. Some of the variations possible for apatite are as follows; CaCa$\Rightarrow$ LnNa; CaP $\Rightarrow$LnSi; CaF $\Rightarrow$LnO.

When apatites are observed with CL, a mottled appearance is often revealed due to variations in luminosity throughout the grain. Under normal light, without electron excitation, the apatites appear clear and homogeneous. The mottled appearance is due to uneven distribution of activators throughout the apatite crystal. Furthermore, it is demonstrated by variations in luminosity and by electron microprobe analyses that the local substitution of lanthanides in apatite is accompanied by an increase in silicon (at the same immediate site) and a decrease in calcium and phosphorus.

Thus the uneven distribution of an activator in a host due to crystal-chemical clustering produces a mottled luminescent effect. During electron microprobe scanning of fluorapatite from the Britholite Zone at Oka, Quebec, it has been observed that increases in rare earth content are accompanied by increased brilliance of blue CL. A correlation is therefore made between intensity of luminescence and lanthanide content of fluorapatite. The mottled effect visible with CL, and revealed also by the microprobe analyses, suggests the existence of clustering in the fluorapatite due to an A:X* substitution of the type 3+:4+ ($Ln^{3+}:Si^{4+}$) where the 3+ and 4+ charges will tend to be associated with the same oxygen ions in order to satisfy Pauling's rule. Then the substituted ions will tend to enter adjacent A and X sites, resulting in the formation of clusters having a different composition and luminescent intensity than that of normal apatite.

8.6 Summary

Many examples have been presented which show that CL observations, combined with the acquisition of emission spectra, can provide important information on trace element composition and trace element oxidation states. This information contributes significantly to the understanding of the nature of the crystallization environment.

The interpretations which have been made in the work described in this chapter have, for the most part, relied only on the major peaks in the spectra. It is clear, however, that the noise level of some of the spectra is sufficiently low that additional recordings could be made on an expanded scale to detect the minor peaks. This may be an important avenue of investigation.

The spectra which have been reproduced in this chapter are reproduced as faithfully as possible from the original charts with no smoothing. In some cases, this results in the noise level being quite noticeable, though not sufficient to change the interpretation. Also, most monochromators have nonuniform transmission functions and this can lead to small differences in spectra between different instruments, unless they are fully corrected. This nonuniform transmission effect appears to be significant in our instrument only in a narrow region centered about 450 nm.

The origins of the broad-band peak in the blue region in some samples, e.g., in Figures 8.13 and 8.15, is still open to question. In some cases I have been able to identify it as due to the Eu^{2+} activator, or to intrinsic luminescence, enhanced by Ti impurities. However, there may also be situations where there is a contribution from Ce^{3+} which can produce a broad-band peak in the ultraviolet–blue region. Further work, possibly with a spectral analysis system using quartz optics, will be needed to resolve this.

Similarly, as was noted in earlier chapters, there is additional information in the infra-red region which remains to be fully explored.

All of these important areas remain to be investigated. I hope that the survey of applications which has been presented in this chapter will encourage these investigations.

* The general formula for apatite may be written as $A_{10}(XO_4)_6F_2$.

Chapter 9

Some other applications of cathodoluminescence

9.1 Lunar samples and meteorites

Link, in 1946, proposed that luminescence occurs on the lunar surface, excited by the existing radiation environment, protons and UV, in the solar wind. Part of the impetus to general luminescence research in the 1960s derived from the interest in investigating this proposal and much of this was supported by NASA.

The results of this work are of general interest, but the excitation conditions chosen emphasize proton beams and UV rather than electron excitation. Protons apparently produce luminescence spectra similar to those produced by electrons, but cause appreciable damage to the specimen. They also are more difficult beams to generate than electrons and are not in common use.

Many of the publications resulting from this work contain excellent discussions on instrumental techniques, including excitation sources and spectral analysis and detection systems. The instrumental discussions are much more extensive than those provided in many of the papers which deal with terrestrial mineralogical and petrological studies using CL. This group of papers also includes useful information on proton bombardment damage and the relative efficiencies of different types of excitation.

Geake, Walker, Derham, and coworkers (Geake & Walker 1966, Geake *et al.* 1970, 1971, Derham *et al.* 1964) have studied the luminescence spectra of meteorites and lunar samples, usually under UV, visible, or proton excitation. Proton energies ranged from 60 to 120 keV. They found that the emission spectra shape, as measured with a grating spectrophotometer, was not a function of proton energy over this range of proton energies. Specimen damage by proton bombardment was appreciable and took place in times of the order of one minute at proton beam intensities of 60 keV and 1 μA. In at least one instance (Derham *et al.* 1964), they observed that electron excitation (5–40 keV) gave about the same color and relative efficiencies as low-energy protons (<100 keV).

Much of their work was done on unseparated lunar 'fines' and dust, and lunar rocks. However, they also investigated separated mineral components including plagioclases and enstatite. Their comments on enstatite have been expanded on in Chapter 4 and their comments on feldspars are included in the discussion in Chapter 5. They also noted that oldhamite (CaS) luminesced yellow (5,800 Å) and osbornite (TiN) did not luminesce (Derham *et al.* 1964).

Geake, Walker and coworkers continued to publish on their work in this area, and the references by Geake *et al.* (1977) and by Walker (1983) will guide the interested reader to their more recent work.

Edgington and Blair (1970) also used proton excitation in their studies of lunar materials. Their studies included measurements at −196 °C. They observed blue and red luminescence and an energy efficiency, for the plagioclase fraction, of approximately 1 percent.

Nash (1966) studied proton-excited luminescence of silicates with an interest in lunar implications. He used a proton beam (5 keV, 10^{14} protons/cm^{2}sec, 0.08 watts/cm^2) and found that several silicates produced a red flash upon initial exposure. He investigated, experimentally, the concept of synergism in connection with possible lunar luminescence (Nash 1973). This was investigated with the combined (simultaneous) use of proton and UV excitation of a lunar sample. The luminescence efficiency was only very slightly larger than would be calculated from the sum of the separate excitations (Fig. 9.1). Thus an important synergistic effect was ruled out, at least for this sample and this particular set of excitation conditions. Additional observations of proton-excited luminescence, darkening under proton irradiation, and absence of synergistic effects by combined UV and proton bombardment were published by Nash *et al.* (1970).

Reid and Cohen (1967) made a comprehensive study of enstatite from enstatite achondrites, using CL as one of their investigative tools. They noted blue, purple, and red luminescence, and a correlation between the intensity of the red component and the amount of manganese. For a more detailed discussion, see Chapter 4.

Grögler and Liener (1968), using 30 keV electrons, observed the same CL colors in the Khor Temiki aubrite as noted by others for proton excitation, i.e., enstatite as reddish-purple or blue and oldhamite as yellow-green.

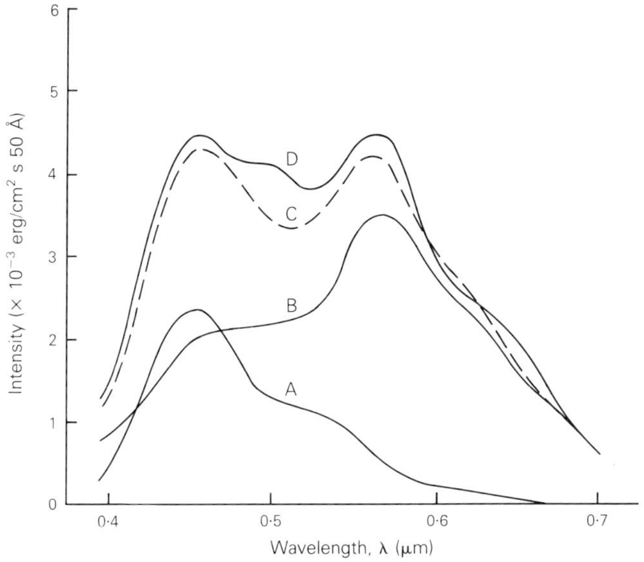

Figure 9.1 Spectra demonstrating the lack of a significant synergistic effect in the luminescence of lunar material. A and B were taken separately under UV (0.25–0.4 μm UV) and proton (2 keV) irradiation. Curve C is the graphical sum of curves A and B and curve D is a spectrum recorded when the UV and proton beams were simultaneously bombarding the sample. The sample was a fine-grained lunar basalt. Measurements were made at ~35 °C. (From Nash 1973.)

Diopside emits a green-blue light. They noted that colors are 'very characteristic' and therefore helpful in the recognition of diopside and oldhamite. One of their microprobe scans across an enstatite crystal, showing the correlation between CL color and Mn content, is reproduced in Figure 9.2.

Greenman and Gross (1970, 1971) also made luminescence measurements on lunar and terrestrial rocks and minerals using UV, X-ray, and proton excitation. In another instance (Greenman & Milton 1968) use was also made of electron excitation, obtaining the electrons from a ^{90}Sr source which provides a flux of 2.3 meV electrons. The total luminescence intensity was measured using a photomultiplier tube with a response between 3,000 and 7,000 Å. Their work is of particular interest, even though their electron source is not a common choice, because they set up a standard and compared the luminescence intensity of other samples to it, deriving information on relative intensities or relative efficiencies. A sample of smoky quartz was chosen as the standard. The relative efficiencies that they found for other minerals are shown in Table 9.1. The results indicate that, for the most part, the relative efficiencies are comparable for the three techniques of excitation. The results also demonstrate, quantitatively, the extreme range of intensities often dealt with when observing CL of natural samples.

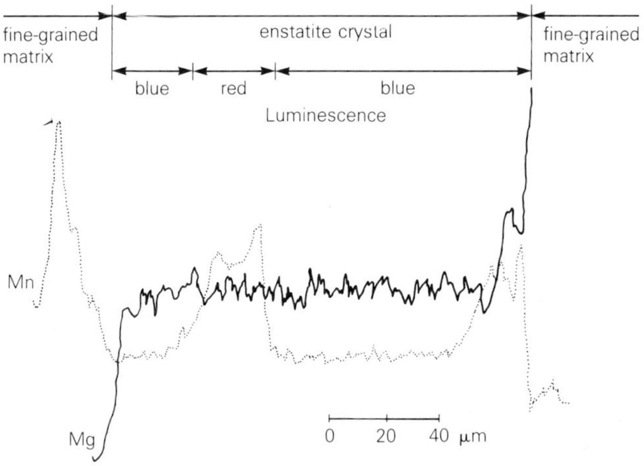

Figure 9.2 Microprobe scanning profile for Mg and Mn across an enstatite crystal from the Khor Temiki aubrite with notes on the observed CL colors. The color is correlated with the Mn content. (From Grögler & Liener 1968.)

Greer *et al.* (1968) have also done extensive CL work on minerals separated from a number of meteorites. Their article contains a good summary discussion of crystal field theory. They used an electron probe for excitation and reported their CL colors by reference to the Munsell color charts (Munsell 1975). Greer and Weber (1969) also presented a comprehensive discussion of the luminescence of enstatite from meteorites.

Of all the investigators who studied luminescence phenomena in lunar-related samples, the work of Sippel and coworkers is most nearly related to the major theme of this text. They used the same equipment as was used in their

Table 9.1 Relative luminescence intensity of the common igneous rock-forming minerals with various types of excitation.* (Value for smoky quartz, 100, taken as standard.)

Mineral	Intensity with excitation by		
	proton	X-ray	electron
quartz			
milky	57	54	250
rose	49	43	120
smoky	100	100	100
K-feldspar			
orthoclase	380	270	260
microcline	720	680	410
plagioclase feldspar			
albite	450	200	400
oligoclase	320	590	730
labradorite	58	19	21
bytownite	74	260	92
mica			
muscovite	1.8	0	2.8
biotite	0.4	0	0.8
amphibole: hornblende	0.1	0	1.6
pyroxene			
enstatite	3.2	0	13
diopside	0.5	57	47
hypersthene	1.4	0	1.5
hedenbergite	0.1	15	3.6
augite	2.8	1.4	6.3
olivine			
olivine I (Arizona)	5.5	0	15
olivine II (North Carolina)	0.6	0	12

* The relative intensities of luminescence for some common minerals under proton, X-ray, or electron beam excitation. The protons were obtained from a neutron generator providing 20 keV energy protons. X-rays were obtained from a tungsten target, bombarded by 200 keV electrons. The electrons were obtained from beta decay from ^{90}Sr and had an energy up to 2.3 meV. The intensities are based on a measurement of the total emission in the 4,000–7,000 Å range, using a photomultiplier (Greenman & Milton 1968).

work on sandstones and carbonates, and the observations, using an electron beam in the energy range of 15–20 keV, are directly of interest and pertinent to the results obtained by others using an electron probe or CMA.

Sippel and Spencer (1970a, b) reported on the luminescence petrography of lunar crystalline rocks and breccias. Their CMA was equipped with a monochromator and they obtained spectral information in the 3,700–8,200 Å range. With a system of apertures to limit the beam size on the specimen, they were able to obtain spectra from individual mineral grains, 40 μm or larger. Their comments on observations of feldspars and other terrestrial minerals have been discussed in earlier chapters. They also observed the CL properties of maskelynite ('plagioclase disordered by shock waves') which exhibited a dull red luminescence. They found peak shifts, peak broadening, and reduced intensity associated with minerals which had undergone shock damage.

Berkley *et al.* (1976) included CL observations in their study of the Kenna Ureilite. The distribution of forsterite ('dark red-orange' CL) and fayalite (nonluminescing) was easily verified with CL, and small grains of, e.g., blue-fluorescing diamond were easily located. Luminescence was also used to examine olivine–carbon vein contacts and to determine the distribution of small diamonds in metallic veins. Some carbon also occurred as lonsdaleite which exhibited blue and green CL. These same authors also used CL as part of their petrological studies of the Allende meteorite. Again, olivine exhibited brilliant red CL of Fe-free (forsteritic) rims, and carbon in the form of diamond and lonsdaleite exhibits 'strong blue and green' CL. Alkali chlorides, present in minute amounts, showed up well. They also found Mg–Al spinels with bright red CL in the Allende. They noted that CL is a valuable technique facilitating mineral separation, both on a micro and a coarse-grained scale.

Hutcheon *et al.* (1978) also used CL as one of the tools in their comprehensive study of inclusions in the Allende meteorite. They noted that 'isolated plagioclase grains were easily recognized by their brilliant blue luminescence, the intensity of which appeared to correlate with Mn content.' (Their study of the correlation of plagioclase CL with Ti, Na, and Mg content has been discussed in Ch. 4.) Spinel showed brilliant red and orange CL, and blue when the spinel is Fe-rich. CL of melilite, grossular, and pyroxene were noted also.

Steele (1986a, b) reported on the cathodoluminescence in forsterite from extraterrestrial samples. Pure Mg olivine is rare in terrestrial samples but common in some chondrites, and, when present, 'CL has revealed significant textural features that probably cannot be seen by any other technique . . .' (Steele 1986a, p. 966). Most of his work was done with an electron probe, equipped with a spectrometer for displaying the CL emission spectrum. Steele saw CL ranging from red to purple to blue in the forsterites. In the forsterites from one group of meteorites, the spectrum is made up of two broad peaks, one at 458 nm and one near 800 nm. In the other group, there is a weak broad blue band and a broad peak at 723 nm. He suggested that the CL spectra may be a possible way to distinguish original sources of single forsterite grains found in deep-sea particles or cosmic dust. He did extensive analytical work for several elements including Al, Ca, Ti, V, Mn, Cr, and Fe and some correlations were made between CL and minor-element composition. Blue CL correlates with high levels of Al, Ca, Ti, and V, and red CL with high amounts of Mn and Fe. He reported that above 2 percent FeO there was no CL.

It is worth noting that many of the measurements made on the CL of materials, with an eye to lunar applications, were made at room temperature. However, some materials have a very strong increase of CL and other forms of luminescence with reduced temperatures. As Hanusiak (1975) has pointed out, this may have an important bearing on the interpretation of the luminescence phenomena observed on the Moon.

9.2 Cathodoluminescence of gems

Many of the gemstones are among the brightest luminescing minerals and Crookes observed their CL properties over 100 years ago. Despite this long history, the amount of serious scientific work which has been published on the CL of gemstones is still relatively small, the most notable exception to this statement being the case of diamond. Several investigators have studied the CL of diamond from several different viewpoints and their work has already been outlined in Chapter 4.

Gaal (1976–77) has presented the results of a survey of CL in gemstones, and reports results for diamonds, jadeite, labradorite, alexandrite, ruby, and benitoite. Brightly luminescent examples, many showing complex structures and/or zonation, were described for each of these and published in full color.

Plates 1a, b, e, and f are examples of the complex appearance often noted in gem materials, in this case for a natural and a synthetic diamond, a natural emerald, and a synthetic corundum, respectively. The complex features revealed by CL in some of these examples are not visible with ordinary light microscopy.

The study of the cathodoluminescence of gemstones contributes to the understanding of both the geological conditions of their formation and also of the underlying solid-state physics processes which play a rôle. Besides these two aspects of cathodoluminescence in gemstones, there are at least two other important questions which the investigators of CL in gemstones have addressed. One question is whether it is possible to distinguish natural from synthetic gems, based on their CL properties. The other question is whether CL properties are similar (or even unique) to each single gemstone, or to gems from the same source. An affirmative answer to any of these questions would mean that the CL of a gemstone could be used as a fingerprint for identification.

Some work has been done in both of these areas but much more remains to be done. Ponahlo and Koroschetz (1985) have been involved in this research. They have found that synthetic and natural stones exhibit significant differences in the slope of the curve of CL intensity versus power of the bombarding electron beam. For both rubies and emeralds, the natural stones show a relatively small increase in CL intensity with electron beam power, whereas their synthetic counterparts show a much greater increase.

Epilogue

Bringing this book to completion is not an easy task. Many of the subjects discussed have 'loose ends' and there is an indication that there will be a rush of new information in the near future. Some of the accepted ideas are undergoing modification as further work is done. We have attempted, throughout the text, to point out some of these areas.

Further advances in the applications of cathodoluminescence to geological problems will require the combined talents of the earth scientist, the solid-state physicist, the mineral synthesizer, and the analyst. If this book catalyzes some of this research, then it will have served one of its intended purposes.

Appendix

Photomicrography systems and photographic procedures used for color plates

Many of the color plate legends contain the available information on photographic and CL conditions. Where an author has furnished a number of plates, the conditions common to all are summarized herein, rather than being duplicated on each legend. Unless otherwise stated, the author referred to in the plate captions has furnished the sample and done the photography.

Buyce, M. R. Luminoscope® mounted on Leitz Dialux monocular microscope with 10× eyepiece and 20× objective.

Dickson, J. A. D. University-made instrument with cold cathode gun. Mounted on Nikon Optiphot, F head, with Nikon 10× objective.

Hagni, R. D. Luminoscope® mounted on Jena monocular microscope. ASA 400 Kodak Ektachrome daylight, pushed to 1,600 in development.

Kopp, O. C. Luminoscope® mounted on Bausch and Lomb monocular microscope with 3.5× objective and 10× hyperplane ocular. Zeiss Ikon camera. ASA 400 Kodak Ektachrome daylight, normal development.

Mariano, A. N. (I). Luminoscope® mounted on a Gamma Scientific Model 700-10-1 monocular microscope system. Objective is Gamma 2.5× with NA variable from 0.03 to 0.25. 10× ocular. Long dimension of 35 mm transparency = 3.8 mm on sample. Kodak Ektachrome. Normal development. Air-supported discharge. Beam focused to about 4 mm.

Mariano, A. N. (II). Luminoscope® mounted on a Leitz, CM monocular. Reflected light objective is 6.3× with NA of 0.2. Ocular is 10×. Kodak Ektachrome. Normal development. Long dimension of 35 mm transparency = 2.2 mm on sample. Air-supported discharge. Beam focused to about 3 mm.

Marshall, D. J. Luminoscope® mounted on an Olympus POS, monocular microscope, with OM-2 series camera and adapters. ASA 400 Kodak Ektachrome, normal development. Air-supported discharge.

Miller, J. Luminoscope® mounted on monocular Olympus POS with 10× objective and 10× ocular. Ektachrome, ASA 200, daylight, normal development.

Owen, M. R. Luminoscope® mounted on Olympus BHT-P microscope with trinocular head, with Zeiss UD-20/0.57 objective and Olympus NFK 2.5× LD ocular. 16.5 kV; 0.2 mA; 2 mm diameter beam; air discharge. Photographic conditions are 30 second exposure, ASA 400 Kodak Ektachrome, normal development. Long dimension of 35 mm transparency = 1 mm.

Powell, T. G., W. T. Thompson and J. B. Thomas. Luminoscope® mounted on Zeiss Binocular Petrographic Microscope, Model 18, with trinocular head. Objective: 2.5× Achromat, 0.1 NA, ASA 400 Kodak Ektachrome, normal development. Magnification on transparency is 63 to 1, i.e., a 0.5 mm grain will essentially be the full width of the long dimension of the transparency.

Bibliography

Adams, A. E. and K. Schofield 1983. Recent submarine aragonite, magnesian calcite, and hematite cements in a gravel from Islay, Scotland. *Journal of Sedimentary Petrology* **53**, 417–21.

Aitken, M. J. 1974. *Physics and archaeology*. Oxford: Clarendon Press.

Ali, A. D. and P. Turner 1982. Authigenic K-feldspar in the Bromsgrove Sandstone Formation (Triassic) of central England. *Journal of Sedimentary Petrology* **52**, 187–97.

Amieux, P. 1982. La cathodoluminescence: méthode d'étude sédimentologique des carbonates. *Bulletin des Centres Recherches Exploration–Production Elf-Aquitaine* **6**, 437–83.

Amieux, P., J. L. Ducreux, E. Jautée and P. Pailleret 1983. Septarias en domaine continental dans le Stampien de la Limagne D'Allier (Issoire, Puy-de-Dôme). *Revue des Sciences Naturelle d'Auvergne* **49**, 7–18.

Bai, G. and Z. Yuan 1985. Carbonatites and related mineral resources. *Bulletin of the Institute of Mineral Deposits, Chinese Academy of Geological Sciences* **13**, 195. (Chinese with English summary.) Abstracted in *Mineralogical Abstracts* **37**, 398.

Bailiff, I. K., D. A. Morris and M. J. Aitken 1977. A rapid-scanning interference spectrometer: applications to low-level thermoluminescence emission. *Journal of Physics E: Scientific Instruments* **10**, 1,156–60.

Bamford, R. W. 1972. The Mount Fubilan (Ok Tedi) porphyry copper deposit, Territory of Papua, New Guinea. *Economic Geology* **67**, 1,019–33.

Barsanov, G. P. and K. K. Sarsembaeva 1962. Luminescence of Iceland spar. *Trudy Mineralogicheskogo Muzeya Akademiya Nauk S.S.S.R.* **13**, 147–52 (in Russian).

Barnes, D. F. 1958. *Infrared luminescence of minerals*. Geological Society of America Bulletin 1,052–C.

Beaman, D. R. and J. A. Isasi 1972. *Electron beam microanalysis*. ASTM Special Publication 506. Philadelphia: American Society for Testing and Materials.

Bencini, A. and A. Turi 1974. Mn distribution in the Mesozoic carbonate rocks from Lima Valley, Northern Apennines. *Journal of Sedimentary Petrology* **44**, 774–82.

Berkley, J. L., H. G. Brown IV, K. Keil, N. L. Carter, J-C. C. Mercier and G. Huss 1976. The Kenna Ureilite: an ultramafic rock with evidence for igneous, metamorphic, and shock origin. *Geochimica et Cosmochimica Acta* **40**, 1,429–37.

Bernhardt, H. 1982. A contribution to high-temperature thermoluminescence investigations in quartz crystals. *Physica Status Solidi A* **74**, 159–65.

Bhalla, R. J. R. S. B. and E. W. White 1970. Polarized cathodoluminescence emission from willemite [$Zn_2SiO_4(Mn)$] single crystals. *Journal of Applied Physics* **41**, 2,267–8.

Bhalla, R. J. R. S. B. and E. W. White 1971. Intrinsic cathodoluminescence emission from willemite single crystals. *Journal of Luminescence* **4**, 194–200.

Bhalla, R. J. R. S. B. and E. W. White 1972. Cathodoluminescence characteristics of Mn^{2+}-activated willemite (Zn_2SiO_4) single crystals. *Journal of the Electrochemical Society* **119**, 740–3.

Blair, I. M. and J. A. Edgington 1968. Luminescent properties of rocks, meteorites and natural glasses under proton bombardment. *Nature* **217**, 157–60.

Blair, I. M. and J. A. Edgington 1970. The luminescent and thermoluminescent properties of granite and its constituent minerals and their relevance to transient lunar phenomena. In *Space science applications of solid state luminescence phenomena*, J. N. Weber and E. White (eds) 22–8, MRL publication 70–101. State College, Pa: Penn. State University.

Blank, H., A. El Goresy, J. Janicke, R. Nobiling and K. Traxel 1984. Trace element determination in lunar rocks – quantitative PIXE analyses on thick samples. *Nuclear Instruments and Methods in Physics Research* **83**, 681–7.

Brand, U. and J. Veizer 1980. Chemical diagenesis of a multicomponent carbonate system – 1: Trace elements. *Journal of Sedimentary Petrology* **50**, 1,219–36.

Bröcker, W. and G. E. Pfefferkorn 1978. Bibliography on cathodoluminescence. In *Scanning electron microscopy/1978/I*, O. Johari (ed.), 333–51. Chicago: SEM.

Bröcker, W. and G. E. Pfefferkorn 1980. Bibliography on cathodoluminescence, Part II. In *Scanning electron microscopy/1980/I*, O. Johari (ed.), 298–302. Chicago: SEM.

Brøgger, W. C. 1921. Die Eruptivgesteine des Kristianiagebietes. IV. Das Fengebiet in Telmarken, Norwegen. *Norske Videnskapsakedimi I–Akad. Oslo Matematisk Naturvidenskapelig klasse* **1920**, no. 9

Buchanan, R. A., K. A. Wickersheim, J. L. Weaver, L. E. Sobon and E. E. Anderson 1967. Cathodoluminescence of rare earth-activated yttrium oxide. In *6th rare earth research conference, Gatlinburg, Tenn.*, 271–82.

Burgner, R. P., B. E. Scheetz and W. B. White 1978. Vibrational structure of the S_2^- luminescence in scapolite. *Physics and Chemistry of Minerals* **2**, 317–24.

Burns, G., E. A. Geiss, B. A. Jenkins and M. I. Nathan 1965. Cr^{3+} fluorescence in garnets and other crystals. *Physics Review* **139**, 1,687–93.

Burns, R. G. 1970. *Mineralogical applications of crystal field theory*. Cambridge: Cambridge University Press.

Buyce, M. R. and G. M. Friedman 1975. Significance of authigenic K-feldspar in Cambrian–Ordovician carbonate rocks of the Proto-Atlantic shelf in North America. *Journal of Sedimentary Petrology* **45**, 808–21.

Carmichael, I. S. E. and J. Nicholls 1967. Iron–titanium oxides and oxygen fugacities in volcanic rocks. *Journal of Geophysical Research* **72**, 4,665–87.

Carpenter, A. B. and T. W. Oglesby 1976. A model for the formation of luminescently zoned calcite cements and its implications. *Abstracts with Programs* **8**, 469–70. Boulder, CO: Geological Society of America.

Carrasco, V. B. 1973. Catodoluminescencia y diagénesis de rudistas dolomitizados y su relación con desarrollos de porosidad. *Revista del Instituto Mexicano del Petroleo* **V**, 5–14.

Carslaw, H. S. and J. C. Jaeger 1959. *Conduction of heat in solids*, 2nd edn. Oxford: Oxford University Press.

Castaing, R. 1952. *Application des sondes électronique à une méthode d'analyse ponctuelle chimique et cristallographique.* Thesis, University of Paris.

Charlet, J. M. 1970. Le photomètre de microthermoluminescence, son intérêt dans les applications de la TL à la géologie. *Annales de la Société Gèologique du Nord* **1,** 23–31.

Charlet, J. M. 1971. Thermoluminescence of detrital rocks used in paleogeographical problems. *Modern Geology* **2,** 265–74.

Charlet, J. M. 1974. Répartition de la thermoluminescence artificielle (TLA) au sein d'un même grain de quartz dans quelques sédiments meubles récent ou ancien. *Annales de la Société Géologique de Belgique* **T 97,** 221–33.

Chinner, G. A., J. V. Smith and C. R. Knowles 1969. Transition-metal contents of Al_2SiO_5 polymorphs. *American Journal of Science* **267–A,** 96–113.

Choquette, P. W. and R. P. Steinen 1980. Mississippian non-supratidal dolomite, Ste. Genevieve limestone, Illinois basin: evidence for mixed-water dolomitization. In *Concepts and models of dolomitization,* D. H. Zenger, J. B. Dunham and R. L. Ethington (eds), 163–96. SEPM Spec. Pub. No. 28. Tulsa: SEPM.

Churnet, H. G. and K. C. Misra 1981. Genetic implications of the trace element distribution pattern in the Upper Knox carbonate rocks, Copper Ridge District, East Tennessee. *Sedimentary Geology* **30,** 173–94.

Claffy, E. W. and R. J. Ginther 1959. Red-luminescing quartz. *Journal of the Optical Society of America* **49,** 412–13.

Collins, A. T. and G. S. Woods 1982. Cathodoluminescence from 'giant' platelets, and of the 2.526 vibronic system in type Ia diamonds. *Philosophical Magazine* **45,** 385–97.

Conzemius, R. J. and H. J. Svec 1978: Scanning laser mass spectrometer milliprobe. *Analytical Chemistry* **50,** 1,854–60.

Coy-Yll, R. 1969/70. Quelques aspects de la cathodoluminescence des mineraux. *Chemical Geology* **5,** 243–54.

Crookes, W. 1879. Contributions to molecular physics in high vacua. *Philosophical Transactions of the Royal Society* **170,** 641–62.

Curie, D. 1963. *Luminescence in crystals.* London: Methuen.

Davey, J. P. 1965. A scanning electron microprobe system for the study of cathodoluminescence. In *Optique des rayons X et microanalyse,* R. Castaing, P. Deschamps and J. Philibert (eds), 566–71. A conference held in Orsay in 1965. Paris: Hermann.

Davies, G. 1974. Vibronic spectra in diamond. *Journal of Physics C: Solid State Physics* **7,** 3,797–809.

Davies, G. 1979. Cathodoluminescence. In *The properties of diamond,* J. E. Field (ed.), 165–81. London: Academic Press.

Dawson, J. B. 1964. Carbonatitic volcanic ashes in northern Tanganyika. *Bulletin of Volcanology* **27,** 1–11.

Deer, W. A., R. A. Howie and J. Zussman 1971. *Rock-forming minerals. 4. Framework silicates.* London: Longman.

Dennen, W. H. 1967. Trace elements in quartz as indicators of provenance. *Geological Society of America Bulletin* **78,** 125–30.

Dennen, W. H., W. H. Blackburn and A. Quesada 1970. Aluminum in quartz as a geothermometer. *Contributions to Mineralogy and Petrology* **27,** 332–42.

Derham, C. J., J. E. Geake and G. Walker 1964. Luminescence of enstatite achondrite meteorites. *Nature* **203,** 134–6.

Dexter, D. L. and J. H. Schulman 1954. Theory of concentration quenching in inorganic phosphors. *Journal of Chemical Physics* **22,** 1,063–70.

Dickson, J. A. D. 1966. Carbonate identification and genesis as revealed by staining. *Journal of Sedimentary Petrology* **36,** 491–505.

Dickson, J. A. D. 1980. Artificial colouration of fluorite by electron bombardment. *Mineralogical Magazine* **43,** 820–2.

Dickson, J. A. D. 1983. Graphical modelling of crystal aggregates and its relevance to cement diagnosis. *Philosophical Transactions of the Royal Society of London A* **309,** 465–502.

Dickson, J. A. D. 1985. Diagenesis of shallow-marine carbonates. In *Sedimentology – recent developments and applied aspects,* P. J. Benchley and B. P. J. Williams (eds), 173–88. Oxford: Blackwell Scientific/Geological Society.

Dickson, J. A. D. and M. L. Coleman 1980. Changes in carbon and oxygen isotope composition during limestone diagenesis. *Sedimentology* **27,** 107–18.

Dieleman, J. 1966. Electron spin resonance in luminescent solids. In *Luminescence of inorganic solids,* P. Goldberg (ed.), 338–84. New York: Academic Press.

Dietrich, D. and P. R. Grant 1985. Cathodoluminescence petrography of syntectonic fibres. *Journal of Structural Geology* **7,** 541–53.

Dravis, J. J. and D. A. Yurewicz 1985. Enhanced carbonate petrography using fluorescence microscopy. *Journal of Sedimentary Petrology* **55,** 795–804.

Dreimanis, A., G. Hütt, A. Raukas and P. W. Whippey 1978. Dating methods of Pleistocene deposits and their problems: 1. Thermoluminescence dating. *Geoscience Canada* **5,** 55–60.

Dudley, R. J. 1976. The use of cathodoluminescence in the identification of soil minerals. *Journal of Soil Science* **27,** 487–94.

Dynasil Corp. 1970. *Fused silica catalog #SP-707-2.5M.* Berlin, New Jersey: Dynasil Corp.

Eastman Kodak Company 1980. *Photography through the microscope.* Kodak Publication No. P-2. Rochester: Eastman Kodak.

Ebers, M. L. and O. C. Kopp 1979. Cathodoluminescent microstratigraphy in gangue dolomite, the Mascot–Jefferson City District, Tennessee. *Economic Geology* **74,** 908–18.

Edgington, J. A. and I. M. Blair 1970. Luminescence and thermoluminescence induced by bombardment with protons of 159 million electron volts. *Science* **167,** 715–17.

Ehrenberg, W. and J. Franks 1953. The penetration of electrons into luminescent material. *Proceedings of the Physics Society of London, Section B* **66,** 1,057–66.

Escobar, R. and A. N. Mariano 1976. *On the origin of Colombian emeralds.* Presented at the 2nd biannual meeting of the Mineralogical Society of America, Friends of Mineralogy symposium, February 15–17, 1976, Tucson (abst.).

Evamy, B. D. 1969. The precipitational environment and correlation of some calcite cements deduced from artificial staining. *Journal of Sedimentary Petrology* **39,** 787–821.

Fairbairn, H. W. *et al.* 1951. *A cooperative investigation of precision and accuracy in chemical, spectrochemical and modal analysis of silicate rocks.* USGS bulletin 980.

Fairchild, I. J. 1980. Stages in a Precambrian dolomitization, Scotland: cementing versus replacement textures. *Sedimentology* **27,** 631–50.

Fairchild, I. J. 1983. Chemical controls of cathodoluminescence of natural dolomites and calcites: new data and review. *Sedimentology* **30,** 579–83.

Faupl, P. and A. Beran 1983. Diagenetische Veränderungen an Radiolarien- und Schwammspicula-führenden Gesteinen der Strubbergschichten (Jura, Nördliche Kalkalpen, Österreich). *Neues Jahrbuch für Geologie und Paläontologie Monatshejte* **5,** 129–40.

Faure, G. and J. L. Powell 1972. *Strontium isotope geology.* Berlin: Springer.

Faye, G. H. 1969. The optical absorption spectrum of tetrahedrally bonded Fe^{3+} in orthoclase. *Canadian Mineralogist* **10,** 112–17.

Ferrigno, K. F. and K. R. Walker 1973. *Ectoproct (Bryozoa) communities in the Middle Ordovician Holston Reefs of Tennessee.* SEGSA meeting, Knoxville, Tenn., April 1973 (abst.).

Field, J. E. (ed.) 1979. *The properties of diamond.* New York: Academic Press.

Finnerty, T. A. 1977. Exchange of Mn, Ca, Mg, and Al between synthetic garnet, orthopyroxene, clinopyroxene, and olivine. In *Annual Report of the Carnegie Institute* 1976–77, issued Dec. 1977, 572–9.

Fleischer, M. 1969. The lanthanide elements in fluorite. *Indian Mineralogist* **10,** 36–9.

Fox, L. A. 1979. *Porosity and permeability reduction in the Nugget Sandstone, southwestern Wyoming*. M. Thesis, University of Missouri-Columbia.

Frank, J. R. 1981. Dedolomitization in the Taum Sauk limestone (upper Cambrian), southeast Missouri. *Journal of Sedimentary Petrology* **51,** 7–17.

Frank, J. R., A. B. Carpenter and T. W. Oglesby 1982. Cathodoluminescence and composition of calcite cement in the Taum Sauk limestone (upper Cambrian), southeast Missouri. *Journal of Sedimentary Petrology* **52,** 631–8.

Freeman, T. 1971. Morphology and composition of an Ordovician vadose cement. *Nature Physical Sciences* **233,** 133–4.

Friedman, G. M. 1975. The making and unmaking of limestones or the downs and ups of porosity. *Journal of Sedimentary Petrology* **45,** 379–98.

Gaal, R. A. P. 1976–77. Cathodoluminescence of gem materials. *Gems and Gemology* **15,** 238–44.

Gallup, J. 1936. The vacuum cell luminescence microscope and its use in the study of luminescent materials. *Journal of the Optics Society of America* **26,** 213–15.

Gamma Scientific 1983. Private communication from E. G. & G. Gamma Scientific, San Diego, California.

Geake, J. E. and G. Walker 1966. The luminescence spectra of meteorites. *Geochimica et Cosmochimica Acta* **30,** 929–37.

Geake, J. E. and G. Walker 1975. Luminescence of minerals in the near-infrared. In *Infrared and Raman spectroscopy of lunar and terrestrial minerals*, C. Karr Jr (ed.), 73–89. New York: Academic Press.

Geake, J. E., G. Walker, D. J. Telfer, A. A. Mills and G. F. J. Garlick 1973. Luminescence of lunar, terrestrial, and synthesized plagioclase, caused by Mn^{2+} and Fe^{3+}. Proc. Fourth Lunar Science Conference, Supplement 4, *Geochimica et Cosmochimica Acta* **3,** 3,181–9.

Geake, J. E., G. Walker and D. J. Telfer 1977. The cause and significance of luminescence in lunar plagioclase. *Philosophical Transactions of the Royal Society of London A* **285,** 403–8.

Geake, J. E., G. Walker, A. A. Mills and G. F. J. Garlick 1971. Luminescence of Apollo lunar samples. In *Proceedings of the Second Lunar Conference*, vol. 3, A. A. Levinson (ed.), 2,265–75. *Geochimica et Cosmochimica Acta*, Supplement 2. Cambridge, Mass.: MIT.

Geake, J. E., A. Dollfus, G. F. J. Garlick, W. Lamb, G. Walker, G. A. Steigmann and C. Titulaer 1970. Electron paramagnetic resonance, and optical properties of lunar material. *Science* **167,** 717–20.

Gentry, R. V. 1973. Radioactive halos. *Annual Review of Nuclear Science* **23,** 347–62.

George, M. 1977. *Carbonate equilibrium in the Hosston Formation, central Mississippi*, M. Thesis, University of Missouri–Columbia.

Gies, H. 1975. Activation possibilities and geochemical correlations of photoluminescing carbonates, particularly calcites. *Mineralogy Deposita* **10,** 216–27.

Gilliland, J. W. 1971. Electron beam aging of $BaSi_2O_5$:Pb. *Journal of Luminescence* **4,** 345–57.

Glover, E. D. 1977. *Cathodoluminescence, iron and manganese content, and the early diagenesis of carbonates*. PhD Thesis, Madison: University of Wisconsin.

Goebel, V. W. and W. J. Patzelt 1976. Cathodoluminescence-investigations of solids with the Leitz contrasting device. *Leitz Scientific and Technical Information* **VI,** 263–7.

Goni, J. and G. Rémond 1969. Localization and distribution of impurities in blende by cathodoluminescence. *Mineralogical Magazine* **37,** 153–5.

Görz, H., R. J. R. S. B. Bhalla and E. W. White 1970. Detailed cathodoluminescence characterization of common silicates. In *Space science applications of solid state luminescence phenomena*, J. N. Weber and E. White (eds), 62–70. MRL pub. 70–101. State College, Pa.: Penn. State University.

Grant, P. 1978. The role of the scanning electron microscope in cathodoluminescence petrology. In *Scanning electron microscopy in the study of sediments*, B. Whalley (ed.), 1–11. Norwich: Geo Abstracts.

Grant, P. R. and S. H. White 1978. Cathodoluminescence and microstructure of quartz overgrowths on quartz. In *Scanning electron microscopy 1978/I*, O. Johari (ed.), 789–94. Chicago: Scanning Electron Microscopy.

Grauert, B. and M. E. Wagner 1975. Age of the granulite-facies metamorphism of the Wilmington Complex, Delaware–Pennsylvania Piedmont. *American Journal of Science* **275,** 683–91.

Greenman, N. N. and H. G. Gross 1970. Luminescence of Apollo 11 lunar samples. *Science* **167,** 720–1.

Greenman, N. N. and H. G. Gross 1971. Luminescence of Apollo 11 and Apollo 12 lunar samples. In *Proceedings of the Second Lunar Science Conference*, vol. 3, A. A. Levinson (ed.), 2,223–33. Cambridge, Mass.: MIT.

Greenman, N. N. and W. B. Milton 1968. Silicate luminescence and remote compositional mapping. In *Proceedings of the Sixth Annual Meeting, Working Group on Extraterrestrial Resources*. NASA SP-177, 55–63.

Greer, R. T. 1970. Thermoluminescence-chemical composition correlations for meteoritic enstatite. In *Space science applications of solid state luminescence phenomena*, J. N. Weber and E. White (eds), 49–55. MRL pub. 70–101. State College, Pa.: Penn. State University.

Greer, R. T. and J. N. Weber 1969. Correlation of mineral luminescent phenomena and its selenological implications. *Icarus* **11,** 55–65.

Greer, R. T., W. G. Staley Jr and V. Vand 1967. Cathodoluminescence device for rapid identification of phase assemblages. *Bulletin of the American Ceramic Society* **46,** 829–32.

Greer, R. T., V. Vand and J. N. Weber 1968. Applications of luminescence techniques to the study of the lunar surface. *Symposium on the interpretation of lunar probe data – II*, Huntington Beach, Ca.

Gregg, J. M. 1985. Regional epigenetic dolomitization in the Bonneterre Dolomite (Cambrian), southeastern Missouri. *Geology* **13,** 503–6.

Grögler, N. and A. Liener 1968. Cathodoluminescence and thermoluminescence observations of Aubrites. In *Thermoluminescence of geological materials*, D. J. McDougall (ed.), 569–78. New York: Academic Press.

Grover, G. Jr and J. F. Read 1983. Paleoaquifer and deep burial related cements defined by regional cathodoluminescent patterns, Middle Ordovician carbonates, Virginia. *American Association of Petroleum Geologists Bulletin* **67,** 1,275–303.

Haggerty, S. E. and A. N. Mariano 1983. Strontian-loparite and strontio-chevkinite: two new minerals in rheomorphic fenites from the Paraná Basin carbonatites, South America. *Contributions to Mineralogy and Petrology* **84,** 365–81.

Hagni, R. D. 1984. Cathodoluminescence microscopy applied to mineral exploration and beneficiation. In *Applied mineralogy* (proceedings of the Second International Congress on Applied Mineralogy), W. C. Park, D. M. Hausen and R. D. Hagni (eds), 41–66. New York: AIME.

Hall, M. R. and P. H. Ribbe 1971. An electron microprobe study of luminescence centers in cassiterite. *American Mineralogist* **56,** 31–44.

Hanley, P. L., I. Kiflawi and A. R. Lang 1977. On topographically identifiable sources of cathodoluminescence in natural diamonds. *Philosophical transactions of the Royal Society of London A Mathematical and Physical Science* **284,** 329–68.

Hanson, R. F., R. Zamora and W. D. Keller 1981. Nacrite, dickite, and kaolinite in one deposit in Nayarit, Mexico. *Clays and Clay Minerals* **29,** 451–3.

Hanusiak, W. M. 1975. *Low temperature cathodoluminescence of crystalline silica for use in the characterization of respirable dusts.* MSc Thesis. State College, Pa.: Penn. State University.

Hanusiak, W. M. and E. W. White 1975. SEM cathodoluminescence for characterization of damaged and undamaged alpha-quartz in respirable dusts. In *Proceedings of the 8th Annual Scanning Electron Microscope Symposium*, O. Johari and I. Corvin (eds), 125–32. Chicago: IIT Research Institute.

Harmer, P. R, S. D. Ford and R. A. Dugdale 1968. Electron bombardment luminescence stage for optical microscopy. *Journal of Scientific Instrument Series 2* **1,** 59–60.

Harshaw Chemical Co. *Vacuum ultraviolet window catalog.* Solon, Ohio: Harshaw Chemical, Crystal Electronics Products Dept.

Harwood, G. 1983. The application of cathodoluminescence in relative dating of barite mineralization in the Lower Magnesian limestone (upper Permian), United Kingdom. *Economic Geology* **78,** 1,022–7.

Hay, R. L. 1983. Natrocarbonatite tephra of Kerimasi volcano, Tanzania. *Geology* **11,** 599–602.

Hearn, P. P. Jr. and J. F. Sutter 1985. Authigenic potassium feldspar in Cambrian carbonates: evidence of Alleghanian brine migration. *Science* **228,** 1,529–31.

Heidel, R. H. 1971a. Precision and detection limits of certain minor and trace elements in silicates by electron microprobe analysis. *Analytical Chemistry* **43,** 1,907–8.

Heidel, R. H. 1971b. Standardization and data reduction for electron probe analysis of geologic materials. *American Laboratory* **Sept.,** 10–14.

Heimlich, R. A., L. B. Shotwell, T. Cookro and M. J. Gawell 1975. Variability of zircons from the Sharon Conglomerate of northeastern Ohio. *Journal of Sedimentary Petrology* **45,** 629–35.

Herzog, L. F., D. J. Marshall and R. F. Babione 1970. The Luminoscope – a new instrument for studying the electron-stimulated luminescence of terrestrial, extra-terrestrial and synthetic materials under the microscope. In *Space science applications of solid state luminescence phenomena*, J. N. Weber and E. White (eds), 79–98. MRL pub. 70–101. State College, Pa.: Penn. State University.

Houseknecht, D. W. 1984. Influence of grain size and temperature on intergranular pressure solution, quartz cementation, and porosity in a quartzose sandstone. *Journal of Sedimentary Petrology* **54,** 348–61.

Hummel, F. A. 1961. Cordierite:Indialite; a new manganese-activated phosphor. *Journal of the Electrochemical Society* **108,** 809–10.

Hummel, F. A. and J. F. Sarver 1964. The cathodoluminescence of Mn^{+2}- and Fe^{+3}-activated magnesium aluminate spinel. *Journal of the Electrochemical Society* **111,** 252–3.

Hutcheon, I. D., I. M. Steele, J. V. Smith and R. N. Clayton 1978. Ion microprobe, electron microprobe and cathodoluminescence data for Allende inclusions with emphasis on plagioclase chemistry. In *Proceedings of the Lunar Planetary Science Conference 9th*, R. B. Merrill, Managing Editor (and many others), 1,345–68, *Geochimica et Cosmochimica Acta*, Supplement 10. Pergamon Press: New York, Oxford.

Iijima, A., M. Utada, R. Matsumoto and R. Tada 1983. Device of cathodoluminescence microscope and its application to sedimentary petrology. *Journal of the Japanese Association of Petroleum Technologists* **48,** 144–51.

James, N. P. and C. F. Klappa 1983. Petrogenesis of early Cambrian reef limestones, Labrador, Canada. *Journal of Sedimentary Petrology* **53,** 1,051–96.

JEOL USA Inc. 1983. Personal communication.

Johannesen, A. 1907. In *Description of the Devil's Tower Quadrangle.* USGS Folio 150. N. H. Darton and C. C. O'Harra (eds). Washington, DC: USGS.

Kastner, M. A. 1971. Authigenic feldspars in carbonate rocks. *American Mineralogist* **56,** 1,403–42.

Kaufman Glass Co. 1984. *Industrial glass handbook.* Wilmington, Del.: Kaufman Glass.

Kiflawi, I. and A. R. Lang 1974. Linearly polarized luminescence from linear defects in natural and synthetic diamond. *Philosophical Magazine* **30,** 219–23.

Klick, C. C. and J. H. Schulman 1957. Luminescence in solids. In *Solid state physics, vol. 5*, F. Seitz and D. Turnbull (eds), 97–192. New York: Academic Press.

Kniseley, R. N., F. C. Laabs and V. A. Fassel 1969. Analysis of rare earth materials by cathodoluminescence spectra excited in an electron microprobe. *Analytical Chemistry* **41,** 50–3.

Kopp, O. C. 1980. Personal communication.

Kopp, O. C. 1981. Cathodoluminescence petrography: a valuable tool for teaching and research. *Journal of Geological Education* **29,** 108–13.

Krinsley, D. H. and P. J. W. Hyde 1971. Cathodoluminescence studies of sediments. In *Scanning Electron Microscopy/1971*, O. M. Johari and I. Corvin (eds), 411–13. Chicago: IIT Research Institute.

Lang, A. R. 1979. Internal structure. In *The properties of diamond*, J. E. Field (ed.), 425–69. London: Academic Press.

Lang, A. R. 1980. Polarized infra-red cathodoluminescence from synthetic diamonds. *Philosophical Magazine* **41,** 689–98.

Lange, H. and G. Kressin 1955. Der Einfluss der Kristallstruktur auf die Lumineszenz des Calciumsilikates (Mn, Pb). *Zeitschrift für Physik* **142,** 380–6.

Larach, S. 1968. Cathode-ray-excited emission spectroscopic analysis of trace rare earths. Part I. Qualitative studies. *Analytica Chimica Acta* **41,** 189–95.

Laud, K. R., E. F. Gibbons, T. Y. Tien and H. L. Stadler 1971. Cathodoluminescence of Ce^{3+}- and Eu^{2+}-activated alkaline earth feldspars. *Journal of the Electrochemical Society* **118,** 918–23.

Lebiedzik, J., E. W. White and R. J. R. S. B. Bhalla 1974. Simple instrumentation for rise and decay time measurement of cathodoluminescence in an electron microprobe. *Review of Scientific Instruments* **45,** 451–2.

Leitch, C. A. and J. V. Smith 1980. Two types of clinoenstatite in Indarch enstatite chondrite. *Nature* **283,** 60–1.

Le Poole, J. B., A. B. Bok and W. J. Boogerd 1968. An electron luminescence microscope. *Geologie en Mijnbouw* **47,** 443–50.

Leverenz, H. W. 1968. *An introduction to luminescence of solids.* New York: Dover.

Lindholm, R. C. and R. B. Finkelman 1972. Calcite staining: semi quantitative determination of ferrous iron. *Journal of Sedimentary Petrology* **42,** 239–42.

Link, F. 1946. Sur la luminescence de la Lune. *Comptes Rendus* **223,** 976–7.

Lohmann, K. C. and W. J. Meyers 1977. Microdolomite inclusions in cloudy prismatic calcites: a proposed criterion for former high-magnesium calcites. *Journal of Sedimentary Petrology* **47,** 1,078–88.

Long, J. V. P. 1963. Recent advances in electron-probe analysis. In

Advances in X-ray analysis, vol. 6, Proceedings of the 11th Annual Conference on Applications of X-ray Analysis. August 1962, W. M. Mueller and M. Fay (eds), 276–90. New York: Plenum.

Long, J. V. P. and S. O. Agrell 1965. The cathodo-luminescence of minerals in thin section. *Mineralogical Magazine* **34,** 318–26.

Lyons, P. C. and H. W. Krueger 1976. Petrology, chemistry, and age of the Rattlesnake pluton and implications for other alkalic granite plutons of southern New England. In *GSA Memoir 146*, P. C. Lyons and A. H. Brownlow (eds), 71–102. Boulder: Geological Society of America.

McCauley, J. W. 1975. Structural and chemical characterization of processed crystalline ceramic materials. In *Characterization of materials in research, ceramics and polymers*, J. Burke (ed.), 175–209. Proceedings of the 20th Sagamore Army Materials Research Conference at Syracuse. Syracuse: Syracuse University Press.

Machel, H. G. 1983. Cathodoluminescence in carbonate petrography: some aspects of geochemical interpretation. (Abst.) *American Association of Petroleum Geologists Bulletin* **67,** 507–8.

Machel, H. G. 1985. Cathodoluminescence in calcite and dolomite and its chemical interpretation. *Geoscience Canada* **12,** 139–47.

McIntire, W. L. 1963. Trace element partition coefficients – a review of theory and applications to geology. *Geochimica et Cosmochimica Acta* **27,** 1,209–64.

Malcuit, R. J. and R. A. Heimlich 1972. Zircons from precambrian gneiss, Southern Bighorn Mountains, Wyoming. *American Mineralogist* **57,** 1,190–209.

Marfunin, A. S. 1979. *Spectroscopy, luminescence, and radiation centers in minerals*. Translated from the Russian publication by V. V. Schiffer. Berlin: Springer Verlag.

Mariano, A. N. 1973. Notes at CL workshop, University of Tennessee, Knoxville.

Mariano, A. N. 1974. UNDP Emerald Project: Colombia.

Mariano, A. N. 1978. The application of cathodoluminescence for carbonatite exploration and characterization. In *Proceedings of the International Symposium on Carbonatites, 1st, Pocos de Caldas, Minas Gerais, Brasil. June 1976*, C. J. Braga (ed.), 39–57. Brasilia: Brasil Departamento Nacional da Producão Mineral.

Mariano, A. N. 1983a. *Macro-photography of cathodoluminescence*. Nuclide pub. 1080/1083.

Mariano, A. N. 1983b. Personal communication.

Mariano, A. N. and P. J. Ring 1975. Europium-activated cathodoluminescence in minerals. *Geochimica et Cosmochimica Acta* **39,** 649–60.

Mariano, A. N. and P. L. Roeder 1983. Kerimasi: a neglected carbonatite volcano. *Journal of Geology* **91,** 449–55.

Mariano, A. N., J. Ito and P. J. Ring 1973. Cathodoluminescence of plagioclase feldspars. *Abstracts with Programs* **5,** 726. Boulder, Co: Geological Society of America.

Marshall, D. J. 1975. *The status of the cathodoluminescence technique in the study of carbonates*. Presented at the IXth International Congress of Sedimentology, Nice, France.

Marshall, D. J. 1978. Suggested standards for the reporting of cathodoluminescence results. *Journal of Sedimentary Petrology* **48,** 651–3.

Marshall, D. J. 1980a. *A cooled stage for cathodoluminescence observations with the Luminoscope®*. Presented at the Edinboro Workshop on Cathodoluminescence, Edinboro, June 1980. Nuclide Corp. Pub. 1007/1180.

Marshall, D. J. 1980b. *The influence of microscope performance on cathodoluminescence observations*. Presented at the Edinboro Workshop on Cathodoluminescence, Edinboro, June 1980. Nuclide Corp. Pub. 1019/0281.

Martin, H. and H. Zeegers 1969. Cathodo-luminescence et distribution du manganèse dans les calcaires et dolomies du Tournaisien supèrieur au Sud de Dinant (Belgique). *Comptes Rendus de l'Académie des Sciences Paris* **269,** 1,922–4.

Mason, B. and L. G. Berry 1968. *Elements of mineralogy*. New York: W. H. Freeman.

Matter, A. and K. Ramseyer 1985. Cathodoluminescence microscopy as a tool for provenance studies of sandstones. In *Provenance of arenites*, G. C. Zuffa (ed.), 191–211. Dordrecht: D. Reidel.

Medlin, W. L. 1959. Thermoluminescent properties of calcite. *Journal of Chemical Physics* **30,** 451–8.

Medlin, W. L. 1960. Comments on 'Thermoluminescent properties of calcite.' *Journal of Chemical Physics* **32,** 943.

Medlin, W. L. 1961a. Thermoluminescence in aragonite and magnesite. *Journal of Physical Chemistry* **65,** 1,172–7.

Medlin, W. L. 1961b. Thermoluminescence in dolomite. *Journal of Chemical Physics* **34,** 672–7.

Medlin, W. L. 1963a. Emission centers in thermoluminescent calcite, dolomite, magnesite, aragonite, and anhydrite. *Journal of the Optical Society of America* **53,** 1,276–85.

Medlin, W. L. 1963b. Thermoluminescence in quartz. *Journal of Chemical Physics* **38,** 1,132–43.

Medlin, W. L. 1968. The nature of traps and emission centers in thermoluminescent rock materials. In *Thermoluminescence of geological materials*, D. J. McDougall (ed.), 193–223. New York: Academic Press.

Mendelssohn, M. J., H. J. Milledge, E. R. Vance, E. Nave and P. A. Woods 1979. Internal radioactive haloes in diamond. In *Industrial diamond review*, 31–6. London: Industrial Diamond.

Merritt, E., E. L. Nichols and C. D. Child 1923. Selected topics in the field of luminescence. *Bulletin of the National Research Council* **5,** (30), 1–126.

Metz, M. C. 1971. *The geology of the Snowbird deposit, Mineral County, Montana*. MS Thesis, Washington State University.

Metz, M. C., D. G. Brookins, P. E. Rosenberg and R. E. Zartman 1985. Geology and geochemistry of the Snowbird deposit, Mineral County, Montana. *Economic Geology* **80,** 394–409.

Meyers, W. J. 1974. Carbonate cement stratigraphy of the Lake Valley formation (Mississippian) Sacramento Mountains, New Mexico. *Journal of Sedimentary Petrology* **44,** 837–61.

Meyers, W. J. 1978. Carbonate cements: their regional distribution and interpretation in Mississippian limestones of southwestern New Mexico. *Sedimentology* **25,** 371–400.

Meyers, W. J. and K. C. Lohmann 1978. Microdolomite-rich syntaxial cements: proposed meteoric–marine mixing zone phreatic cements from Mississippian limestones, New Mexico. *Journal of Sedimentary Petrology* **48,** 475–88.

Meyers, W. J. and K. C. Lohmann 1979. Reply: microdolomite-rich syntaxial cements: proposed meteoric–marine mixing zone phreatic cements from Mississippian limestones, New Mexico. *Journal of Sedimentary Petrology* **49,** 674–6.

Meyers, W. J. and Lohmann, K. C. 1985. Isotope geochemistry of regionally extensive calcite cement zones and marine components in Mississippian limestones, New Mexico. In *Carbonate cements*, N. Schneidermann and P. M. Harris (eds), 223–39. Tulsa: SEPM.

Michard, M. G. 1968. Coprécipitation de l'ion manganeux avec le carbonate de calcium. *Comptes Rendus de l'Académie des Sciences Paris* **267,** 1,685–8.

Miller, J. 1983. *Introduction to luminescence. Notes for sedimentologists*. Edinburgh: Edinburgh University.

Miller, J. and E. N. K. Clarkson 1980. The post-ecdysial development of the cuticle and the eye of the Devonian trilobite *Phacops rana milleri* Stewart 1927. *Philosophical Transactions* **288,** 461–80.

Moldovanyi, E. P. and K. C. Lohmann 1984. Isotopic and petrographic record of phreatic diagenesis: lower Cretaceous Sligo and Cupido formations. *Journal of Sedimentary Petrology* **54,** 972–85.

Moore, C. E. 1949. *Atomic energy levels, Vol. I*. National Bureau of Standards, circular 467. Washington, DC: US Govt Printing Office.

Moore, C. E. 1952. *Atomic energy levels, Vol. II*. National Bureau of Standards, circular 467. Washington, DC: US Govt Printing Office.

Morgan, J. W. and G. A. Wandless 1980. Rare earth element distribution in some hydrothermal minerals: evidence for crystallographic control. *Geochimica et Cosmochimica Acta* **44,** 973–80.

Morozov, A. M., L. G. Morozova, A. K. Trofimov and P. P. Feofilov 1970. Spectral and luminescent characteristics of fluorapatite single crystals activated by rare earth ions. *Optika i Spektroskopiya* **29** (6), 1,106–18 (in Russian).

Morse, S. A. and K. M. Nolan 1985. Kiglapait geochemistry VII: Yttrium and the rare earth elements. *Geochimica et Cosmochimica Acta* **49,** 1,621–44.

Morton, R. D. 1978. Cathodoluminescence applied to uranium exploration. *Nuclide Spectra* **XI,** no. 1. Acton, Mass.: Nuclide.

Mou, D. C. and R. L. Brenner 1982. Control of reservoir properties of Tensleep Sandstone by depositional and diagenetic facies: Lost Soldier Field, Wyoming. *Journal of Sedimentary Petrology* **52,** 367–81.

Mugridge, S. J. and H. R. Young 1984. Rapid preparation of polished thin sections for cathodoluminescence study of carbonate rocks. *Canadian Mineralogist* **22,** 513–15.

Muir, M. D., P. R. Grant, G. Hubbard and J. Mundell 1971. Cathodoluminescence spectra. In *Scanning electron microscopy/ 1971/I*, O. Johari and I. Corvin (eds), 401–8. Chicago: IIT Research Institute.

Munsell, A. H. 1975. *A color notation*. Baltimore: Munsell Color.

Murata, K. J. and R. L. Smith 1946. Manganese and lead as coactivators of red fluorescence in halite. *American Mineralogist* **31,** 527–38.

Nash, D. B. 1966. Proton-excited luminescence of silicates: experimental results and lunar implications. *Journal of Geophysical Research* **71,** 2,517–34.

Nash, D. B. 1973. Experimental results on combined ultraviolet–proton excitation of moon rock luminescence. *Journal of Geophysical Research* **78,** 3,512–14.

Nash, D. B., J. E. Conel and R. T. Greer 1970. Luminescence and reflectance of tranquility samples: effects of irradiation and vitrification. *Science* **167,** 721–4.

Nassau, K. 1978. The origins of color in minerals. *American Mineralogist* **63,** 219–29.

Nichols, E. L. and E. Merritt 1909. Studies of luminescence IX, spectrophotometer study of certain cases of Kathodoluminescence. *Physical Review* **28,** 349–60.

Nichols, E. L., H. L. Howes and D. T. Wilber 1928. *Cathodoluminescence and the luminescence of incandescent solids*. Publication no. 384. Washington, DC: Carnegie Institution.

Nickel, E. 1978. The present status of cathode luminescence as a tool in sedimentology. *Minerals Science and Engineering* **10,** 73–100.

Niemann, J. C. 1984. *Regional cementation from unconformity-sourced aquifers and burial fluids, Mississippian Newman Limestone, Kentucky*. MS Thesis, Virginia Polytechnic Institute.

Oglesby, T. W. 1976. *A model for the distribution of manganese, iron, and magnesium in authigenic calcite and dolomite cements in the Upper Smackover formation in Eastern Mississippi*. MA Thesis, University of Missouri–Columbia.

Oglesby, T. W. 1979. Discussion: microdolomite-rich syntaxial cements: proposed meteoric–marine mixing zone phreatic cements from Mississippian limestones, New Mexico. *Journal of Sedimentary Petrology* **49,** 670–3.

Olaussen, S., A. Dalland, T. G. Gloppen and E. Johannessen 1984. Depositional environment and diagenesis of Jurassic reservoir sandstones in the eastern part of Troms I area. In *Petroleum geology of the North European Margin, Proceedings of North European Margin Symposium, Trondheim, May 1983*, A. M. Spencer *et al.* (eds), 61–79. Norwegian Petroleum Society. London: Graham & Trotman.

Ono, A. 1976. Chemistry and zoning of zircon from some Japanese granitic rocks. *Journal of the Japanese Association of Mineralogists, Petrologists and Economic Geologists* **71,** 6–17.

Orgel, L. E. 1955a. Phosphorescence of solids containing the manganous or ferric ions. *Journal of Chemical Physics* **23,** 1,958.

Orgel, L. E. 1955b. Spectra of transition-metal complexes. *Journal of Chemical Physics* **23,** 1,004–14.

Orgel, L. E. 1966. *An introduction to transition-metal chemistry. Ligand-field theory*. Methuen: London.

Oriel Corp. 1979. *Optical systems & components catalog*. Stamford, Conn.: Oriel.

Ovcharenko, V. K. and L. D. Yur'ev 1971. Luminescence of Ukrainian apatites. *Dopovidi Akademii Nauk Ukrains'koi RSR, ser. B* **33,** (11), 974–7 (in Russian).

Owen, M. R. 1984a. *Sedimentary petrology and provenance of the Upper Jackfork Sandstone (Morrowan), Ouachita Mountains, Arkansas, USA*. PhD Thesis, Urbana: University of Illinois.

Owen, M. R. 1984b. *Some phenomena associated with melting detrital quartz grains in the Luminoscope*. Nuclide Corp. Pub. 1090/0184.

Owen, M. R. 1984c. Southern source of the Jackfork Sandstone determined by petrography, cathodoluminescence of quartz, and hafnium content of zircons. *Abstracts with Programs* **16,** 616. Boulder, Co: Geological Society of America.

Owen, M. R. and A. V. Carozzi 1983. Cathodoluminescence applied to provenance study of a quartz arenite: the Jackfork Sandstone, Ouachita Mountains, Arkansas. *Abstracts with Programs* **15,** 657. Boulder, Co: Geological Society of America.

Owen, M. R. and A. V. Carozzi 1986. Southern provenance of upper Jackfork Sandstone, southern Ouachita Mountains: cathodoluminescence petrology. *Geological Society of America Bulletin* **97,** 110–15.

Pauling, L. 1970. *General chemistry*, 3rd edn. New York: W. H. Freeman.

Pearce, T. H. 1984. Multiple frequency laser interference microscopy: a new technique. *The Microscope* **32,** 69–81.

Pearce, T. H. and M. Thomas 1984. Laser-induced fluorescence in minerals: preliminary results. *Canadian Mineralogist* **22,** 379–82.

Peters, H. 1930. On the fluorescence of quartz under the influence of cathode rays of low voltage. *Physical Review 2nd Series* **36,** 1,631–5.

Pierson, B. J. 1981. The control of cathodoluminescence in dolomite by iron and manganese. *Sedimentology* **28,** 601–10.

Pingitore, N. E. Jr 1976. Vadose and phreatic diagenesis: processes, products and their recognition in corals. *Journal of Sedimentary Petrology* **46,** 985–1,006.

Pingitore, N. E. Jr 1978. The behaviour of Zn^{++} and Mn^{++} during carbonate diagenesis: theory and applications. *Journal of Sedimentary Petrology* **48,** 799–814.

Pingitore, N. E. Jr 1982. The role of diffusion during carbonate diagenesis. *Journal of Sedimentary Petrology* **52,** 27–39.

Pirsson, L. V. 1894. On some phonolitic rocks from the Black Hills. *American Journal of Science*, CXLVII, 341–6.

Polaroid Corp. 1978. Polarized light brochure.

Ponahlo, J. 1983. Personal communication.

Ponahlo, J. and T. Koroschetz 1985. Quantitative Kathodolumineszenz – ein neues Verfahren zur Unterscheidung echter von synthetischen Smaragden und Rubinen. *Zeitschrift der Deutschen Gemmologischen Gesellschaft* **34,** 132–42.

Popp, B. N., T. F. Anderson and P. A. Sandberg 1983. Importance of coordinating compositional and textural analyses of limestone constituents. *Abstracts with Programs* **15,** 662. Boulder, Co: Geological Society of America.

Popp, B. N., T. F. Anderson and P. A. Sandberg 1986a. Textural, elemental, and isotopic variations among constituents in Middle Devonian Limestones, North America. *Journal of Sedimentary Petrology* **56,** 715–27.

Popp, B. N., F. A. Podosek, J. C. Brannon, T. F. Anderson and J. Pier 1986b. $^{87}Sr/^{86}Sr$ ratios in Permo-Carboniferous sea water from the analyses of well-preserved brachiopod shells. *Geochimica et Cosmochimica Acta* **50,** 1,321–8.

Portnov, A. M. and B. S. Gorobets 1969. Luminescence of apatite from different rock types. *Doklady Akademii Nauk SSSR* **184,** 110–13 (in Russian).

Potosky, R. A. and O. C. Kopp 1970. Application of cathodoluminescence to petrographic studies. *The Compass* **47,** 63–9.

Prezbindowski, D. 1980. Microsampling technique for stable isotopic analysis of carbonates. *Journal of Sedimentary Petrology* **50,** 643–4.

Raman, C. V. 1966. Luminescence of fluorspar. *Review Roumaine Physics* **11,** (9–10), 769–72.

Ramseyer, K. 1982a. *A cathodoluminescence microscope for the low intensity luminescence of quartz – application to diagenetic and source rock studies.* Presented at the workshop – conference on the geological applications of cathodoluminescence, Manchester, UK, January 1982.

Ramseyer, K. 1982b. A new cathodoluminescence microscope and its application to sandstone diagenesis. In *11th International Congress on Sedimentology, (Abst.)*, 120. Meeting held in 1982 at McMaster University.

Ramseyer, K. 1982c. Cathodoluminescence microscopy, a powerful tool in sandstone petrology. In *11th International Congress on Sedimentology, (Abst.)*, 85. Meeting held in 1982 at McMaster University.

Randall, J. T. 1939. Some recent experiments in luminescence. *Transactions of the Faraday Society* **35,** 2–14.

Reeder, R. J. and J. L. Prosky 1986. Compositional sector zoning in dolomite. *Journal of Sedimentary Petrology* **56,** 237–47.

Reid, A. M. and A. J. Cohen 1967. Some characteristics of enstatite from enstatite achondrites. *Geochimica et Cosmochimica Acta* **31,** 661–72.

Reid, A. M., T. E. Bunch and A. J. Cohen 1964. Luminescence of orthopyroxenes. *Nature* **204,** 1,292–3.

Reid, L. R. and J. P. Smith 1969. Versatile cathodoluminescence attachment for electron microprobe analysis. *Review of Scientific Instruments* **40,** 316–19.

Rémond, G. 1977. Applications of cathodoluminescence in mineralogy. *Journal of Luminescence* **15,** 121–55.

Renfrew, C. and J. S. Peacey 1968. Aegean Marble: a petrological study. *Annual British School Athens* **63,** 45–66.

Richter, D. K. and U. Zinkernagel 1975. Petrographie des 'Permoskyth' der Jaggl-Einheit (Südtirol) und Diskussion der Detritusherkunft mit Hilfe von Kathoden-Lumineszenz-Untersuchungen. *Geologische Rundschau* **64,** 783–807.

Richter, D. K. and U. Zinkernagel 1981. Zur Anwendung der Kathodolumineszenz in der Karbonatpetrographie. *Geologische Rundschau* **70,** 1,276–302.

Rindone, G. E. 1966. Luminescence in the glassy state. In *Luminescence of inorganic solids*, P. Goldberg (ed.), 419–64. New York: Academic Press.

Robbins, M. 1983. *The collector's book of fluorescent minerals.* New York: Van Nostrand Reinhold.

Robinson, R. F. and A. Cook 1966. The Safford copper deposit Lone Star Mining District, Graham County, Arizona. In *Geology of the porphyry copper deposits: southwestern North America*, S. R. Titley and C. L. Hicks (eds), 251–66. Tucson: University of Arizona Press.

Roeder, P. L. 1974. Activity of iron and olivine solubility in basaltic liquids. *Earth and Planetary Science Letters* **23,** 397–410.

Roeder, P. L., J. D. MacArthur, X. P. Ma, G. R. Palmer and A. N. Mariano 1987. Cathodoluminescence and microprobe study of rare earth elements in apatite. *American Mineralogist* **72,** 801–11.

Rolyn Optics Co. 1981. *Optics catalog*. Arcadia, Calif.: Rolyn Optics.

Ross, M., J. J. Papike and P. W. Weiblen 1968. Exsolution in clinoamphiboles. *Science* **159,** 1,099–102.

Ruppert, L. F., C. B. Cecil and R. W. Stanton 1985. Authigenic quartz in the Upper Freeport coal bed, west-central Pennsylvania. *Journal of Sedimentary Petrology* **55,** 334–9.

Ryan, D. E. and J. P. Szabo 1981. Cathodoluminescence of detrital sands: a technique for rapid determination of the light minerals of detrital sands. *Journal of Sedimentary Petrology* **51,** 669–70.

Saether, E. 1957. The alkaline rock province of the Fen area in southern Norway. *Det Kongelige Norske Vidensk abers Selskabs* 1–150.

Saksena, B. D. and L. M. Pant 1951. Cathode luminescence of crystalline quartz. *Journal of Chemical Physics* **19,** 134–5.

Saksena, B. D. and L. M. Pant 1954. Cathodo-luminescence of MgO. *Proceedings of the Physics Society of London, Section B* **67b,** 811–16.

Scatena-Wachel, D. E., A. P. Jones, A. N. Mariano, R. W. Hinton and R. N. Clayton 1986. Strontium isotopes of kimberlites and carbonatites. *Terra Cognita* **6,** 202 (Abst.).

Schindler, P. and S. Ghose 1970. Electron paramagnetic resonance of Mn^{++} in dolomite, and magnesite, and Mn^{++} distribution in dolomites. *American Mineralogist* **55,** 1,889–96.

Schluger, P. R. 1976. Petrology and origin of the red beds of the Perry Formation New Brunswick, Canada, and Maine, USA. *Journal of Sedimentary Petrology* **46,** 22–37.

Schrank, J. A. and G. M. Friedman 1975. Lithified layers from subbottom carbonate sediments of the Red Sea: Deep Sea Drilling Project Leg 23B. Presented at *IXth International Congress of Sedimentology,* Nice, France.

Schulman, J. H. 1946. Luminescence of $(Zn,Be)_2SiO_4$:Mn and other manganese-activated phosphors. *Journal of Applied Physics* **17,** 902–8.

Schulman, J. H., L. W. Evans, R. J. Ginther and K. J. Murata 1947. The sensitized luminescence of manganese-activated calcite. *Journal of Applied Physics* **18,** 732–9.

Scotford, D. M. 1975. A test of aluminum in quartz as a geothermometer. *American Mineralogist* **60,** 139–42.

Semenov, Y. I. 1958. Relation between composition of the rare earths and composition and structure of minerals. *Geokhimiya* no. 5.

Shankai, Z., R. J. Conzemius and H. J. Svec 1984. Determination of carbon, nitrogen, and oxygen in solids by laser mass spectrometry. *Analytical Chemistry* **56,** 382–5.

Shapiro, J. 1972. *Radiation protection. A guide for scientists and physicians.* Cambridge: Harvard University Press.

Shockley, W. 1948. Electronic polarizability of O^{--} in presence of Ti^{+4} (Abstract). *Bulletin of the American Physical Society* **23,** no. 2. Annual Meeting – Supplementary Programme, Series 1, p. 46.

Sibley, D. F. 1975. Personal communication.

Sibley, D. F. 1982. The origin of common dolomite fabrics: clues from the Pliocene. *Journal of Sedimentary Petrology* **52,** 1,087–100.

Sibley, D. F. and H. Blatt 1976. Intergranular pressure solution and cementation of the Tuscarora orthoquartzite. *Journal of Sedimentary Petrology* **46,** 881–96.

Sippel, R. F. 1965. Simple device for luminescence petrography. *Review of Scientific Instruments* **36,** 1,556–8.

Sippel, R. F. 1968. Sandstone petrology, evidence from luminescence petrography. *Journal of Sedimentary Petrology* **38,** 530–54.

Sippel, R. F. 1971. Luminescence petrography of the Apollo 12 rocks and comparative features in terrestrial rocks and meteorites. In *Proceedings of the Second Lunar Conference,* Vol. 1, A. A. Levinson (ed.), 247–63. Cambridge: Mass.: MIT.

Sippel, R. F. and E. D. Glover 1965. Structures in carbonate rocks made visible by luminescence petrography. *Science* **150,** 1,283–7.

Sippel, R. F. and A. B. Spencer 1970a. Cathodoluminescence properties of lunar rocks. *Science* **167,** 677–9.

Sippel, R. F. and A. B. Spencer 1970b. Luminescence petrography and properties of lunar crystalline rocks and breccias. *Proceedings of the Apollo 11 Lunar Science Conference,* Vol. 3. *Geochimica et Cosmochimica Acta,* Supplement 1, A. A. Levinson (ed.), 2,413–26. New York: Pergamon.

Smith, A. L. 1949. New complex silicate phosphors containing calcium, magnesium, and beryllium. *Journal of the Electrochemical Society* **96,** 287–96.

Smith, J. V. 1974a. *Feldspar minerals,* Vol. 1. Berlin: Springer.

Smith, J. V. 1974b. *Feldspar minerals,* Vol. 2. Berlin: Springer.

Smith, J. V. and R. C. Stenstrom 1965. Electron-excited luminescence as a petrologic tool. *Journal of Geology* **73,** 627–35.

Sommer, S. E. 1969. *Characterization and application of cathodoluminescence from manganese activated carbonate minerals.* PhD Thesis, Penn. State University.

Sommer, S. E. 1972a. Cathodoluminescence of carbonates, 1. Characterization of cathodoluminescence from carbonate solid solutions. *Chemical Geology* **9,** 257–73.

Sommer, S. E. 1972b. Cathodoluminescence of carbonates, 2. Geological applications. *Chemical Geology* **9,** 275–84.

Sommer, S. E. 1972c. Selected area X-ray luminescence spectroscopy with the X-ray milliprobe. *Applied Spectroscopy* **26,** 557–8.

Sprunt, E. S. 1981. Causes of quartz cathodoluminescence colors. In *Scanning electron microscopy/I,* O. Johari (ed.), 525–35. Chicago: SEM.

Sprunt, E. S. and A. Nur 1977. Destruction of porosity through pressure solution. *Geophysics* **42,** 726–41.

Sprunt, E. S. and A. Nur 1979. Microcracking and healing in granites: new evidence from cathodoluminescence. *Science* **205,** 495–7.

Sprunt, E. S., L. A. Dengler and D. Sloan 1978. Effects of metamorphism on quartz cathodoluminescence. *Geology* **6,** 305–8.

Steele, I. M. 1986a. Cathodoluminescence and minor elements in forsterites from extraterrestrial samples. *American Mineralogist* **71,** 966–70.

Steele, I. M. 1986b. Compositions and textures of relic forsterite in carbonaceous and unequilibrated ordinary chondrites. *Geochimica et Cosmochimica Acta* **50,** 1,379–95.

Stel, H. 1981. Crystal growth in cataclasites: diagnostic microstructures and implications. *Tectonophysics* **78,** 585–600.

Stenstrom, R. C. and J. V. Smith 1964. *Electron-excited luminescence as a petrologic tool* (Abst.): Geological Society of America Special Paper no. 76, 158. Meeting held in New York, 1963.

Steyn, J. B., P. Giles and D. B. Holt 1976. An efficient spectroscopic detection system for cathodoluminescence mode scanning electron microscopy (SEM). *Journal of Microscopy* **107,** 107–28.

Stow, D. A. V. and J. Miller 1984. Mineralogy, petrology, and diagenesis of sediments at site 530, southeast Angola basin. *Initial reports of the deep sea drilling project,* vol. LXXV, W. W. Hay, J. C. Sibuet *et al.* (eds), 857–73. Washington: US Govt Printing Office.

Sutton, S. R. 1978. Personal communication.

Taftø, J. and J. C. H. Spence 1982. Crystal site location of iron and trace elements in a magnesium–iron olivine by a new crystallographic technique. *Science* **218,** 49–51.

Tarashchan, A. N. 1978. *Luminescence of minerals.* Kiev: Naukova Dumka (in Russian).

Tarashchan, A. N., A. I. Serebrennikov and A. N. Platanov 1973. Features of lead ions luminescence in amazonite. In *Constitution and properties of minerals,* vol. 8, 106–11 (in Russian).

Teegarden, K. 1966. Halide lattices. In *Luminescence of inorganic solids,* P. Goldberg (ed.), 53–118. New York: Academic Press.

Telfer, D. J. and G. Walker 1975. Optical detection of Fe^{3+} in lunar plagioclase. *Nature* **258,** 694–5.

Telfer, D. J. and G. Walker 1978. Ligand field bands of Mn^{2+} and Fe^{3+} luminescence centres and their site occupancy in plagioclase feldspars. *Modern Geology* **6,** 199–210.

ten Have, T. and W. J. Heijnen 1985. Cathodoluminescence activation and zonation in carbonate rocks: an experimental approach. *Geologie en Mijnbouw* **64,** 297–310.

Thomas, J. B. 1974. *Cathodoluminescence as applied to sandstone petrology: sediment– source rock relationships.* Presented at Short course on geological applications of cathodoluminescence, NEGSA, Baltimore, Md., March 1974.

Thornber, C. R., P. L. Roeder and J. R. Foster 1980. The effect of composition on the ferric–ferrous ratio in basaltic liquids at atmospheric pressure. *Geochimica et Cosmochimica Acta* **44,** 525–32.

Tovey, N. K. and D. H. Krinsley 1980. A cathodoluminescent study of quartz sand grains. *Journal of Microscopy* **120,** 279–89.

Turner, P., J. R. Ashworth and F. A. Iskenderian 1982. Authigenic albite in the Pendleside Formation (Dinantian) of northern England. *Geology Magazine* **119** (4), 395–404.

van Doorn, C. Z., D. J. Schipper and P. T. Balwijn 1972. Optical investigation of cathodochromic sodalite. *Journal of the Electrochemical Society, Solid-state Science and Technology,* **119,** 85–92.

Vlam, C. C. 1954. The shape of the emission bands of luminescent solids. *British Journal of Applied Physics* **5,** 443–6.

Vlasov, K. A. (ed.) 1966. *Geochemistry and mineralogy of rare elements and genetic types of their deposits. Vol. II, mineralogy of rare elements.* Israel Program for Scientific Translations, Jerusalem.

Vocke, R. D. Jr and G. N. Hanson 1981. U–Pb zircon ages and petrogenetic implications for two basement units from Victoria Valley, Antarctica. In *Antarctic Research Series, Dry Valley Drilling Project,* Vol. 33, L. D. McGinnis (ed.) 247–55. Washington: American Geophysical Union.

Waff, H. S. 1977. The structural role of ferric iron in silicate melts. *Canadian Mineralogist* **15,** 198–9.

Walker, G. 1983. Luminescence centres in minerals. *Chemistry in Britain* **19,** 824–31.

Walker, G. 1985. Mineralogical applications of luminescence technique. In *Chemical bonding and spectroscopy in mineral chemistry,* F. J. Berry and D. J. Vaughan (eds), 103–40. London: Chapman & Hall.

Wang, K. 1981. Distribution characteristics of the rare earth elements in Bayan Obo iron ore deposit. *Scientia Geologica Sinica,* 360–7 (Chinese with English abstract).

Weber, B. B. 1963. Luminescent phenomena and zirconia research. *Interceram* **2,** 90–6.

Weber, J. N. and E. White (eds) 1970. *Space science applications of solid state luminescent phenomena.* MRL Special Pub. 70–101. State College, Pa.: Penn. State University.

Weber, J. N., R. T. Greer and V. Vand 1967. Electron-excited fluorescence of serpentines. *Planetary Space Science* **15,** 633–42.

Weiblen, P. W. 1965. Investigation of cathodo-luminescence with the petrographic microscope. In *Developments in applied spectroscopy*, vol. 4, 1964 meeting, E. N. Davis (ed.), 245–51. New York: Plenum.

White, E. W. and W. B. White 1967. Electron microprobe and optical absorption study of colored kyanites. *Science* **158,** 915–17.

White, W. B. and K. L. Keester 1966. Optical absorption spectra of iron in the rock-forming silicates. *American Mineralogist* **51,** 774–91.

Wildeman, T. R. 1969/70. The distribution of Mn^{2+} in some carbonates by electron paramagnetic resonance. *Chemical Geology* **5,** 167–77.

Wilkinson, B. H., A. L. Smith and K. C. Lohmann 1985. Sparry calcite marine cement in upper Jurassic limestones of southeastern Wyoming. In *Carbonate cements*, N. Schneidermann and P. M. Harris (eds), 169–84. Tulsa: SEPM.

Williams, F. 1966. Theoretical basis for solid-state luminescence. In *Luminescence of inorganic solids*, P. Goldberg (ed.), 1–52. New York: Academic Press.

Winkler, E. M. 1983. Ultra-violet luminescence, a new tool in geology. *Bulletin of the Association of Engineering Geologists* **XX,** 227–30.

Wintle, A. G. 1975. Thermal quenching of thermoluminescence in quartz. *Geophysical Journal of the Royal Astronomical Society* **41,** 107–13.

Woods, G. S. and A. R. Lang 1975. Cathodoluminescence, optical absorption and X-ray topographic studies of synthetic diamonds. *Journal of Crystal Growth* **28,** 215–26.

Yanguas, J. E. and J. J. Dravis 1985. Blue fluorescent dye technique for recognition of microporosity in sedimentary rocks. *Journal of Sedimentary Petrology* **55,** 600–2.

Zinkernagel, U. 1978. Cathodoluminescence of quartz and its application to sandstone petrology. *Contributions to Sedimentology*, No. 8. Stuttgart: E. Schweizerbart'sche Verlagsbuchhandlung.

Additional Literature

The following articles are those which have been brought to the author's attention since the time of preparing this book:

Amieux, P. 1987. Description pètrographique de foraminifères par combinaison d'images en lumière naturelle et en cathodoluminescence. *Comptes Rendu de l'Academie des Sciences de Paris* **304,** Sèrie II, 741–4.

Lottermoser, B. G. 1986. Cathodoluminescence of phenakite. *Mineralogical Magazine* **50,** 733–4.

Machel, H. G. 1987. Comment: Cathodoluminescence activation and zonation in carbonate rocks: an experimental approach. *Geologie en Mijnbouw* **66,** 67–74.

Martini, R., P. Amieux, A. Gandin and L. Zaninetti 1987. Triassic foraminifers from Punta Tonnara (SW Sardinia) observed in cathodoluminescence. *Revue de Palèobiologie* **6,** 23–7.

McLemore, V. T. and J. M. Barker 1987. Some geological applications of cathodoluminescence: examples from the Lemitar Mountains and Riley travertine, Socorro County, New Mexico. *New Mexico Geology* **9,** 37–40.

Ponahlo, J. and T. Koroschetz 1986. Quantitative cathodoluminescence of gemstones. *The Australian Gemmologist* **16,** 64–71.

White, W. B., M. Matsumura, D. G. Linnehan, T. Furukawa and B. K. Chandrasekhar 1986. Absorption and luminescence of Fe^{3+} in single-crystal orthoclase. *American Mineralogist* **71,** 1,415–19.

White, W. B. 1975. Luminescent materials. *Transactions of the American Crystallographic Association* **11,** 31–49.

Subject Index

Note: Figure numbers are *underlined*. Section numbers are in bold print.

Author Index

Notes: One and two author papers are referenced to the text citation. Multiple author papers are referenced only to the bibliography page. Figure numbers are printed in italics.

INDEX